Plantation Forests and Biodiversity: Oxymoron or Opportunity?

TOPICS IN BIODIVERSITY AND CONSERVATION

Volume 9

http://www.springer.com/series/7488

Plantation Forests and Biodiversity: Oxymoron or Opportunity?

Edited by

Eckehard G. Brockerhoff
Hervé Jactel
John A. Parrotta
Christopher P. Quine
Jeffrey Sayer

and

David L. Hawksworth

Reprinted from *Biodiversity and Conservation,* volume 17:5 (2008)

Editors
Eckehard G. Brockerhoff
Scion
(New Zealand Forest Research Institute)
PO Box 29237
Christchurch 8540
New Zealand
eckehard.brockerhoff@scionresearch.com

Hervé Jactel
INRA
69 route d'Arcachon
33612 Cestas Cedex
France
herve.jactel@pierroton.inra.fr

John A. Parrotta
U.S. Forest Service
Research & Development
1601 N. Kent Street
Arlington VA 22209
USA
jparrotta@fs.fed.us

Christopher P. Quine
Forest Research
Northern Research Station
Roslin, Midlothian
United Kingdom EH25 9SY
chris.quine@forestry.gsi.gov.uk

Jeffrey Sayer
IUCN The World Conservation Union
Forest Conservation
rue Mauverney 28
1196 Gland
Switzerland
jeff.sayer@iucn.org

David L. Hawksworth
Universidad Complutense
Fac. Farmacia
Dept. Biología Vegetal II
Plaza Ramon y Cajal
Ciudad Universitaria
28040 Madrid
Spain
d.hawksworth@nhm.ac.uk

ISBN: 978-94-007-3685-6 e-ISBN: 978-90-481-2807-5

DOI: 10.1007/978-90-481-2807-5

Softcover reprint of the hardcover 1st edition 2009

Printed on acid-free paper

springer.com

Contents

Plantation Forests and Biodiversity: Oxymoron or Opportunity?

Foreword

Plantation forests and biodiversity: Oxymoron[1] or opportunity?

Forests form the natural vegetation over much of the Earth's land, and they are critical for the survival of innumerable organisms. The ongoing loss of natural forests, which in some regions may have taken many millennia to develop, is one of the main reasons for the decline of biodiversity. Preventing the further destruction of forests and protecting species and ecosystems within forests have become central issues for environmental agencies, forest managers, and governments. In this difficult task science has an important role in informing policy and management as to how to go about this. So how do industrial and other plantation forests fit into this?

Plantation forests, comprised of rows of planted trees that may be destined for pulp or sawmills after only a few years of growth, appear to have little to contribute to the conservation of biodiversity. Yet there is more to this than meets the eye (of the casual observer), and there are indeed numerous opportunities, and often untapped potential, for biodiversity conservation in plantation forestry. With plantation forests expanding at a rate of approximately three million hectares per year, it is crucial to understand how plantations can make a positive contribution to biodiversity conservation and how the potentially negative impacts of this land use can be minimised. That is the topic of this book.

In some countries, expansion of plantation forests represent a threat to natural forests, along with many other pressures on natural land cover from our rapidly growing population and our thirst for natural resources. Clearly, new plantation forests should be established on land that does not have important conservation value. However, plantation forests replacing agricultural and other 'un-natural' land uses often benefit conservation by providing new or expanded habitat for forest species of all kinds, from microorganisms, fungi, and insects to birds and mammals. This is particularly so when the use of native trees leads to a forest that shares key attributes with the former natural land cover, but even exotic trees offer opportunities for restoration of forest habitat. Regardless of the origin or

[1] An 'oxymoron' is a figure of speech using an intended combination of two apparently contradictory terms.

history of a plantation forest, much can be done in terms of forest design and management to improve their value for biodiversity. This applies to both the planted area as well as natural forest remnants and other reserve areas maintained within the plantation forest landscape.

This book is the result of ongoing collaboration among forest scientists from numerous countries who have joined efforts under the umbrella of the International Union of Forest Research Organisations (IUFRO), with contributions from the World Conservation Union (IUCN), the WWF-World Wide Fund For Nature, and many others. The contents of this volume are derived from papers presented at three conferences that took place in Europe and Australia between 2005 and 2006, with the aim of increasing our understanding of conservation issues and opportunities around plantation forestry. The research presented here covers a wide range of taxa living in forests, from lichens to primates, from various temperate and tropical regions around the world. The findings are equally of interest to the scientific community, policy makers and forest managers. This work can assist with the improvement of best-practice guidelines for the establishment and management of plantation forests. The topical examples of applied conservation issues will make the volume also highly valuable for use in conservation biology courses.

The 14 contributions presented here were first published in Biodiversity and Conservation 17(5):925–1211 (2008). They are being released again now in book form in view of the great interest shown by the scientific community in the compilation. Since the conception of this project there has been an increasing momentum towards planting more forests as carbon sinks to combat climate change. Much of the contents of this volume are relevant to such "carbon forests", and its appearance is timely. As with all types of planted forests, there are significant win-win opportunities for multiple benefits, including biodiversity conservation.

ECKEHARD G. BROCKERHOFF
HERVÉ JACTEL
JOHN A. PARROTTA
CHRISTOPHER P. QUINE
JEFFREY SAYER
DAVID L. HAWKSWORTH

Plantation forests and biodiversity: oxymoron or opportunity?

Eckehard G. Brockerhoff · Hervé Jactel · John A. Parrotta · Christopher P. Quine · Jeffrey Sayer

Originally published in the journal Biodiversity and Conservation, Volume 17, No 5, 925–951.
DOI: 10.1007/s10531-008-9380-x

Abstract Losses of natural and semi-natural forests, mostly to agriculture, are a significant concern for biodiversity. Against this trend, the area of intensively managed plantation forests increases, and there is much debate about the implications for biodiversity. We provide a comprehensive review of the function of plantation forests as habitat compared with other land cover, examine the effects on biodiversity at the landscape scale, and synthesise context-specific effects of plantation forestry on biodiversity. Natural forests are usually more suitable as habitat for a wider range of native forest species than plantation forests but there is abundant evidence that plantation forests can provide valuable habitat, even for some threatened and endangered species, and may contribute to the conservation of biodiversity by various mechanisms. In landscapes where forest is the natural land cover, plantation forests may represent a low-contrast matrix, and afforestation of agricultural land can assist conservation by providing complementary forest habitat, buffering edge effects, and

An 'oxymoron' is a figure of speech using an intended combination of two apparently contradictory terms.

E. G. Brockerhoff (✉)
Scion (New Zealand Forest Research Institute), P.O. Box 29237, Christchurch 8540, New Zealand
e-mail: Eckehard.Brockerhoff@scionresearch.com

H. Jactel
INRA, UMR1202 Biodiversity, Genes & Communities, Laboratory of Forest Entomology and Biodiversity, 69 Route d'Arcachon, 33612 Cestas Cedex, France

J. A. Parrotta
U.S. Forest Service, Research & Development, 4th floor, RP-C, 1601 North Kent Street, Arlington, VA 22209, USA

C. P. Quine
Forest Research, Northern Research Station, Roslin, Midlothian EH25 9SY, UK

J. Sayer
The World Conservation Union (IUCN), Forest Conservation Programme, 28 rue Mauverney, 1196 Gland, Switzerland

increasing connectivity. In contrast, conversion of natural forests and afforestation of natural non-forest land is detrimental. However, regional deforestation pressure for agricultural development may render plantation forestry a 'lesser evil' if forest managers protect indigenous vegetation remnants. We provide numerous context-specific examples and case studies to assist impact assessments of plantation forestry, and we offer a range of management recommendations. This paper also serves as an introduction and background paper to this special issue on the effects of plantation forests on biodiversity.

Keywords Afforestation · Biodiversity conservation · Certification · Context · Deforestation · Forest management · Impact assessment · Land use change · Landscape ecology

Introduction

Deforestation is a major cause of the loss of biological diversity and a significant global concern (e.g., Wilson 1988; Brook et al. 2003; Laurance 2007) as it is estimated that more than half of the known terrestrial plant and animal species live in forests (Millenium Ecosystem Assessment 2005). Globally, the area of natural and semi-natural forests decreases by some 13 million ha annually (ca. 0.3%), mostly due to conversion to agriculture (FAO 2006a, 2007). Plantation forests constitute only about 3.5% of the total forest area (ca. 140 million ha) but the area of plantation forests is increasing by about 2–3 million ha (ca. 2%) annually, against the trend of a globally falling forest cover (FAO 2006a, Table 1). According to the current Food and Agriculture Organization (FAO) and International Union of Forest Research Organisations (IUFRO) definitions (e.g., FAO 2006a), plantation forests are established through planting or seeding of one or more indigenous or introduced tree species in the process of afforestation or reforestation. Particularly in the first rotation after establishment, stands are typically of an even-aged structure with an even spacing of trees. Their main objective is often the production of timber or fuel wood (plantations provided about 35% of the global wood supply in 2000) but some are established to reduce erosion, fix carbon, or provide other environmental, economic, or social benefits. Many plantations are intensively managed including the use of improved tree varieties and silvicultural operations that may involve site preparation (e.g., ploughing, harrowing, use of fertilizers, and herbicides), thinning, and clear-cut harvesting, often with short rotations (e.g., <30 years between planting and harvesting, or as little as 5–10 years for poplars and some tropical species). Apart from such plantations, the FAO definition for "planted forests" (FAO 2006b) also includes some types of semi-natural forests that were established through planting or seeding by human intervention. In reality, it is often difficult to categorise planted forests such as those that have been established as pure stands by planting or sowing centuries ago, and have since become more diverse by natural processes, which is common in much of Europe.

The implications for the conservation of forest biodiversity of plantation forests and their continuing expansion are being debated vigorously. Although plantation forest managers increasingly recognise the need to conserve biodiversity, and many adhere to sustainable management guidelines such as those of the Forest Stewardship Council (FSC Forest Stewardship Council 2007a) or the Programme for the Endorsement of Forest Certification schemes (PEFC 2007), certification does not always benefit biodiversity conservation (Gullison 2003), and criticism of plantation forestry from some stakeholders remains strong (e.g., Cossalter and Pye-Smith 2003, and see below). Plantation forests are the focus of

Table 1 Forest cover, proportion of plantation forests, and approximate area of principle tree species used in plantation forests in selected countries with different species composition

Country	Total forest cover (million ha) and annual % change	Percentage of land area forested	Area of plantation forests (million ha) and annual % change	Percentage of forest cover in plantation forest	Approximate area of principle plantation tree species in million ha[a] Conifers	Broadleaved trees
World	3952.0 −0.2%	30.3	139.8 +2.0%	3.5%		
Examples of countries with mostly exotic plantation species						
Brazil	477.7 −0.6%	57.2	5.4 +0.4%	1.1%	1.8 *Pinus spp.** (*P. taeda**, *P. caribaea**, *P. elliotii**) 0.08 *Araucaria angustifolia*	3.0 *Eucalyptus* spp.* (*E. grandis**, *E. urophylla**, *E. urograndis**)
Indonesia	88.5 −2.0%	48.8	3.4 +2.3%	3.8%	0.8 *Pinus merkusii*	3.5 *Hevea* spp.* 1.5 *Tectona* spp.* 3.4 other broadleaves
New Zealand	8.3 +0.2%	31.0	1.8 +0.9%	22.3%	1.6 *Pinus radiata** 0.1 *Pseudotsuga menziesii**	
UK	2.8 +0.4%	11.8	1.9 −0.1%	67.6%	0.8 *Picea* spp.* (*P. sitchensis**, *P. abies**) 0.3 *Pinus* spp. (*P. sylvestris*, *P. contorta**, *P. nigra**) 0.13 *Larix* spp.* 0.05 *Pseudotsuga menziesii**	
Examples of countries with mostly native plantation species						
China	197.3 +2.2%	21.2	31.4 +4.7%	15.9%	12.9 *Pinus* spp. (*P. massoniana*, *P. tabulaeformis*, *P. elliottii**) 17.2 *Cunninghamia lanceolata*/*Larix* spp.	1.3 *Eucalyptus* spp.* 11.5 *Populus* spp. and other broadleaves
France	15.6 +0.3%	28.3	2.0 +0.3%	12.7%	1.3 *Pinus* spp. (1.1 *P. pinaster*, 0.2 *P. nigra*) 0.45 *Picea abies* 0.3 *Pseudotsuga menziesii**	0.2 *Populus* spp.
USA	303.1 +0.1%	33.1	17.1 +0.9%	5.6%	15.0 *Pinus* spp. (*P. taeda*, *P. elliottii*, *P. echinata*, *P. palustris*) 1.2 *Pseudotsuga menziesii*/*Larix* spp./*Picea* spp.	

[a] Exotic species are indicated by an asterisk. Compiled from various sources, primarily FAO (2001, 2006a). Additional information, mostly about the species composition of plantation forests, for Brazil from www.itto.or.jp/newsletter/v7n2/08industrial.html (Reis MS, Industrial planted forests in tropical Latin America, accessed January 2002); for France from IFN database, Inventaire Forestier National, France (accessed January 2002 by HJ); for New Zealand from Anon. (2001a) New Zealand forest industry facts and figures. Forest Owners Association, Wellington, New Zealand; for the United Kingdom from Anon. (2001b) Forestry statistics 2001. Forestry Commission, Edinburgh, UK; and Simon Gillam (Forestry Commission, UK, pers. comm.); for the USA from W. Brad Smith (U.S. Forest Service, Washington, DC, pers. comm.). Note, for Indonesia the area of tree species in plantations is based on FAO (2001). Indonesia reported a significant drop in the total plantation area for the 2005 Forest Resources Assessment according to FAO (2006a)

widespread opposition amongst several environmental organisations (e.g., Carrere and Lohmann 1996; World Rainforest Movement 2007). The industrial scale of many plantations, their common structure as monocultures and particularly the fact that they are sometimes established on land previously covered in natural forest all serve to place them in the forefront of the concerns of environmental lobbies. FERN (2007) and the World Rainforest Movement (2007) have also expressed concern about the likely impacts of expanded use of genetically modified trees in some plantations. One recent focus of these concerns about plantations has been the debate over the extent to which, and conditions under which, plantations might be eligible for certification under the FSC. In the past, plantations were certified if they met the same basic conditions of good management that were applied to natural and semi-natural forests, along with some additional criteria specifically for plantations. Some environmental groups argued that certification by FSC implied in the minds of purchasers a natural, green product and that therefore certification should never be given to 'monocultures'. These concerns of NGOs have recently triggered a review of the eligibility of plantation forests for certification under the Forest Stewardship Council (2004, 2007b, c).

The scientific community is equally divided over these issues (e.g., Kanowski et al. 2005), and despite an expanding body of literature on the effects of plantation forestry on biodiversity, there is no simple answer to the question of whether or not plantation forestry is compatible with biodiversity conservation goals. To answer this question, and to determine whether 'plantation forests and biodiversity' are indeed an oxymoron or an opportunity, it is necessary to consider the wider context of a plantation forest and to take numerous factors into account that vary substantially among locations and countries, and that ultimately determine the likely effects on biodiversity. For example, it is essential to know what kind of land use preceded the establishment of a plantation, what alternative land uses are probable at a given location, what tree species are involved, and how and for what purpose a plantation is being managed. While the majority of plantation forests are managed primarily for production purposes, substantial areas serve primarily for environmental protection and conservation, and many plantations have multiple purposes. Although these characteristics vary widely among plantation forests, assessments of their environmental effects often do not consider such factors.

The International Union of Forest Research Organizations (IUFRO), the World-Wide Fund for Nature (WWF), and several other organisations recently sponsored three conferences[1] to facilitate scientific debate on these issues. This special issue of Biodiversity and Conservation represents an account of some of these contributions of recent research that is relevant to this debate. This article serves as a background document to the topic and to give an overview of some of the key issues that need to be considered for an informed debate about plantation forestry and biodiversity. The specific objectives of this paper are:

- to provide a brief review of the value of *plantation forests as habitat*, compared with natural forests and other, mainly agricultural, land uses,
- to examine effects of plantation forests on biodiversity at the *landscape scale*,
- to *place in context the different types of plantation forests* and thereby clarify the situations in which there are positive and negative impacts of plantation forests on biodiversity, and to examine examples of various plantation forests in different countries, and
- to offer suggestions how plantations can be *managed to enhance biodiversity*.

[1] "Biodiversity and Conservation Biology in Plantation Forests." Bordeaux, France, 27–29 April 2005. "Biodiversity and Plantation Forests—Oxymoron or Opportunity", Technical Session at the XXII IUFRO World Congress—Brisbane, Australia, 8 August 2005. "Ecosystem Goods and Services from Planted Forests." Bilbao, Spain, 3–7 October 2006.

Habitat or non-habitat? Is there biodiversity in plantation forests?

A common perception of plantation forests is that they are ecological deserts that do not provide habitat for valued organisms. However, numerous studies in many countries have documented that plantation forests can provide habitat for a wide range of native forest plants, animals, and fungi (Parrotta et al. 1997a; Oberhauser 1997; Humphrey et al. 2000; Brockerhoff et al. 2003; Barbaro et al. 2005; Carnus et al. 2006, and papers in this issue). Even uncommon and threatened species are increasingly recorded in plantations as more targeted surveys are being undertaken. For example, the largest population in Europe of the locally threatened hoopoe, *Upupa epops*, occurs in plantation forests in the Landes region in France (Barbaro et al. 2008—this issue). The flightless cassowary, *Casuarius casuarius*, has been recorded in *Araucaria cunninghamii* plantations in Queensland (Keenan et al. 1997). Substantial populations of the endangered brown kiwi, *Apteryx mantelli*, occur in exotic pine plantations in New Zealand (e.g., Kleinpaste 1990). The critically endangered ground beetle, *Holcaspis brevicula*, a locally endemic species, is thought to depend on a plantation forest as its only remaining habitat (Brockerhoff et al. 2005; Berndt et al. 2008—this issue).

Thus, there is abundant evidence that plantation forests themselves can be valuable as habitat. However, plantation forests are commonly being compared with biodiversity in more natural forests, often without consideration of the circumstances that define whether such comparisons are appropriate (see below). Appropriate or not, it is usually true that natural forests offer superior habitat for native forest species than plantation forests (Armstrong and van Hensbergen 1996; Moore and Allen 1999; Lindenmayer and Hobbs 2004 and references therein, du Bus de Warnaffe and Deconchat 2008—this issue). But the extent of this difference varies considerably across the range of management intensities and the degree to which plantations deviate from the tree species composition and structure of natural forests in the same area (Fig. 1). Plantation forests usually have less habitat diversity and complexity. For example, some forest bird species may not find their required food sources in plantations, or there may be a lack of overmature trees suitable for nesting (Clout and Gaze 1984). The species richness of forest specialists is often lower in plantations than in semi-natural forest, whereas the difference is less strong for generalist species (Magura et al. 2000; Raman 2006). In particular, plants and animals that are old forest specialists may not be able to colonise or reproduce in plantations with comparatively short rotations (e.g., 7–21 years for eucalypts in Brazil, ca. 27 years for *Pinus radiata* in New Zealand). In a study in California that compared species assemblages in exotic eucalypt and native *Quercus agrifolia* woodlands, Sax (2002) reported very similar species diversity for amphibians, birds, mammals, and leaf-litter invertebrates, although species composition was often dissimilar. On the other hand, longer-rotation plantation forests, especially those managed with conservation objectives, may differ little in habitat value from managed natural forests (Keenan et al. 1997; Humphrey et al. 2003; Suzuki and Olson 2008—this issue).

Nevertheless, plantations compare favorably with most other economically productive land uses. For example, in New Zealand far fewer native species are found in pastoral grasslands than in plantation forests (Brockerhoff et al. 2001; Ecroyd and Brockerhoff 2005; Pawson et al. 2008—this issue). Across the scale of management intensity and conservation value, probably all types of plantation forests have a higher conservation value than intensive agriculture land uses (Fig. 1). In the case of plantation forests that were established on agricultural land, this comparison is more appropriate than evaluating plantation forests against what would be found in a natural forest.

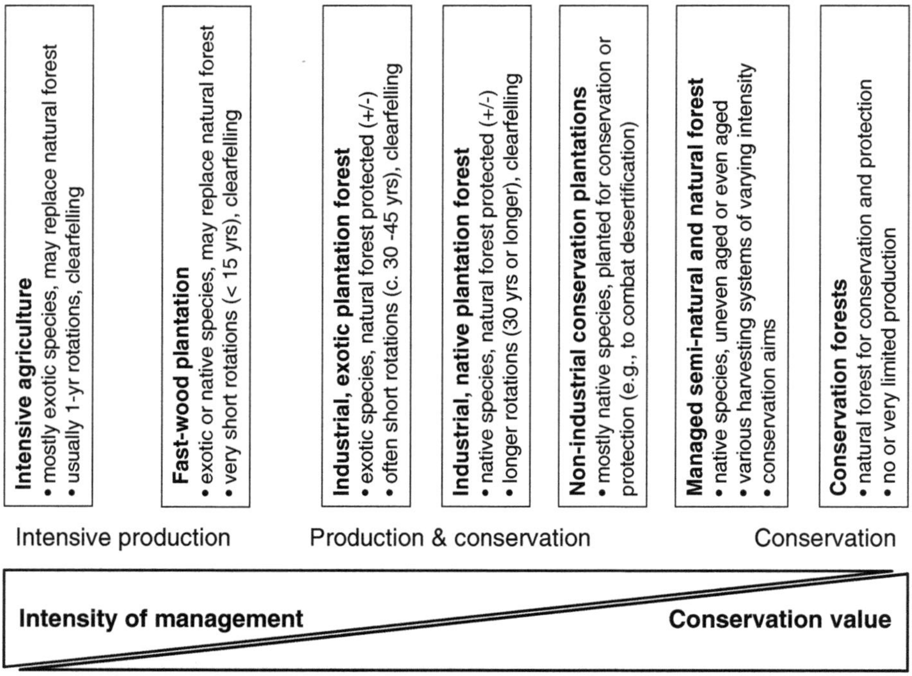

Fig. 1 Conceptual model of the relative conservation value of planted forests relative to conservation forests and agricultural land uses. Note that many plantation forests cannot be clearly assigned to one of the main categories outlined here. Some plantation forests serve multiple purposes including production, protection, and conservation on the same land. Categorisation is also difficult for some forests in Europe that have been established as pure stands by planting or sowing centuries ago, and have since become more diverse by natural processes. "Close-to-nature forests" are included in our "managed semi-natural and natural forest" category. For more details refer to the text and the case study examples provided

Successional processes strongly influence the species assemblages that occur in plantation forests and biodiversity varies considerably with stand age (i.e., time since planting). Older stands provide better habitat for forest species than young stands because of increased spatial and vertical heterogeneity, well-developed soil organic layers and associated fungal floras, increased dead wood on the forest floor, a better light environment, and inter-specific facilitation. For example, the understorey vegetation of pine plantations can show a clear successional trend toward increasing dominance by native shade-tolerant species that are typical of natural forest understories (Allen et al. 1995; Brockerhoff et al. 2003). Similar patterns have been observed for other taxonomic groups including epiphytes (see Coote et al. 2008—this issue), birds (Clout and Gaze 1984; Lopez and Moro 1997; Donald et al. 1998), and insects (Fahy and Gormally 1998; Jukes et al. 2001; Lindenmayer and Hobbs 2004; Barbaro et al. 2005; Pawson et al. 2008—this issue; du Bus de Warnaffe and Deconchat 2008—this issue). These successional processes also facilitate forest restoration on sites that were previously deforested, provided that there is a local source of propagules, dispersal agents, and a favorable climate.

There is compelling evidence that plantation forests can accelerate forest succession on previously deforested sites and abandoned agricultural areas where persistent ecological barriers to succession might otherwise preclude re-establishment of native species (see references below). This is due to the influence of the planted trees on understory microclimate conditions, vegetation structural complexity, and development of litter and humus layers during the early years of plantation growth. These changes lead to increased seed inputs from neighboring native forests by seed dispersing wildlife attracted to the plantations, suppression of grasses or other light-demanding species that normally prevent tree seed germination or seedling survival, and improved light, temperature, and moisture conditions for seedling growth. In the absence of intensive silvicultural management aimed at eliminating woody understory regeneration even mono-specific plantations are replaced by a mixed forest comprised of the planted species and an increasing number of early and late successional tree species and other floristic elements drawn from surrounding forest areas. Examples of this 'catalytic' effect of plantations of both native and exotic species, have been reported in many tropical and subtropical regions of the world (Parrotta 1993, 1999; Armstrong and van Hensbergen 1996; Fang and Peng 1997; Geldenhuys 1997; Keenan et al. 1997; Loumeto and Huttel 1997; Oberhauser 1997; Parrotta et al. 1997a, b; Zuang 1997; Yirdaw 2001; Carnevale and Montagnini 2002). These findings also suggest that populations of numerous native species that occur in plantations are viable over successive rotations.

Effects of plantation forests at the landscape scale

The loss and fragmentation of natural forests remains one of the main causes of biodiversity loss (Hunter 1990; Murcia 1995; Wigley and Roberts 1997; Didham et al. 1998; Magura 2002; Henle et al. 2004). Fragmentation reduces the available area of forest habitat (Watson et al. 2004; Benedick et al. 2006), increases the isolation of forest patches (van der Ree et al. 2004) and edge effects in these patches (Yates et al. 2004), all of which contribute to a higher risk of species extinction (Fahrig 2001; Kupfer et al. 2006). In the past, forest fragments were viewed as islands of habitat embedded in an inhospitable matrix of non-habitat. However, a growing body of evidence, referring to the "continuum model" (Fischer and Lindenmayer 2006), suggests that suitable food, shelter, or climatic conditions may be found along gradients in the matrix, allowing dispersal and survival of fragment-dwelling biota. It is now known that some matrix types can mitigate fragmentation effects (Ewers and Didham 2006; Kupfer et al. 2006). The landscape matrix can (1) supplement or complement species habitat or resources, (2) allow or even facilitate dispersal between isolated patches, and (3) its properties or configuration may dampen the effects of disturbance regimes, such as the provision of buffer zones around fragments against adverse edge effects. In contrast, some matrix habitats may act as ecological traps for native species or as sources of invasive species that can spread into remnants. Therefore, besides the conservation of large patches of native forest, there is increasing consensus that more consideration has to be given to managing the complexity of the matrix, as another important objective of biodiversity conservation in forest landscapes. As a type of forest habitat, plantation forests can greatly contribute to improve the quality of the matrix where native forest remnants are embedded (Lindenmayer and Franklin 2002; Kanowski et al. 2005; Fischer and Lindenmayer 2006), more so than alternative land uses such as intensive agriculture.

Plantations can contribute to biodiversity within landscapes through the following three mechanisms:

Habitat supplementation or complementation to forest species

Some species that survive in forest fragments may compensate for habitat loss by using resources in the matrix (Wunderle 1997; Ewers and Didham 2006; Kupfer et al. 2006,

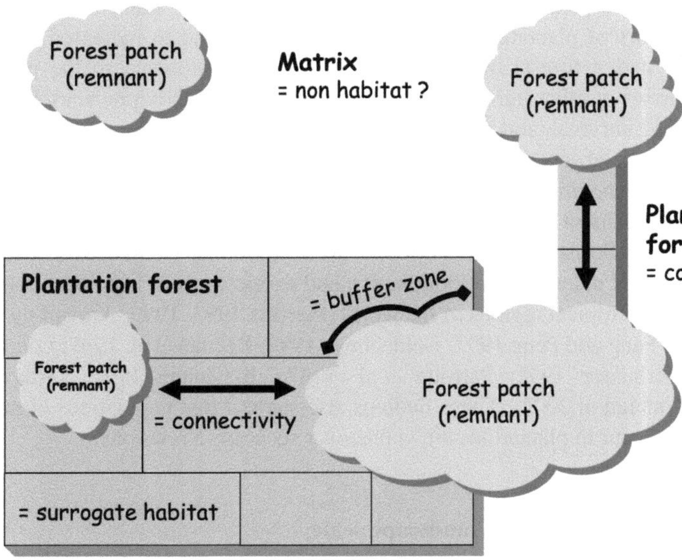

Fig. 2 The 'corridor-patch-matrix' landscape model showing a highly fragmented landscape example with ca. 85% loss of natural forest and ca. 20% plantation forest. Modified after Forman (1995) and Lindenmayer and Franklin (2002)

Fig. 2). Plantation forests can provide suitable habitats for numerous forest species. In addition to the studies mentioned above, comprehensive reviews with relevant examples are those in Gascon et al. (1999) on birds, frogs, mammals, and ants in Amazonia and those in Bernhard-Reversat (2001) on understory plants, birds, mammals, and soil invertebrates in plantations of *Acacia auriculiformis* and *Pinus caribaea* in the Congo.

Connectivity

The presence of plantation forests can enhance indigenous biodiversity by improving connectivity between indigenous forest remnants (Hampson and Peterken 1998; Norton 1998, Fig. 2). This has been demonstrated by studies on a wide range of taxa (e.g., Innes et al. 1991; Parrotta et al. 1997b; Lindenmayer et al. 1999; Wethered and Lawes 2005). For example, plantation forests facilitate the dispersal across of forest dwelling mammals such as the endangered Iberian lynx (Ferreras 2001) and various marsupials (Lindenmayer et al. 1999). Likewise, the maintenance of a network of natural forest remnants, for example along riparian areas, may assist the survival of species for which the plantation matrix is less suitable (Lamb 1998; Carnus et al 2006; Nasi et al. 2008—this issue). However, corridors of insufficient width may not be used by species that avoid edge habitats (Ewers and Didham 2007).

Buffering effects

Native forest remnant edges are characterised by altered microclimates, with edges typically experiencing higher irradiance, temperature, vapor pressure deficit, and wind speed than forest interiors, with consequential changes in biodiversity (Murcia 1995). Plantation forests may enhance the value of indigenous forest remnants by buffering remnant edges from these influences (Renjifo 2001; Fischer et al. 2006, Fig. 2). For example, Denyer et al. (2006) found that microclimate changes across native forest edges adjacent to pine plantations were half those that occurred between native forest and pasture. Furthermore, the vegetation of native forest edges was more similar to the forest interior when the edge was adjacent to plantation forest than when it was adjacent to pasture. This buffering effect is, however, disrupted by the harvesting of the plantation trees, exposing the edge temporarily to the aforementioned negative external influences (Norton 1998).

Plantations forests may also have negative effects on adjacent natural and modified land cover. Planted forests often consist of fast-growing pioneer tree species that may spread beyond the plantation and invade neighboring habitats, particularly open or disturbed habitats. Such invasive trees are also referred to as "wilding trees" (Ledgard 2001). Similar invasion processes can occur with species associated with plantations such as weeds and feral animals (Kanowski et al. 2005; Kupfer et al. 2006). Grazing animals and seed predators may also use plantations where food resources are not limited to build up their population and then cause damage in neighboring remnants (Curran et al. 1999; Lindenmayer and Hobbs 2004; Kanowski et al. 2005). Finally, responses to landscape features are often species-specific or at least dependent on particular traits of species. Generalists and species that are active dispersers are predicted to benefit more from plantation forests in the matrix than rare forest specialists, which could lead to an impoverishment of forest biota compared with native forest communities (Ewers and Didham 2006).

Plantation forests—good or bad for biodiversity? It depends on the context

To determine objectively whether plantation forests are detrimental or give net benefits for conservation is not trivial because this is context-specific and depends on multiple factors (e.g., Brockerhoff et al. 2001; Hartley 2002; Carnus et al. 2006). Essential points that need to be considered include:

1. Whether plantation forestry leads to reduced harvesting and thus improved protection of natural forests, and at what scale,
2. What was the land use or vegetation that preceded the establishment of plantation forests, and how well can the plantation forest provide substitute habitat for species of the former natural land cover (and thus what the appropriate comparison is),
3. How much time has passed since plantation establishment and thus, for example, how long have local species been able to colonise and adapt to the new habitat,
4. Whether the planted area is being managed with conservation goals in mind, whether remnants areas of natural habitat are being protected, and whether conservation goals across the wider landscape are being considered,
5. How plantation forestry compares relative to other alternative land uses that are likely to be practised on a particular piece of land.

Does plantation forestry lead to reduced harvesting and improved protection of natural forests?

Plantation forestry can benefit biodiversity (at a larger scale), if it leads to reduced harvesting of natural forests (Shepherd 1993; Hartley 2002), although this is not necessarily always the case (Clapp 2001). Intensively managed plantations can provide forest products more efficiently than natural forests, and therefore require less land, which may allow greater protection of natural forests (Carle et al. 2002). This is the case in New Zealand where the debate about a national forestry and forest conservation strategy has led to a spatial separation of production and conservation, with the agreement of all stakeholders (Shaw 1997). Over 99% of domestic forest products are obtained from plantation forests (that occupy about 7% of the land base) while there is negligible production in natural forests (that cover about 20% of the land base). This has allowed the majority of New Zealand's natural forests to be managed by the Department of Conservation for conservation and recreation. While this strategy is reasonably successful in New Zealand, a comparatively orderly and corruption-free country, this may not be the case in countries where plantation forestry is still expanding, driven by opportunities for increasing export of forest products, and where natural forests are perhaps not as well protected from conversion to plantations (e.g., Cossalter and Pye-Smith 2003). The paradigm of spatially separating production (in plantation forests) and conservation (in protected areas) is at odds with a strengthening movement in various countries, particularly in Europe, that accepts that production and conservation can occur on the same land. For example, there are efforts to introduce natural features into plantation forests, convert some plantation forests of exotic trees to semi-natural forests and to restore natural forests (e.g., Anderson 2001; Quine et al. 2004). Some of these areas are managed for both production and conservation on the same land and in some cases management is integral to maintaining their conservation value (Fuller et al. 2007). Seymour and Hunter (1999) argued for a hybrid with elements of both these approaches: ecological forestry—containing components of a landscape triad—non-intervention reserves, ecological forestry, and intensive plantations.

What land use or vegetation preceded the establishment of plantation forests, and are there affinities to the natural vegetation and fauna of the area?

The net effects of plantation forests on biodiversity conservation also strongly depend on the land cover that was or is being replaced. It is critical to distinguish if plantations forests replace or replaced natural forests or modified, agricultural land. Clearly, conversion of natural forests into plantations is detrimental to biodiversity conservation, unless deforestation is inevitable and plantation forestry is a 'lesser evil' (see below). Similarly, afforestation of non-forest ecosystems is not desirable where these represent the natural vegetation. For example, afforestation could threaten several rare bird species of open habitats that inhabit South Africa's grasslands (Allan et al. 1997). In maritime pine plantations in the Landes Forest (SW France) few forest specialist species occur even though almost 200 years and three rotations have passed since their establishment in a landscape that was originally dominated by open, moorland habitats. However, rich assemblages of open-habitat beetles, spider and bird species, including several species of conservation concern (e.g., *Harpalus rufipalpis, Carathus erratus, Lullula arborea*), occur in clear-cuts and young pine stands. This suggests that the colonization of clear-felled sites is a key process maintaining the diversity of open-habitat species in this afforested area (Barbaro et al. 2005; Van Halder et al. 2008—this issue).

On the other hand, many plantation forests were established in areas that were originally forested but have lost their natural plant and animal communities long before the plantation was established. Afforestation of intensively managed agricultural land, which is typically inhabited by a highly impoverished flora and fauna (below), usually brings conservation gains (but see Buscardo et al. 2008—this issue). This is particularly true in regions that have experienced significant losses of natural forests. In such situations plantation forests often facilitate the restoration of natural forest elements by natural succession, as outlined below (Sect. "How much time has passed since plantation establishment, and has colonisation by native species occurred?"). Plantation forests can be expected to be better equivalents of natural forests if they are composed of locally occurring native tree species, and in some cases it may be difficult to distinguish older stands from natural forests. However, even plantations of exotic tree species may have an understorey of indigenous plants and a fauna that resemble those of natural forests (e.g., Parrotta and Turnbull 1997 and references therein; Brockerhoff et al. 2003; Humphrey et al. 2003; Pawson et al. 2008—this issue). Given that land use often changes over time, it is also worth considering that plantation forests probably represent a better starting point than agriculture if restoration of natural forest becomes an objective at a later time. A good example of this are efforts in parts of the UK where some Sitka spruce plantations are gradually being restored to Atlantic oak forests and other natural forests, particularly at ancient woodland sites (Humphrey et al. 2006).

How much time has passed since plantation establishment, and has colonisation by native species occurred?

Where plantations replaced natural forests or other natural vegetation it is important to distinguish whether this happened a long time ago or whether this is still an ongoing activity. The FSC currently uses a cut-off point of 1994, and plantation establishment prior to that year is not an impediment, whereas more recent conversion of natural vegetation is not permitted (Forest Stewardship Council 2007a). An obvious benefit of this rule is that it discourages the destruction of natural vegetation. In addition, in older plantations on sites formerly occupied by natural vegetation, certification is likely to lead to improved protection of remaining natural vegetation within such plantations (see below) and it may also encourage the restoration of natural habitats in a proportion of the afforested area.

Plantation forests that were established a long time ago are also more likely to be valuable habitat for biodiversity. Plantation forest habitats become more complex over time which benefits forest species (e.g., Barlow et al. 2008—this issue). Furthermore, colonisation by forest species will have progressed more in an old plantation than in one that was established only recently, if the original vegetation was not forest. Thus, an old plantation forest is likely to be more valuable as habitat in its own right.

Are the planted area and embedded remnants of natural vegetation managed for biodiversity conservation?

The management of plantation forests increasingly meets sustainable forest management goals, particularly in the growing proportion of certified forests in many countries (Forest Stewardship Council 2007c) although many forests that are not certified may also be well managed from a biodiversity point of view. To comply with FSC criteria (Forest Stewardship Council 2007a) concerned with the various aspects of biodiversity conservation, plantation forests have to be managed in accordance with a management plan that specifies conservation goals and resulting actions, including surveys and measures for the protection of rare, threatened and endangered species, the protection of high conservation value forests and other valuable habitats (e.g., wetlands, riparian areas, natural grasslands), and critical examination of the use of pesticides and other potentially detrimental practices (Forest Stewardship Council 2007a). Consultation of a wide range of stakeholders, including

NGOs, and annual re-evaluations of certified 'forest management units' ensure that there is a mechanism that scrutinises whether these criteria are being met. While such processes do not transform plantation forests into biodiversity havens, in many countries FSC-certification has contributed significantly to raising the standards for consideration of biodiversity conservation goals as such issues are among the most frequently issued corrective action requests (Paulsen 2004). Forests managed in accordance with such principles, whether certified or not, clearly contribute more than others that are still managed with little or no special regard for biodiversity. It is important to note that this applies not only to the planted area but also to the often substantial areas of natural habitats that are embedded in plantation forest estates. For example, some holdings in New Zealand include as much as 30% or more of their area in natural forest remnants that are being protected and managed for conservation purposes (Hock and Hay 2003).

Some new plantations include set-aside areas of natural vegetation that are designed to maintain connectivity between these remnants. For example, in Sumatra, some *Acacia mangium* plantations retain up to 26% of the area in natural forest, and, if appropriately designed and managed, these areas can assist the conservation of primates and other species (Nasi et al. 2008—this issue). Similarly, new pine plantations in Patagonia are designed such that connectivity of forest habitats and open steppe habitats are maintained (Lantschner et al. 2008—this issue). Forest management can also contribute to the achievement of conservation goals across the wider landscape. In regions where fragmentation effects are important, plantation forests can increase connectivity between distant remnants of natural vegetation and provide additional forest habitat (Fig. 2). This will be most beneficial in cases where plantations were established in agricultural areas.

Plantation forestry compared with other 'productive' land uses—a 'lesser evil'?

Illegal logging as well as the conversion of natural forests to plantation forests are undoubtedly causing the continued loss of natural vegetation (e.g., Cyranoski 2007; Nasi et al. 2008—this issue). However, land clearance for agriculture is a more significant driver of forest loss. According to the Global Forest Resources Assessment 2000 (FAO 2001) 142 million ha of natural tropical forest were lost from 1990 to 2000, and of these, 132 million ha (93%) were converted to other land uses (i.e., deforestation), whereas only 10 million ha (7%) were converted into plantation forest. Furthermore, plantation forests will provide more suitable habitat for most forest species than agriculture, as outlined above. Many plantation forests also contain substantial areas of natural vegetation in reserve areas that may not be retained if they are embedded in agricultural areas. For these reasons and because forestry companies increasingly make concessions to demands from environmental lobby groups, there is an emerging trend among such groups to accept plantation forestry as a 'lesser evil', and to 'make peace with the enemy' (Cyranoski 2007). Some ecologists believe that working with forestry companies and influencing management will ultimately provide better conservation outcomes than simply opposing plantation forestry. However, afforestation can potentially be more detrimental for biodiversity than agriculture in landscapes where the natural vegetation was not forest but a type of open vegetation, such as grassland, open shrubland, or wetland. Under such circumstances agricultural land uses may be preferable, provided that some natural elements are maintained within the landscape. However, the world wide intensification of agricultural production makes sustainability challenging (Tilman et al. 2002). If plantation forests are established in such areas their impact can be mitigated by protecting adequate areas of open natural habitats (e.g., Lantschner et al. 2008—this issue).

Plantation forests in different contexts—seven countries as case study examples

Exotic tree species are often prevalent in plantations, although in some countries plantation forests consist primarily of native species (Table 1). The desire to maximise timber production and profitability led to the widespread planting of relatively few, mostly fast-growing, tree species. Worldwide, several pine species (*Pinus* spp.) are the most widely used plantation species (ca. 20% of the total plantation area, FAO 2001). Other common plantation genera include spruces (*Picea* spp.), and poplars (*Populus* spp.) in temperate regions, eucalypts (*Eucalyptus* spp.), and rubber (*Hevea* spp.), *Acacia* spp., and teak (*Tectona* spp.) in tropical regions (Cossalter and Pye-Smith 2003; FAO 2006a). For biodiversity conservation, the use of native plantation species is generally preferable because of their higher value as habitat for native species and because of the risk of the planted species becoming invasive. However, in many situations even exotic tree species can make a considerable contribution to biodiversity conservation (below).

Regardless of the identity of the tree species, the relevance of plantation forests for biodiversity conservation in a given country or region needs to be assessed in relation to the relative forest cover and its composition. Some countries that historically had large forest areas still have a considerable cover of mostly natural forests, whereas others have little remaining forest and manage plantations also for biodiversity conservation. The following case studies were chosen as representative examples of countries where various kinds of plantation forests are significant as a land use.

Brazil

With about 478 million ha Brazil has the second largest forest area in the world (after Russia) and the most primary forest of all countries (31% of the global total), but a significant proportion of global deforestation also occurs there (FAO 2006a). Well over half of Brazil's land area is still covered in natural forests (Table 1). Plantation forests, though extensive (5.4 million ha), represent only a small proportion (Table 1) and are of comparatively minor significance for biodiversity conservation. Most of these plantations are of exotic tree species (Table 1) that are managed on very short rotations. Such plantations are also referred to as 'fast-wood plantations', and they are common in many tropical and subtropical countries (Cossalter and Pye-Smith 2003). While plantation forests sometimes replaced natural forests, particularly from the 1960s until the 1980s, their total area is small (1.6%) compared to the total forest area cleared for agriculture (FAO 2006a; IBGE 2007). In recent decades plantation establishment is increasingly occurring on lands that were deforested decades earlier for large-scale agricultural development, particularly in southern and southeastern Brazil. Although the area of plantation forests represents only about 1% of the total forest cover, plantations provide most of Brazil's forest products, according to FAO statistics. For example, 62% of Brazil's industrial roundwood comes from plantations (Carle et al. 2002). This indicates that there is much potential for substitution of wood production in natural forest by plantation forestry which may enable the protection of natural forests. Plantation forests are more valuable for biodiversity conservation than agricultural land uses because most species of conservation concern in Brazil are forest species of which some can use plantation forests as habitat (see also Barlow et al. 2008—this issue). As of mid-2007, certification of forest management under FSC covered an area of 4.8 million ha in Brazil, of which plantation forests (including mixed plantation and other forest types) contribute about 2.1 million ha (Romona Anton, FSC, pers. comm., August 2007). PEFC is also common, presently covering about 0.76 million ha (PEFC 2007).

Indonesia

Officially Indonesia has just under half of its land area in forests but much of this land has long been modified and the extent of near-natural forest is now probably below 20% of total land area (Table 1) (FAO 2006a; Indonesian Ministry of Forestry 2007). Deforestation is thought to continue at a rate of 2% annually (FAO 2006a). Teak has been extensively planted in Java and parts of Sulawezi for hundreds of years. Since the 1960s these plantations have been greatly expanded. Tropical pines have long been planted in Sumatra and throughout Indonesia vast areas are covered in smallholder managed agroforests. In the past two decades there has been a dramatic increase in establishment of fast-growing tree plantations to supply large industrial pulp mills, mainly in Sumatra. There are also significant plantations of *A. mangium* in Kalimantan (the Indonesian part of Borneo) although many of them are poorly managed and some are now abandoned. In addition, large areas have been converted to oil palm plantations which provide fewer conservation benefits than less intensively managed plantation forests. Until recently biodiversity conservation measures have focussed almost exclusively on protected areas but the potential of set-asides within industrial plantations now receives much attention. One reason for this is the realisation that deforestation rates are higher in protected areas than in managed forests—at least in Kalimantan (Curran et al 2004; Meijaard et al. 2005; Meijaard and Sheil 2007). The vast majority of plantations now being established are devoted to *A. mangium* with smaller areas of *A. auriculiformis, Paraserianthes* sp., *Pinus* spp., and *Eucalyptus* spp. Little is known of the within-stand biodiversity value of these plantations but it is likely to be low (but see Nasi et al. 2008—this volume). Despite efforts to improve the protection of natural forest habitats in Indonesia, there are reports that much plantation establishment by conversion of secondary natural forests is ongoing (Cossalter and Pye-Smith 2003), with negative impacts on biodiversity. Indonesian law requires that industrial plantation operators allocate 30% of their concessions to retaining sample areas of natural forests within the plantation matrix. Some of these areas are said to still support populations of elephants and other wildlife of conservation concern (Nasi et al. 2008—this volume) although this is not always well documented. The set aside areas are rarely given proper protection and are often subject to illegal logging or even used to supply raw material for pulp mills. However, some of these areas are extensive and under proper management would undoubtedly make a significant contribution to biodiversity conservation (Zuidema et al. 1997). Recently environmental groups have been putting pressure on these companies to observe higher levels of corporate social and environmental responsibility and there appears to have been a greater effort to give more rigorous protection to these forest enclaves within the plantation estate.

Over 0.5 million ha of mostly natural forest is FSC certified but none of the large-scale industrial pulp plantations are as yet certified, although some of the companies have stated their intention of seeking certification. WWF has recently signed an agreement with a large pulp company in Sumatra to collaborate on measures to protect biodiversity in and around the companies' concession, including the maintenance of natural forest set-asides within the plantation estate. A recent initiative, the Grand Perfect plantations, in a large-scale pulp plantation in Sarawak (a state in the Malaysian part of Borneo) has made even greater investments in maintaining biodiversity both within the planted forest and in the set asides within the plantation estate (Cyranoski 2007). Grand Perfect is sponsoring research efforts which are beginning to provide evidence that the forest mosaic and the planted forests are supporting populations of important components of the lowland forest biodiversity of Borneo.

United Kingdom

The United Kingdom is an example of a country that has lost most of its natural forests but has comparatively large areas of planted forests. Perhaps as much as 80% of the United Kingdom was once covered in natural forests but after many centuries of deforestation due to demand for timber and agricultural land, few semi-natural forest remnants were left and at the start of the 20th century only 5% of the land area was covered with trees. An extensive afforestation program during the 20th century has increased the woodland cover to 12% (Mason 2007); much of this expansion was on marginal agricultural land (especially pasture) but some (particularly pre-1980) involved the conversion of semi-natural woodland. Plantations now contribute almost 70% of the total forest area (Table 1). Because of the scarcity of natural or semi-natural forest, plantations play an important part in the conservation of forest biodiversity in the United Kingdom (e.g., Humphrey et al. 2000), despite the fact that they consist mostly of exotic species (Table 1). Recently there have been considerable efforts to improve the value of plantation forests for biodiversity and other non-wood values, and to restore natural forests (Quine et al. 2004; Humphrey 2005). On sites where plantations of the 20th century replaced former native woodland (ancient woodland), there are now substantial efforts to restore the woodland cover to native habitats. In addition, where afforestation occurred in open habitats such as blanket and raised bogs that are particularly valued today, there are activities to restore these original habitats (Anderson 2001). In both cases, the survival of elements of the former vegetation (or propagules of it, e.g., Eycott et al. 2006), makes restoration a more attractive proposition than trying to recreate such habitats from neighboring sections converted to intensive agriculture. However, there are many plantations that were established on marginal agricultural land (often upland pasture) which had not held tree cover for hundreds of years. On these sites, the benefits of forest cover, and the qualities of the new habitats are increasingly being appreciated. Several rare species are now found within these plantations of exotic tree species (Humphrey et al. 2003). Much effort is being expended on diversifying the structure (across landscapes but also within stands), to provide some of the missing structural elements that are required by native biodiversity (Humphrey 2005; Quine et al 2007). Substantial parts of the plantation area of the UK are FSC certified (Bills 2001).

New Zealand

The biodiversity of New Zealand's forests is very rich for a temperate region and characterised by a high proportion of endemic species, due to its long isolation from other land masses. New Zealand has experienced extensive loss of native forests following the colonisation by Polynesians (about 1000 years ago) and Europeans (from about 150 years ago) but natural forests remain on over 20% of the total land area (Table 1). Plantation forests represent ca. 22% of total forest cover, and the Californian *P. radiata* (radiata or Monterey pine) is the principal tree species (Table 1), managed with rotations of about 27 years and clearfell harvesting. Historically, some plantations have replaced native forests, but today almost all new plantations are on land that was previously in pasture or 'degraded land'. The Forest Accord 1991, an agreement between plantation forest managers and NGOs in effect since 1991, ensures that plantation forests are not established at the expense of natural forests or in areas recommended for protection (see also Shaw 1997), but some conversion of regenerating shrubland and native grasslands has still occurred. Until recently conservation efforts focussed almost exclusively on the publicly owned and largely protected native forests and other natural areas, which are inhabited by a unique and mostly endemic biota. Today there is a growing awareness about the value of plantations as additional habitat for native biodiversity, including several threatened species that can occur in plantations (see above). Some plantations provide particularly valuable habitat in low-lying areas where losses of natural forest were most severe. Some of these areas are now being converted into agricultural land, causing further loss of forest habitat. Approximately 700,000 ha, about 40% of the plantation area, are being managed with FSC certification (Goulding 2006; Romona Anton, FSC, pers. comm., August 2007). Certification has led to widespread biodiversity surveys in plantations, improved management of plantations and embedded remnants of natural vegetation, and it improved the general awareness of biodiversity issues among forest managers (Hock and Hay 2003; Goulding 2006). There is no PEFC certification in New Zealand (PEFC 2007).

China

China has the largest area of plantation forest of all countries, and these plantations consist mostly of native tree species (Table 1). Currently, there are massive ongoing afforestation programs with new plantings between 2000 and 2005 amounting to nearly 1.5 million ha per year, the most of any country (FAO 2006a). These programs were initiated to mitigate environmental problems resulting from the substantial loss and degradation of China's forests, in addition to increasing efforts in forest conservation and restoration of degraded forest ecosystems (Wenhua 2004). With the growth of China's population it was difficult to meet demands for wood and wood products, and this caused the overexploitation of forests and losses of biodiversity, particularly in those densely populated regions where much forest had already been lost (Wenhua 2004). The consequences for biodiversity of this forest loss are severe because China has a very rich biota; for example, there are 27,000 species of higher plants including more than 7,000 woody species. In a global assessment of biodiversity hotspots that are rich in endemic species and where threats to biodiversity are important (Myers et al. 2000), several Chinese regions were identified, along with several other regions in most of the countries covered in the case studies here. The Chinese Government has embarked on a plan to conserve biodiversity and to establish new nature reserves (Wenhua 2004). Although mixed forests are being encouraged, it appears that there are only limited efforts to integrate the expanding plantation forest estate into these biodiversity conservation activities. However, even though afforestation programmes focus on timber production and environmental benefits such as soil and water protection, several studies have shown that it also enhances the restoration of forest biodiversity (e.g., Fang and Peng 1997). Until now there has been limited uptake of FSC certification in China with only six certificates covering a total of about 0.75 million ha of plantations and other forest types (Romona Anton, FSC, pers. comm., August 2007). There is no PEFC certification of forests in China (PEFC 2007).

United States of America

The United States have the fourth largest total forest area and the second largest area of production plantation forest worldwide (FAO 2006a). Most of the plantation forests in the US consist of native tree species (Table 1). The principal areas are the intensively managed plantations of native loblolly pine (*Pinus taeda*) and slash pine (*P. elliotii*) in the south-east. These are the result of intensified management and improvement (particularly since the 1950s) of degraded natural stands that had occupied the region prior to extensive logging and large-scale conversion for agriculture (Stanturf et al. 2003). All forests in the U.S. are subject to national and state environmental regulations and the Endangered Species Act which effectively prohibit management actions that threaten listed species and their habitats. Biodiversity conservation is an important objective in the management of natural and semi-natural forests in the U.S. but less so in plantation forests. In the northwestern U.S., biodiversity conservation issues have been more prominent, and the efforts to protect spotted owls have strongly influenced forestry policy (Lindenmayer and Franklin 2002). The forestry debate has at times been polarized, whereby 'timber or biodiversity' were considered mutually exclusive. However, a compromise was reached with the Northwest Forest Plan, which set out areas assigned for forestry and others for conservation, and there is an ongoing debate about how the management of forests can be improved (e.g., Suzuki and Olson 2008—this issue). There has been considerable uptake of FSC certification in the U.S., and retail policy of some do-it-yourself chains appears to have contributed substantially to the demand for FSC-certified forest products (Hock and Hay 2003). The total area of forests with FSC-certified management exceeds 9.1 million ha although the majority of this area comprises natural forests and little plantation forest (Romona Anton, FSC, pers. comm., August 2007). Certification by Sustainable Forestry Initiative (SFI), a member system of PEFC (2007), is more widespread in the U.S. than FSC certification, and some forests are certified under both systems. In 2007, there were nearly 22 million ha of SFI-certified forests in the U.S. which included substantial areas of plantation forest (SFI 2007).

France

With forests covering over 15 million ha, France is one of the most forested countries in Europe. France has forest plantations covering 2 million ha (ranking 9th in the world in terms of plantation area) that consist mostly of native tree species and contribute ca. 13% of the total forest area (Table 1). The Landes Forest represents the largest continuous plantation forest in Europe with ca. 1 million ha of maritime pine (*Pinus pinaster*). This resulted from an afforestation program that was launched by Napoleon III in the 19th century to develop the economy of the Landes region, at that time a moorland with some deciduous trees. Seeds from the natural pine forests of the adjacent coastal area were sowed in drained soils. Although these stands have some of the lowest tree species diversity in France (Ministry of Agriculture and Fishery 2005), there are embedded semi-natural riparian forests and remnants of broadleaved forest of high biodiversity value (Barbaro et al. 2005; 2008—this issue; Van Halder et al. 2008—this issue). Contrary to some of the neighbouring countries, there is little uptake of FSC certification in France (currently only about 15,500 ha; Romona Anton, FSC, pers. comm., August 2007) but PEFC is gaining importance in plantation forests (currently about 4.3 million ha, PEFC 2007). PEFC certification of some plantation forests has led to efforts to improve both conservation and restoration of deciduous patches and hedgerows within the pine plantation matrix.

Enhancing biodiversity in plantation forests

As discussed above, many factors influence biodiversity in plantation forests and the landscapes in which they occur, and all these offer opportunities to improve forest management for the benefit of biodiversity. In a recent working paper on 'voluntary guidelines for the responsible management of planted forests', FAO (2006b) provides a comprehensive bibliography on plantation issues and some general recommendations for planners and managers to conserve biodiversity. These include "preparing baseline studies to monitor the impact of planted forest management on the maintenance of plants and animals and the conservation of genetic resources." The use of indicator species and other biodiversity indicators has been advocated for this purpose (e.g., Larsson 2001) because the assessment of a wide range of taxa is often too time consuming and expensive. In this issue a comprehensive set of such indicators is being proposed for use with plantation forest management in Ireland (Smith et al. 2008—this issue). However, such indicators have their limitations, and they have been criticised as being potentially too simplistic (e.g., Lindenmayer and Franklin 2002). Furthermore, results often vary among taxa (Barlow et al. 2007), and hence, the effects of stand and landscape-level management should ideally be examined with a wide range of taxa, if resources permit.

Management actions can broadly be divided between those that are concerned with stand-level management and those that are concerned with the spatial and landscape aspects of the entire plantation and its surroundings. Many of the following recommendations are also reflected in the criteria used for Forest Stewardship Certification of forest management.

Stand-level recommendations

A recent summary of recommendations for management at the stand level has been given by Hartley (2002). Biodiversity can be enhanced through appropriate management choices regarding composition and structure. The first approach is to consider the tree species that are being planted (e.g., du Bus de Warnaffe and Deconchat 2008—this issue). Several studies have shown that the establishment of a greater diversity of tree species will increase the range of habitat types available for native species (Lamb 1998; Norton 1998; Hartley 2002 and references therein). Planting a larger number of tree species will result in a greater diversity of habitats and thus of dependent species (Spellerberg and Sayer 1996). Moreover, mixed plantation being more resistant and resilient to natural and human disturbances (Scherer-Lorenzen et al. 2005; Jactel and Brockerhoff 2007) may provide a more stable environment for native species. Careful selection of species for these plantings could considerably improve habitat for native species, particularly if they provide food resources such as nectar and fruit and help to create understorey microclimate, soil conditions and stand structures that would favor native species (Parrotta et al. 1997a; Hartley 2002; Lindenmayer and Franklin 2002; Carnus et al. 2006). Although native species are more likely to meet these criteria, some exotics can fulfil the same role.

The amount and quality of available habitats can be influenced by a variety of stand management practices (Decocq et al. 2005; Quine et al. 2007). If possible, intensive site preparation should be avoided if the previous land cover has conservation value as it may destroy herbaceous vegetation and coarse woody debris which provide resources for many native forest species (Hartley 2002; Lindenmayer and Franklin 2002; Carnus et al. 2006). Similarly, wider tree spacing at plantation establishment and heavy pre-commercial thinning may help to maintain understorey vegetation (Moore and Allen 1999; Hartley 2002; Lindenmayer and Hobbs 2004). The age at which plantations are harvested is also often seen as a key issue for native biodiversity (Lindenmayer and Hobbs 2004). Native biodiversity is often greatest in the oldest stands, although conservation value is not always correlated with stand age. However, the trend of decreasing rotation length in many plantation areas (e.g., in New Zealand radiata pine from 40–50 years in the 1970s to 25–30 years in the 1990s) is usually a concern. An increase in rotation length has been widely advocated as a means to enhance native biodiversity in plantations (Rosoman 1994; Humphrey 2005); however, this is usually considered uneconomical because financial profitability begins to fall above a certain stand age or because of increasing environmental risks (such as wind damage) with increasing stand age. But as Peterken et al. (1992) suggested for British plantations, there can be a trade-off for increasing rotation lengths in some areas by reducing rotation lengths in other areas thus maintaining financial returns from the forest. The use of single-tree, group selection or small-coupe harvesting will result in the continued presence of mature forest at a site and this has been suggested as beneficial for native biodiversity and provides a useful alternative to the traditional forest clear-cut. Maintaining some stand structural attributes such as old trees or snags within stands will also enhance the value of plantations for biodiversity (Lindenmayer and Franklin 2002; Humphrey et al. 2006). Various stand-level management recommendations based on thinning, weed control, burning and other methods are given by Cummings and Reid (2008—this issue). These were aimed mainly at the restoration of plantations to a more natural vegetation, but many of their findings are also useful to improve the biodiversity value of production plantations.

Landscape level recommendations

General guidelines and management recommendations to increase the value of plantation forests for biodiversity have in the past focussed mostly at the stand-level. Less attention has been given to management issues at the landscape level (but see Wigley and Roberts 1997; Lamb 1998; Norton 1998; Humphrey et al. 2000; Lindenmayer and Franklin 2002). There is increasing evidence that the complexity of the landscape matrix is of great importance in maintaining biodiversity at the landscape level and that plantations forests can contribute to this complexity (Norton 1998; Lindenmayer and Franklin 2002; Fischer et al. 2006; Kupfer et al. 2006; Barbaro et al 2007).

The *structural complexity* throughout the landscape may be enhanced by juxtaposition of different plantations types, size and shapes which will in turn increase the probability of providing suitable alternative habitats to native forest species (Lamb 1998; Lamb et al. 2005). Landscape-level biodiversity issues can also be addressed by considering the spatial arrangement of different-aged plantation stands with respect to other landscape components, especially native forest remnants. Plantations in fragmented landscapes can contribute to the connectivity of native remnants, particularly when they provide corridors or stepping stones for forest specialist species (Norton 1998; Fischer et al. 2006; Nasi et al. 2008—this issue). They may also be placed side by side with native remnants to buffer adverse edge effects (Harper et al. 2005). For example, special-purpose exotic or native plantations are likely to be more beneficial when located adjacent to native forest remnants than when located distant from them (Lamb et al. 1997; Parrotta et al. 1997a; Lindenmayer and Hobbs 2004). Many forestry companies make considerable use of amenity plantings, for example along roads and around recreational amenities.

Changes in the *spatial and temporal pattern of plantation forests harvesting* offer another avenue to improve biodiversity conservation in plantation-dominated landscapes (Lamb et al. 1997; Lindenmayer and Franklin 2002; Carnus et al. 2006, see also Suzuki and Olson 2008—this issue). In Australia, Lindenmayer and Pope (2000) suggested that some advanced regrowth radiata pine plantations in the matrix should always link eucalyptus remnants to maintain connectivity for native birds. Rotational harvesting, where a core old growth remnant is surrounded by a series of managed stands that have a sufficiently long gap between harvesting to ensure that at any one time the old forest remnant is surrounded by a large proportion of mature forest, has been advocated for managing old growth Pacific northwest forests of North America (Harris 1984). A similar system has been proposed for managing upland conifer plantations in Britain (Peterken et al. 1992) involving assigning 15–20% of the plantation to long rotations surrounding permanently uncut cores. In New Zealand, Norton (1998) has suggested that a similar approach could be used for managing plantation forests around indigenous forest remnants or between remnants. For example, the native biodiversity values of plantations would be enhanced by ensuring that there is always a large area of mature forest present adjacent to the remnant, and that a continuous sequence of older plantation stands occurred between remnants (e.g., Nasi et al. 2008—this issue). Variable retention harvesting has been advocated to mitigate detrimental impacts of clear-cutting on biodiversity in large harvested areas. Residual tree patches can function as valuable refugia, at least in the short-term, for frogs (Chan-McLeod and Moy 2007), spiders and carabids (Hyvarinen et al. 2005; Matveinen-Huju et al. 2006) and birds (Vergara and Schlatter 2006). The beneficial effects of green-tree retention are expected to increase with patch size (Chan-McLeod and Moy 2007) and decrease with distance from undisturbed areas (Deans et al. 2005; Vergara and Schlatter 2006).

Plantation forests and certification

Several aspects of the current debate about the eligibility of plantation forests to be certified under the FSC involve biodiversity issues. Many of the concerns expressed by environmental NGOs are valid in some situations but generalisations about the impact of plantation forests on biodiversity are not doing justice to this complex issue (see above). It is also difficult to draw a clear line between a plantation and an intensively managed natural forest. For example, the forests covering much of Europe have all been intensively managed for centuries, many of the trees have been established by planting or sowing in pure stands (although they often became more diverse by natural succession). Should these forests therefore really be classed alongside the old growth forests of the tropics or should they be lumped together with the plantations? The different stakeholders in the FSC have been struggling with these issues for several years and do appear to be reaching a compromise that would allow certification of some plantations but under very strict conditions (Forest Stewardship Council 2006). FAO has also contributed to this debate and have recently published a voluntary code of conduct which gives a comprehensive and balanced view of the issues surrounding plantation establishment (FAO 2006b, see above).

Another complicating factor is that much criticism that has been expressed against plantation forests is concerned about social impacts of plantation forests, which are beyond the scope of the present paper. However, there are forestry companies that take longer term views of sustainability, including social sustainability (Porter and Kramer 2006) concerning the people living in the areas where these companies operate. The challenge for environmental groups is to distinguish between companies that are performing better, those that are just window dressing, their operations and those that will grab short-term profits and move on. Because certification leads to increased scrutiny of forest management, it can be expected that there are benefits for social aspects and for biodiversity conservation within plantation forests. Conversely, without certification from organisations such as FSC, there would be fewer incentives to address such concerns in the management of plantation forests. Special problems may occur in situations where plantations are established in countries with weak institutions or where corruption is widespread. In any case, blanket disapproval of plantation forests appears inappropriate given the wide range of issues and context-specific impacts of plantation forests with regard to biodiversity and other criteria.

Conclusions

Plantations can make an important contribution to the conservation of native biodiversity, but not if their establishment involves the replacement of native natural or semi-natural ecosystems—should they do so, there will indeed be a contradiction (oxymoron) in the juxtaposition of the terms plantation and biodiversity. While a plantation stand will usually support fewer native species than a native forest at the same site, plantations are increasingly replacing other human-modified ecosystems (e.g., degraded pasture) and will almost always support a greater diversity of native species. As such, plantations can play an important role in sustaining native biodiversity in production landscapes—and indeed be an opportunity for biodiversity. As well as providing habitat in their own right, plantations play particularly important roles in buffering native forest remnants and in enhancing connectivity between areas of native ecosystems, including patches of primary forests, riparian strips, and amenity plantings.

The opportunities afforded by plantations can be realised when particular attention to biodiversity informs management choices, and the objectives become multi-purpose (sustainable forest management). So, to sustain native biodiversity within plantations forest managers need to consider using a greater diversity of planted species, extending rotation lengths in some stands, and adopting a variety of harvesting approaches. Managers also have to consider plantations from a landscape perspective and the contribution that can be made by planning the spatial array of individual stands or compartments of different age and species composition as well as natural or semi-natural conservation areas. Although our understanding of such approaches is improving, there is still a need for further research on the specific requirements for the protection of biodiversity in regions that are not yet well studied. Another question that has not yet been adequately addressed is whether plantation forests composed of locally occurring native tree species are in fact providing better habitat for biodiversity than plantations of exotic tree species, and if so, how the use of native trees in plantations could be encouraged.

Tensions remain between the objectives of biodiversity conservation and plantation productivity (Lindenmayer and Hobbs 2004). The goal of higher ecosystem complexity may conflict with current trends in forest management towards increasing intensification and simplification; this is another area that requires more research. Furthermore local people, particularly in developing countries, may view biodiversity conservation as a luxury as they struggle to meet their basic food and fuel needs. Trade-offs between biodiversity conservation and improvement in human well-being are probably easier to achieve at the landscape scale where a spatial partition of forest objectives can be made, for instance by the juxtaposition of natural reserves and a productive matrix (Lamb et al. 2005). Exploration of different harvesting scenarios can be used to identify harvesting plans that provide improved biodiversity outcomes without unduly affecting economic objectives. In North America, spatial modelling tools have been used to optimise timber harvesting in native forests to meet biodiversity conservation goals, including "adjacency requirements" (Bettinger et al. 1997; Snyder and ReVelle 1997; Van Deusen 2001). Similar modelling could be used to maximise timber production and biodiversity conservation as well as ecosystem stability. The key feature of this approach is that it considers biodiversity conservation at the landscape scale rather than at the stand scale and thus removes the direct conflict between biodiversity conservation and timber production at any individual site. Thus, we suggest that the role of plantations in biodiversity conservation can be enhanced if plantations are managed in a manner in which they can contribute to biodiversity conservation across the whole landscape, rather than focusing only on the values within the plantations themselves.

Acknowledgments We would like to thank all authors for their contributions to this special issue and the three conferences or conference sessions that spawned this initiative. Many thanks also to David Hawksworth (Editor of Biodiversity and Conservation), Raphael Didham (Associate Editor), and all the staff in the Springer journal office who provided invaluable support with this special issue. This article was partly funded by the New Zealand Foundation for Research Science and Technology (contract C04X0214).

References

Allan DG, Harrison JA, Navarro RA, van Wilgen BW, Thompson MW (1997) The impact of commercial afforestation on bird populations in Mpumalanga Province, South Africa—insights from bird-atlas data. Biol Conserv 79:173–185. doi:10.1016/S0006-3207(96)00098-5

Allen RB, Platt KH, Coker REJ (1995) Understorey species composition patterns in a *Pinus radiata* D. Don plantation on the central North Island volcanic plateau, New Zealand. N Z J For Sci 25:301–317

Anderson AR (2001) Deforesting and restoring peat bogs: a review. Technical paper 32, Forestry Commission, Edinburgh, UK

Armstrong AJ, van Hensbergen HJ (1996) Impacts of afforestation with pines on assemblages of native biota in South Africa. South Afr For J 175:35–42

Barbaro L, Pontcharraud L, Vetillard F, Guyon D, Jactel H (2005) Comparative responses of bird, carabid, and spider assemblages to stand and landscape diversity in maritime pine plantation forests. Ecoscience 12:110–121. doi:10.2980/i1195-6860-12-1-110.1

Barbaro L, Rossi JP, Vetillard F, Nezan J, Jactel H (2007) The spatial distribution of birds and carabid beetles in pine plantation forests: the role of landscape composition and structure. J Biogeog 34:652–664

Barbaro L, Couzi L, Bretagnolle V, Nezan J, Vetillard F (2008) Landscape complementation for breeding and foraging in the declining Eurasian hoopoe (*Upupa epops*). Biodivers Conserv (this issue)

Barlow J, Gardner TA, Araujo IS, Bonaldo AB, Costa JE, Esposito MC, Ferreira LV, Hawes J, Hernandez MIM, Leite RN, Lo-Man-Hung NF, Malcolm JR, Martins MB, Mestre LAM, Nunes-Gutjahr AL, Overal WL, Parry L, Peters SL, Ribeiro-Junior MA, da Silva Motta C, da Silva MNF, Peres CA (2007) Quantifying the biodiversity value of tropical primary, secondary and plantation forests. Proc Natl Acad Sci USA 104:18555–18560. doi:10.1073/pnas.0703333104

Barlow J, Araujo IS, Overal WL, Gardner TA, Da Silva Mendes F, Lake I, Peres CA (2008) Diversity and composition of fruit-feeding butterflies in tropical *Eucalyptus* plantations. Biodivers Conserv. doi:10.1007/s10531-007-9240-0

Benedick H, Hill K, Mustaffa N, Chey VK, Maryati M, Searle B, Schilthuizen M, Hamer KC (2006) Impacts of rain forest fragmentation on butterflies in northern Borneo: species richness, turnover and the value of small fragments. J Appl Ecol 43:967–977. doi:10.1111/j.1365-2664.2006.01209.x

Berndt L, Brockerhoff E, Jactel H (2008) Can exotic plantation forests provide surrogate habitat for carabid beetles where native forest is rare? Biodivers Conserv (this issue)

Bernhard-Reversat F (ed) (2001) Effect of exotic tree plantations on plant diversity and biological soil fertility in the Congo savanna: with special reference to eucalypts. Center for International Forestry Research, Bogor, p 71

Bettinger P, Sessions J, Boston K (1997) Using Tabu search to schedule timber harvests subject to spatial wildlife goals for big game. Ecol Model 94:111–123. doi:10.1016/S0304-3800(96)00007-5

Bills D (2001) The UK Government and certification. Int For Rev 3:323–326

Brockerhoff EG, Ecroyd CE, Langer ER (2001) Biodiversity in New Zealand plantation forests: policy trends, incentives, and the state of our knowledge. N Z J For 46:31–37

Brockerhoff EG, Ecroyd CE, Leckie AC, Kimberley MO (2003) Diversity and succession of adventive and indigenous vascular understorey plants in *Pinus radiata* plantation forests in New Zealand. For Ecol Manage 185:307–326

Brockerhoff EG, Berndt LA, Jactel H (2005) Role of exotic pine forests in the conservation of the critically endangered ground beetle *Holcaspis brevicula* (Coleoptera: Carabidae). N Z J Ecol 29:37–43

Brook BW, Sodhi NS, Ng PKL (2003) Catastrophic extinctions follow deforestation in Singapore. Nature 424:420–423. doi:10.1038/nature01795

Buscardo E, Smith GF, Kelly DL, Freitas H, Iremomger S, Mitchell F, O'Donoghue S, Mckee A-M (2008) The early effects of afforestation on biodiversity of grasslands in Ireland. Biodivers Conserv. doi:10.1007/s10531-007-9275-2
Carle J, Vuorinen P, Del Lungo A (2002) Status and trends in global forest plantation development. For Prod J 52(7/8):12–23
Carnevale NJ, Montagnini F (2002) Facilitating regeneration of secondary forests with the use of mixed and pure plantations of indigenous tree species. For Ecol Manage 163:217–227
Carnus J-M, Parrotta J, Brockerhoff EG, Arbez M, Jactel H, Kremer A, Lamb D, O'Hara K, Walters B (2006) Planted forests and biodiversity. J For 104(2):65–77
Carrere R, Lohmann L (1996) Pulping the south: industrial tree plantations and the world paper economy. WRM and Zed Books Ltd, London
Chan-McLeod ACA, Moy A (2007) Evaluating residual tree patches as stepping stones and short-term refugia for red-legged frogs. J Wildlife Manage 71:1836–1844. doi:10.2193/2006-309
Clapp RA (2001) Tree farming and forest conservation in Chile: do replacement forests leave any originals behind? Soc Nat Resour 14:341–356
Clout MN, Gaze PD (1984) Effects of plantation forestry on birds in New Zealand. J Appl Ecol 21:795–815. doi:10.2307/2405048
Coote L, Kelly DL, Dowding P, Smith GF (2008) Epiphytes of Sitka spruce (*Picea sitchensis*) plantations in Ireland and the effects of open spaces. Biodivers Conserv. doi:10.1007/s10531-007-9302-3
Cossalter C, Pye-Smith C (2003) Fast-wood forestry—myths and realities. Center for International Forestry Research, Jakarta
Cummings J, Reid N (2008) Stand-level management of plantations to improve biodiversity values. Biodivers Conserv. doi:10.1007/s10531-008-9362-z
Curran LM, Caniago I, Paoli GD, Astianti D, Kusneti M, Leighton M, Nirarita CE, Haeruman H (1999) Impact of El Nino and logging on canopy tree recruitment in Borneo. Science 286:2184–2188. doi:10.1126/science.286.5447.2184
Curran LM, Trigg SN, McDonald AK, Astiani D, Hardiono YM, Siregar P, Caniago I, Kasischke E (2004) Lowland forest loss in protected areas of Indonesian Borneo. Science 303:1000–1003. doi:10.1126/science.1091714
Cyranoski D (2007) Logging: the new conservation. Nature 446:608–610. doi:10.1038/446608a
Deans AM, Malcolm JR, Smith SM, Bellocq MI (2005) Edge effects and the responses of aerial insect assemblages to structural-retention harvesting in Canadian boreal peatland forests. For Ecol Manage 204:249–266
Decocq G, Aubert M, Dupont F, Bardat J, Wattez-Franger A, Saguez R, De Foucault B, Alard D, Delelis-Dusoliier A (2005) Silviculture-driven vegetation change in a European temperate deciduous forest. Ann For Sci 62:313–323. doi:10.1051/forest:2005026
Denyer K, Burns B, Ogden J (2006) Buffering of native forest edge microclimate by adjoining tree plantations. Austral Ecol 31:478–489. doi:10.1111/j.1442-9993.2006.01609.x
Didham RK, Hammond PM, Lawton JH, Eggleton P, Stork NE (1998) Beetle species responses to tropical forest fragmentation. Ecol Monogr 68:295–323
Donald PF, Fuller RJ, Evans AD, Gough SJ (1998) Effects of forest management and grazing on breeding bird communities in plantations of broadleaved and coniferous trees in western England. Biol Conserv 85:183–197. doi:10.1016/S0006-3207(97)00114-6
Du Bus de Warnaffe G, Deconchat M (2008) Impact of four silvicultural systems on birds in the Belgian Ardenne: implications for plantation management. Biodivers Conserv. doi:10.1007/s10531-008-9364-x
Ecroyd CE, Brockerhoff EG (2005) Floristic changes over 30 years in a Canterbury Plains kanuka forest remnant, and comparison with adjacent vegetation types. N Z J Ecol 29:279–290
Ewers RM, Didham RK (2006) Confounding factors in the detection of species responses to habitat fragmentation. Biol Rev 81:117–142. doi:10.1017/S1464793105006949
Ewers RM, Didham RK (2007) The effect of fragment shape and species' sensitivity to habitat edges on animal population size. Conserv Biol 21:926–936. doi:10.1111/j.1523-1739.2007.00720.x
Eycott AE, Watkinson AR, Dolman PM (2006) Ecological patterns of plant diversity in a plantation forest managed by clearfelling. J Appl Ecol 43:1160–1171. doi:10.1111/j.1365-2664.2006.01235.x
Fahy O, Gormally M (1998) A comparison of plant and carabid beetle communities in an Irish oak woodland with a nearby conifer plantation and clearfelled site. For Ecol Manage 110:263–273
Fang W, Peng SL (1997) Development of species diversity in the restoration process of establishing a tropical man-made forest ecosystem in China. For Ecol Manage 99:185–196
Fahrig L (2001) How much habitat is enough? Biol Conserv 100:65–74. doi:10.1016/S0006-3207(00)00208-1
FAO (2001) Global Forest Resources Assessment 2000. Main report. FAO Forestry Paper 140, Food and Agriculture Organization of the United Nations, Rome, Italy

FAO (2006a) Global Forest Resources Assessment 2005—Progress towards sustainable forest management. FAO Forestry Paper 147, Food and Agriculture Organization of the United Nations, Rome, Italy
FAO (2006b) Planted Forests and Trees Working Paper FP37E. Responsible Management of Planted Forests, Voluntary Guidelines. FAO, Rome, Italy
FAO (2007) The State of the World's Forests. ftp.fao.org/docrep/fao/009. FAO, Rome, Italy
FERN (2007) www.fern.org/publications. Accessed 22 April 2007
Ferreras P (2001) Landscape structure and asymmetrical inter-patch connectivity in a metapopulation of the endangered Iberian lynx. Biol Conserv 100:125–136. doi:10.1016/S0006-3207(00)00213-5
Fischer J, Lindenmayer DB (2006) Beyond fragmentation: the continuum model for fauna research and conservation in human-modified landscapes. Oikos 112:473–480. doi:10.1111/j.0030-1299.2006.14148.x
Fischer J, Lindenmayer DB, Manning AD (2006) Biodiversity, ecosystem function, and resilience: ten guiding principles for commodity production landscapes. Front Ecol Environ 4:80–86. doi:10.1890/1540-9295(2006)004[0080:BEFART]2.0.CO;2
Forman RTT (1995) Land mosaics: the ecology of landscapes and regions. Cambridge University Press, Cambridge
Forest Stewardship Council (2004) Perspectives on plantations. A review of the issues facing plantation management. Background paper to the FSC plantations review. FSC, Bonn, Germany. www.fsc.org/plantations/content/en/home.htm. Accessed 19 June 2007
Forest Stewardship Council (2006) The FSC plantations review policy working group final report. FSC, Bonn, Germany. www.fsc.org/plantations/content/en/home.htm. Accessed 19 June 2007
Forest Stewardship Council (2007a) FSC principles and criteria of forest stewardship. http://www.fsc.org/en/about/policy_standards/princ_criteria. Accessed 19 June 2007
Forest Stewardship Council (2007b) FSC and Plantations: the link with responsible forest management. FSC, Bonn, Germany.www.fscus.org . Accessed 22 April 2007
Forest Stewardship Council (2007c) Plantations review. FSC, Bonn, Germany. www.fsc.org/plantations/content/en/home.htm. Accessed 19 June 2007
Fuller RJ, Smith KW, Grice PV, Currie FA, Quine CP (2007) Habitat change and woodland birds in Britain: implications for management and future research. Ibis 149(Suppl. 2):261–268. doi:10.1111/j.1474-919X.2007.00775.x
Gascon C, Lovejoy TE, Bierregaard RO Jr, Malcolm JR, Stouffer PC, Vasconcelos HL, Laurance WF, Zimmerman B, Tocher M, Borges S (1999) Matrix habitat and species richness in tropical forest remnants. Biol Conserv 91:223–229. doi:10.1016/S0006-3207(99)00080-4
Geldenhuys CJ (1997) Native forest regeneration in pine and eucalypt plantations in Northern Province, South Africa. For Ecol Manage 99:101–115
Goulding CJ (2006) Forest Stewardship Council certification of industrial plantation forests. Allg Forst-Jagdzeitung 177(3–4):42–47
Gullison RE (2003) Does forest certification conserve biodiversity? Oryx 37(2):153–165. doi:10.1017/S0030605303000346
Hampson AM, Peterken GF (1998) Enhancing the biodiversity of Scotland's forest resource through the development of a network of forest habitats. Biodivers Conserv 7:179–192. doi:10.1023/A:1008832401731
Harper KA, Macdonald SE, Burton PJ, Chen J, Brosofske KD, Saunders SC, Euskirchen ES, Roberts D, Jaiteh MS, Esseen P-A (2005) Edge influence on forest structure and composition in fragmented landscapes. Conserv Biol 19:768–782. doi:10.1111/j.1523-1739.2005.00045.x
Harris LD (1984) The fragmented forest: Island biogeography theory and the preservation of biotic diversity. University of Chicago Press, Chicago
Hartley MJ (2002) Rationale and methods for conserving biodiversity in plantation forests. For Ecol Manage 155:81–95
Henle K, Davies KF, Kleyer M, Margules C, Settele J (2004) Predictors of species sensitivity to fragmentation. Biodivers Conserv 13:207–251
Hock B, Hay E (2003) Forest certification in New Zealand: how are we going? N Z J For 47:17–23
Humphrey JW (2005) Benefits to biodiversity from developing old-growth conditions in British upland spruce plantations: a review and recommendations. Forestry 78:33–53. doi:10.1093/forestry/cpi004
Humphrey JW, Newton AC, Peace AJ, Holden E (2000) The importance of conifer plantations in northern Britain as a habitat for native fungi. Biol Conserv 96:241–252. doi:10.1016/S0006-3207(00)00077-X
Humphrey JW, Ferris R, Quine CP (eds) (2003) Biodiversity in Britain's planted forests: results from the Forestry Commission's Biodiversity assessment project. Forestry Commission, Edinburgh
Humphrey JW, Quine CP, Watts K (2006) The influence of forest and woodland management on biodiversity in Scotland: recent findings and future prospects. In: Davison R, Galbraith CA (eds) Farming, forestry

and the natural heritage: towards a more integrated approach. Scottish Natural Heritage, Edinburgh, pp 59–75

Hunter M (1990) Wildlife, forest and forestry: principles of managing for biological diversity. Prentice Hill, Engelwood Cliffs, New Jersey

Hyvarinen E, Kouki J, Martikainen P, Lappalainen H (2005) Short-term effects of controlled burning and green-tree retention on beetle (Coleoptera) assemblages in managed boreal forests. For Ecol Manage 212:315–332

IBGE (2007) Instituto Brasileiro de Geographia e Estatistica. http://www.ibge.gov.br. Accessed 18 July 2007

Indonesian Ministry of Forestry (2007) Program Kehutanan, Nasional, 2005–2025. http://www.dephut.go.id/content.php?id=229. Accessed 15 August 2007

Innes J, Calder B, Williams D (1991) Native meets exotic—kokako and pine forest. What's new in forest research, no. 209. Forest Research Institute, Rotorua

Jactel H, Brockerhoff EG (2007) Tree diversity reduces herbivory by forest insects. Ecol Lett 10:835–848. doi:10.1111/j.1461-0248.2007.01073.x

Jukes MR, Peace AJ, Ferris R (2001) Carabid beetle communities associated with coniferous plantations in Britain: the influence of site, ground vegetation and stand structure. For Ecol Manage 148:271–286

Kanowski J, Catterall CP, Wardell-Johnson GW (2005) Consequences of broadscale timber plantations for biodiversity in cleared rainforest landscapes of tropical and subtropical Australia. For Ecol Manage 208:359–372

Keenan R, Lamb D, Woldring O, Irvine T, Jensen R (1997) Restoration of plant biodiversity beneath tropical tree plantations in Northern Australia. For Ecol Manage 99:117–131

Kleinpaste R (1990) Kiwis in a pine forest habitat. In: Fuller E (ed) Kiwis, a monograph of the family Apterygidae. SeTo Publishing, Auckland, pp 97–138

Kupfer JA, Malanson GP, Franklin SB (2006) Not seeing the ocean for the islands: the mediating influence of matrix-based processes on forest fragmentation processes. Glob Ecol Biogeogr 15:8–20. doi:10.1111/j.1466-822X.2006.00204.x

Lamb D (1998) Large-scale ecological restoration of degraded tropical forest lands: the potential role of timber plantations. Restor Ecol 6:271–279. doi:10.1046/j.1526-100X.1998.00632.x

Lamb D, Parrotta JA, Keenan R, Tucker NIJ (1997) Rejoining habitat remnants: restoration of degraded tropical landscapes. In: Laurance WF, Bierregaard RO Jr (eds) Tropical forest remnants: ecology, management and conservation of fragmented communities. University of Chicago Press, Chicago, pp 366–385

Lamb D, Erskine P, Parrotta JA (2005) Restoration of degraded tropical forest landscapes. Science 310:1628–1632. doi:10.1126/science.1111773

Lantschner MV, Rusch V, Peyrou C (2008) Bird assemblages in pine plantations replacing native ecosystems of N.W. Patagonia, Argentina. Biodivers Conserv. doi:10.1007/s10531-007-9243-x

Larsson T-B (ed) (2001) Biodiversity evaluation tools for European forests. Ecological Bulletins 50, Blackwell Science, Oxford

Laurance WF (2007) Have we overstated the tropical biodiversity crisis? Trends Ecol Evol 22:65–70. doi:10.1016/j.tree.2006.09.014

Ledgard N (2001) The spread of lodgepole pine (*Pinus contorta* Dougl.) in New Zealand. For Ecol Manage 141:43–57

Lindenmayer DB, Franklin JF (2002) Conserving forest biodiversity: a comprehensive multiscaled approach. Island Press, Washington

Lindenmayer DB, Hobbs RJ (2004) Fauna conservation in Australian plantation forests—a review. Biol Conserv 119:151–168. doi:10.1016/j.biocon.2003.10.028

Lindenmayer DB, Pope ML (2000) The design of exotic softwood plantations to enhance wildlife conservation: preliminary lessons from the Tumut fragmentation experiment. In: Craig J, Mitchell N, Saunders DA (eds) Nature conservation. 5: Nature conservation in production environments: managing the matrix. Surrey Beatty and Sons, Sydney, pp 44–49

Lindenmayer DB, Cunningham RB, Pope ML (1999) A large-scale "experiment" to examine the effects of landscape context and habitat fragmentation on mammals. Biol Conserv 88:387–403. doi:10.1016/S0006-3207(98)00111-6

Lopez G, Moro MJ (1997) Birds of Aleppo Pine plantations in south-east Spain in relation to vegetation composition and structure. J Appl Ecol 34:1257–1272. doi:10.2307/2405236

Loumeto JJ, Huttel C (1997) Understory vegetation in fast-growing tree plantations on savanna soils in Congo. For Ecol Manage 99:65–81

Magura T (2002) Carabids and forest edge: spatial pattern and edge effect. For Ecol Manage 157:23–37

Magura T, Tothmeresz B, Bordan Z (2000) Effects of nature management practice on carabid assemblages (Coleoptera: Carabidae) in a non-native plantation. Biol Conserv 93:95–102. doi:10.1016/S0006-3207(99)00073-7

Matveinen-Huju K, Niemela J, Rita H, O'Hara RB (2006) Retention-tree groups in clear-cuts: do they constitute 'life-boats' for spiders and carabids? For Ecol Manage 230:119–135

Mason WL (2007) Changes in the management of British forests between 1945 and 2000 and possible future trends. Ibis 149(Suppl. 2):41–52. doi:10.1111/j.1474-919X.2007.00696.x

Meijaard E, Sheil D, Nasi R, Augeri D, Rosenbaum B, Iskandar D, Setyawati T, Lammertink M, Rachmatika I, Wong A, Soehartono T, Stanley S, O'Brien T (2005) Life after logging. Reconciling wildlife conservation and production forestry in Indonesian Borneo. CIFOR and UNESCO, Jakarta

Meijaard E, Sheil D (2007) A logged forest in Borneo is better than no forest at all. Nature 446:974. doi:10.1038/446974a

Millenium Ecosystem Assessment (2005) Ecosystems and human well-being: current state and trends. Findings of the condition and trends working group. In: Hassan R, Scholes R, Ash N (eds) Millenium ecosystem assessment series. Island Press, Washington

Ministry of Agriculture, Fishery (2005) ISMFF—indicators for the sustainable management of French forests. Paris, France

Moore SE, Allen HL (1999) Plantation forestry. In: Hunter ML Jr (ed) Maintaining biodiversity in forest ecosystems. Cambridge University Press, New York, pp 400–433

Murcia C (1995) Edge effects in fragmented forests: implications for conservation. Trends Ecol Evol 10:58–62. doi:10.1016/S0169-5347(00)88977-6

Myers N, Mittermeier RA, Mittermeier CG, da Fonseca GAB, Kent J (2000) Biodiversity hotspots for conservation priorities. Nature 403:853–858. doi:10.1038/35002501

Nasi R, Koponen P, Poulsen JG, Buitenzorgy B, Rusmantoro W, Ladjar LN, Efendy R (2008) Impact of landscape and corridor design on primates in a large-scale industrial tropical plantation landscape. Biodivers Conserv. doi:10.1007/s10531-007-9237-8

Norton DA (1998) Indigenous biodiversity conservation and plantation forestry: options for the future. N Z For 43(2):34–39

Oberhauser U (1997) Secondary forest regeneration beneath pine (*Pinus kesiya*) plantations in the northern Thai highlands: a chronosequence study. For Ecol Manage 99:171–183

Parrotta JA (1993) Secondary forest regeneration on degraded tropical lands: the role of plantations as 'foster ecosystems'. In: Lieth H, Lohmann M (eds) Restoration of tropical forest ecosystems. Kluwer Academic Publishers, Dordrecht, pp 63–73

Parrotta JA (1999) Productivity, nutrient cycling and succession in single- and mixed-species plantations of *Casuarina equisetifolia*, *Eucalyptus robusta* and *Leucaena leucocephala* in Puerto Rico. For Ecol Manage 124:45–77

Parrotta JA, Turnbull J (eds) (1997) Catalyzing native forest regeneration on degraded tropical lands. For Ecol Manage (special issue) 99:1–290

Parrotta JA, Turnbull J, Jones N (1997a) Catalyzing native forest regeneration on degraded tropical lands. For Ecol Manage 99:1–8

Parrotta JA, Knowles OH, Wunderle JM (1997b) Development of floristic diversity in 10-year-old restoration forests on a bauxite mined site in Amazonia. For Ecol Manage 99:21–42

Paulsen A (2004) Analysis of corrective actions issued to existing plantation certificates. FSC, Bonn, Germany. www.fsc.org/plantations/content/en/home.htm. Accessed 19 June 2007

Pawson S, Brockerhoff E, Meenken E, Didham R (2008) Non-native plantation forests as alternative habitat for native forest beetles in a heavily modified landscape. Biodivers Conserv. doi:10.1007/s10531-008-9363-y

PEFC (2007) www.pefc.org. Accessed 31 July 2007

Peterken GF, Ausherman D, Buchenau M, Forman RTT (1992) Old-growth conservation within British upland conifer plantations. Forestry 65:127–144. doi:10.1093/forestry/65.2.127

Porter M, Kramer MR (2006) Strategy and society: the link between competitive advantage and corporate social responsibility. Harvard Business Review (December 2006. www.hbr.org)

Quine CP, Humphrey JW, Watts K (2004) Biodiversity in the UK's forests—recent policy developments and future research challenges. In: Paivanen R, Franc A (eds) Towards the sustainable use of Europe's forests—forest ecosystem and landscape research: scientific challenges and opportunities. European Forest Institute, Joensuu, pp 237–248

Quine CP, Fuller RJ, Smith KW, Grice PV (2007) Stand management: a threat or opportunity for birds in British woodland? Ibis 149(Suppl. 2):161–174. doi:10.1111/j.1474-919X.2007.00742.x

Raman TRS (2006) Effects of habitat structure and adjacent habitats on birds in tropical rainforest fragments and shaded plantations in the Western Ghats, India. Biodivers Conserv 15:1577–1607. doi:10.1007/s10531-005-2352-5

Renjifo LM (2001) Effect of natural and anthropogenic landscape matrices on the abundance of sub-Andean bird species. Ecol Appl 11:14–31. doi:10.1890/1051-0761(2001)011[0014:EONAAL]2.0.CO;2

Rosoman G (1994) The plantation effect. Greenpeace, Auckland
Sax DF (2002) Equal diversity in disparate species assemblages: a comparison of native and exotic woodlands in California. Global Ecol Biogeogr 11:49–57. doi:10.1046/j.1466-822X.2001.00262.x
Scherer-Lorenzen M, Körner C, Schulze E-D (2005) The functional significance of forest diversity: a synthesis. In: Scherer-Lorenzen M, Körner C, Schulze E-D (eds) Forest diversity and function: temperate and boreal systems. Ecological Studies, vol 176. Springer, Heidelberg, pp 377–389
Seymour RS, Hunter ML Jr (1999) Principles of ecological forestry. In: Hunter ML Jr (ed) Maintaining biodiversity in forested ecosystems. Cambridge University Press, Cambridge, pp 22–61
SFI (2007) Sustainable Forestry Initiative. www.aboutsfi.org. Accessed 31 July 2007
Shaw K (1997) Biodiversity position statement. N Z For 42(3):5–6
Shepherd KR (1993) Significance of plantations in a global forestry strategy. Austral For 56:237–335
Smith GF, Gittings T, Wilson M, French L, Oxbrough A, O'Donoghue S, O'Halloran J, Kelly DL, Mitchell FJG, Kelly T, Iremonger S, O'Donnell V, McKee A-M, Giller P (2008) Identifying practical indicators of biodiversity for stand-level management of plantation forests. Biodivers Conserv. doi:10.1007/s10531-007-9274-3
Snyder S, ReVelle C (1997) Dynamic selection of harvests with adjacency restrictions: the SHARe model. For Sci 43:213–222
Spellerberg IF, Sayer JWD (1996) Standards for biodiversity: a proposal based on biodiversity standards for forest plantations. Biodivers Conserv 5:447–459. doi:10.1007/BF00056390
Stanturf JA, Kellison RC, Broerman ES, Jones SB (2003) Innovation and forest industry: domesticating the pine forests of the southern United States, 1920–1999. For Pol Econ 5:407–419
Suzuki N, Olson DH (2008) Options for biodiversity conservation in managed forest landscapes of multiple ownerships in Oregon and Washington, USA. Biodivers Conserv. doi:10.1007/s10531-007-9301-4
Tilman D, Cassman KG, Matson PA, Naylor R, Polasky S (2002) Agricultural sustainability and intensive production practices. Nature 418:671–677. doi:10.1038/nature01014
Van der Ree R, Bennett AF, Gilmore DC (2004) Gap-crossing by gliding marsupials: thresholds for use of isolated woodland patches in an agricultural landscape. Biol Conserv 115:241–249. doi:10.1016/S0006-3207(03)00142-3
Van Deusen P (2001) Scheduling spatial arrangement and harvest simultaneously. Silva Fenn 35:85–92
Van Halder I, Barbaro L, Corcket E, Jactel H (2008) Importance of semi-natural habitats for the conservation of butterfly communities in landscapes dominated by pine plantations. Biodivers Conserv. doi:10.1007/s10531-007-9264-5
Vergara PM, Schlatter RP (2006) Aggregate retention in two Tierra del Fuego Nothofagus forests: short-term effects on bird abundance. For Ecol Manage 225:213–224
Watson JEM, Whittaker RJ, Dawson TP (2004) Habitat structure and proximity to forest edge affect the abundance of forest-dependent birds in tropical coastal forests of southeastern Madagascar. Biol Conserv 120:311–327. doi:10.1016/j.biocon.2004.03.004
Wenhua L (2004) Degradation and restoration of forest ecosystems in China. For Ecol Manage 201:33–41
Wethered R, Lawes MJ (2005) Nestedness of bird assemblages in fragmented Afromontane forest: the effect of plantation forestry in the matrix. Biol Conserv 123:125–137. doi:10.1016/j.biocon.2004.10.013
Wigley TB, Roberts TH (1997) Landscape-level effects of forest management on faunal diversity in bottomland hardwoods. For Ecol Manage 90:141–154
Wilson EO (1988) Biodiversity. National Academy Press, Washington
World Rainforest Movement (2007) www.wrm.org.uy. Accessed 23 April 2007
Wunderle JM Jr (1997) The role of animal seed dispersal in accelerating native forest regeneration on degraded tropical lands. For Ecol Manage 99:223–235
Yates ED, Levia DF, Williams CL (2004) Recruitment of three non-native invasive plants into a fragmented forest in southern Illinois. For Ecol Manage 190:119–130
Yirdaw E (2001) Diversity of naturally regenerated native woody species in forest plantations in the Ethiopian highlands. New For 22:159–177. doi:10.1023/A:1015629327039
Zuang X (1997) Rehabilitation and development of forests on degraded hills of Hong Kong. For Ecol Manage 99:197–201
Zuidema P, Sayer JA, Dijkman W (1997) Forest fragmentation and biodiversity: the case for intermediate sized conservation areas. Environ Conserv 23(4):290–297

Epiphytes of Sitka spruce (*Picea sitchensis*) plantations in Ireland and the effects of open spaces

Linda Coote · George F. Smith · Daniel L. Kelly · Saoirse O'Donoghue · Paul Dowding · Susan Iremonger · Fraser J. G. Mitchell

Originally published in the journal Biodiversity and Conservation, Volume 17, No 5, 953–968.
DOI: 10.1007/s10531-007-9302-3

Abstract The epiphytes of the trunks and branches of mature Sitka spruce (*Picea sitchensis*) trees were studied in twelve plantations containing open spaces (glades, rides and roads) in the east and southwest of Ireland. A pair of trees was studied at each site: one tree at the south-facing edge of an open space and one in the forest interior. Spruce trees were found to support a moderately diverse range of bryophytes and lichens, including two relatively rare bryophyte species. Clear patterns in vertical distribution were identified, with bryophyte richness and cover decreasing and lichen richness and cover increasing from the tree base to the upper trunk. The open spaces themselves did not appear to affect overall epiphyte diversity, with no significant differences in any of the diversity measures between edge and interior trees. The main effect of open spaces was on the epiphyte cover of the edge trees. This was related to increased light levels combined with the presence of live branches from close to ground level on the south sides of the edge trees, which produced optimum conditions for bryophytes at the tree base and lichens in the upper plots. However, this dense side-canopy negatively affected epiphyte diversity on the north sides of the edge trees. Further research is required to assess the effects of open spaces within forestry plantations on epiphyte diversity.

Keywords Bryophyte · Epiphyte · Forest biodiversity · Glade · Lichen · Sitka spruce

Introduction

In Europe, forests and other wooded land cover 47% of the total land area, with plantations accounting for 3% of the total wooded area (Anon. 2003). However, the Republic of Ireland is currently one of the least wooded countries in Europe, with forest cover standing at approximately 10% of the total land area (EPA 2004). In Ireland, as in Britain, the vast

L. Coote (✉) · G. F. Smith · D. L. Kelly · S. O'Donoghue · P. Dowding · S. Iremonger · F. J. G. Mitchell
Department of Botany, School of Natural Sciences, Trinity College Dublin, Dublin 2, Ireland
e-mail: cootel@gmail.com

E.G. Brockerhoff et al. (eds.), *Plantation Forests and Biodiversity: Oxymoron or Opportunity?*
DOI: 10.1007/978-90-481-2807-5_2

majority of forests are even-aged commercial plantations of exotic conifers, with Sitka spruce (*Picea sitchensis*) plantations making up more than 50% of the forest estate (EPA 2004). This means that, in these regions, plantations are potentially of greater importance for woodland biodiversity than they are in regions containing large areas of semi-natural forests. Baseline information on the biodiversity of forestry plantations is required if they are to be effectively managed, particularly in the light of Ireland's commitment to Sustainable Forest Management (SFM) (Forest Service 2000).

Epiphytic bryophytes and lichens are an important component of biological diversity in natural tropical, temperate and boreal forests (Rhoades 1981; Cornelissen and ter Steege 1989; Lesica et al. 1991; Boudreault et al. 2002), although they are known to be decreasing worldwide because of habitat destruction, forest operations and air pollution (Mitchell et al. 2002). They also form a major component of the total botanical diversity of semi-natural woodland in Ireland (Kelly 2000). However, there have been few studies on epiphytes in Ireland. A large proportion of these relate to the study of lichens as indicators of pollution (cf. Richardson 1987). The remainder come from semi-natural woodland and scrub habitats (Richards 1938; Phillips 1959; Mitchell 1964; Dickinson and Thorp 1968; Folan and Mitchell 1970; McCarthy 1980; Kelly 1981; Kirby and O'Connell 1982; McCarthy et al. 1986; Fox et al. 2001), which, compared to forestry plantations, are a minority habitat in Ireland. We know of no published work on the epiphytes of Irish forestry plantations, save a note on pendulous *Hypnum jutlandicum* growing in a plantation in the west of Ireland (Doyle 1987). Any study of forest plant diversity that excludes epiphytes runs the risk of overlooking what could be a significant component of the flora, particularly considering the often poor diversity of the ground flora in closed canopy plantations (Hill 1979; French 2005; Smith et al. 2005).

Open areas have been recognised as being important for maintaining high levels of biodiversity in the natural forests of northwest Europe and North America, where the main Irish forest plantation species originated (Ratcliffe and Peterken 1995; Peterken 1996). In Ireland, where most plantations consist of one or more even-aged stands (Joyce and O Carroll 2002), light levels can decrease dramatically as the forest matures and the canopy closes (French 2005; Smith et al. 2005). Open spaces may be more important in these plantations for the survival of species of flora and fauna excluded from dark, closed-canopy forests (Sparks et al. 1996; Mullen et al. 2003). Neitlich and McCune (1997) found that gaps containing broadleaved trees and shrubs were important for epiphytic lichen diversity in managed conifer stands in Oregon. Rose (1974) has also noted that many epiphytic lichens and a significant amount of bryophytes will occur at or near the edges of woodlands, or along rides or glades, since most are light-demanding. It is therefore important to evaluate the effects of open spaces on the epiphyte diversity of the adjacent stand. The aims of this study are:

1. To examine the diversity of epiphytes in Sitka spruce plantations
2. To assess the effects of open spaces on epiphyte diversity in these plantations

This study is part of a wide-ranging survey investigating plant and animal biodiversity in Irish forestry plantations (the BIOFOREST project) (Iremonger et al. 2006).

Methods

Study sites

To reduce the effects of large-scale environmental variation, twelve study sites were chosen in two geographic clusters in the east and southwest of the country (Table 1, Fig. 1). All

Table 1 The Irish Grid References for the twelve study sites shown in Fig. 1

Site name	Grid Ref.
Athdown	O 076 158
Ballinastoe	O 180 084
Ballycurragh	T 061 831
Ballysmuttan	O 047 145
Carrigagulla	W 384 835
Cleanglass	R 244 218
Glannaharee West	W 444 887
Knocknagoum	Q 958 217
Lugg	O 031 242
Meentinny	R 245 135
Mucklagh One	T 083 864
Toureenmacauliffe	R 256 200

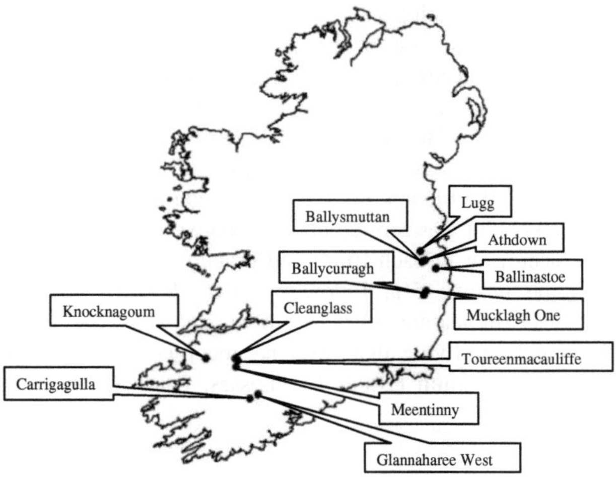

Fig. 1 Location of the twelve study sites

sites were Sitka spruce plantations approaching commercial maturity (28–43 years after planting) and containing glades, rides or roads. Sites containing glades with south-facing edges were preferred, although, in three cases, where such glades were absent or unsuitable, east-west running rides or roads were studied. The sites in the east were mainly well-drained upland sites on mineral soils with an elevation range of 290–540 m, mean 422 m (SE = 27). Those in the southwest were typically poorly drained sites on blanket peat with an elevation range of 180–380 m, mean 288 m (SE = 23). The climate of Ireland is typically Oceanic with mild winters, cool summers and abundant year-round precipitation (Met Éireann 2006). The mean annual rainfall at the study sites fell within rainfall ranges from 800–1,000 mm yr^{-1} up to 1,600–2,000 mm yr^{-1} (Met Éireann 2006); the lower values being for low elevation eastern sites and the higher values for high elevation eastern sites and sites in the southwest. For the purposes of this paper the two geographic clusters are not considered separately in the data analysis.

Epiphyte sampling

Fieldwork was carried out between July and November 2003. All epiphyte surveying took place on the north side (i.e., south-facing side) of each open space. Epiphytes were studied on a pair of trees at each site: one tree at the edge of an open space and one tree in the forest interior. The forest interior trees studied were located at a distance from the edge of the open space greater than or equal to the height of the edge trees at that point.

Epiphytes were studied in plots located on the trunk and branches in 4 different height zones on the tree–tree base (B), lower (L), middle (M) and upper (U) (Fig. 2). The tree base zone began at the point where the trunk emerged from the soil or needle litter and reached to 0.5 m above this point, the lower zone was centred on breast height (1.3 m) and the middle and upper zones were centred on 1/3 and 2/3 of the height of the tree respectively. Trees were climbed with rope, harness and climbing spurs in order to study the middle and upper zones. Trunk plots were located on the side of the trunk facing the open space and the opposite side (approximately south and north facing; edge aspect ranged 138–211°). Plots were 50 cm long and the width varied from a maximum of 25 cm to that required to sample a half cylinder of the trunk. In the middle and upper zones each trunk plot was centred vertically on a branch whorl. A branch from the north side and a branch from the south side of the whorl were removed for study on the ground. Three plots, 25 cm long by 50 cm wide, were studied on each branch. These plots ran perpendicular to the main axis of the branch and included the main axis and the side branches on that side of the branch with the most cover of side branches (Fig. 2). The upper half cylinders of branches were studied (i.e., the undersides were excluded). The first plot was placed at the base of the branch; the third plot was placed near the tip of the branch, but did not include the last 2 years' growth on the main axis; and the second plot was centred half way between the previous two plots. A list of epiphyte species occurring in each trunk and branch plot was made with percentage cover estimated to the nearest 5%. Below 5% two different cover-abundance units were distinguished: '3%' (indicating cover of 1–5%), and '0.5%' (indicating cover <1%). Non-lichenised fungi and free-living algae were not recorded. Nomenclature follows Stace (1997) for vascular plants, Smith (2004) for mosses, Paton (1999) for liverworts and Coppins (2002) for lichens.

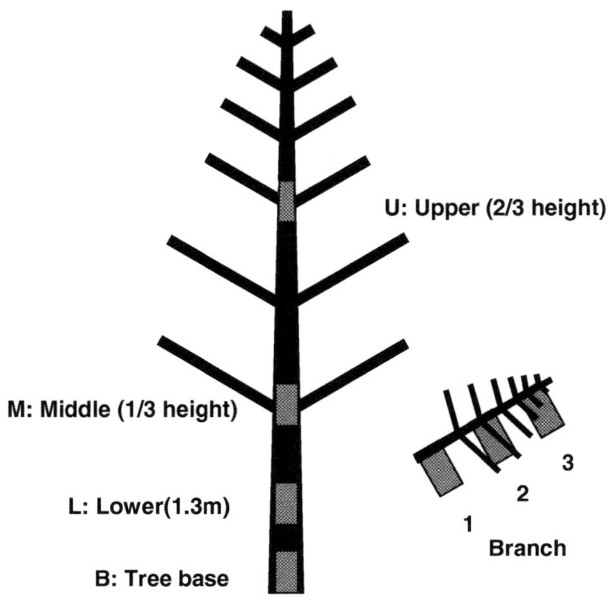

Fig. 2 The plot sampling design used. All plots are 25 cm × 50 cm

Data analysis

Total epiphyte, bryophyte and lichen species richness were calculated for each tree based on the data derived from all plots on each tree. Within-plot diversity was calculated using Simpson's reciprocal diversity index (1/D) (Simpson 1949; Magurran 2003) and the mean diversity of all plots ($n = 20$) was used to give an overall mean diversity value for each tree. Whittaker's beta diversity index (β_w) (Whittaker 1972) was used to examine the compositional variation among the eight trunk plots studied on each individual tree. Indicator species analysis (Dufrêne and Legendre 1997) was carried out using PC-ORD for Windows version 4.01 (McCune and Mefford 1999), in order to identify species characteristic of the edge and interior trees.

Paired samples *t*-tests, or the non-parametric equivalent Wilcoxon signed ranks tests, were used to compare edge and interior trees and the north and south sides of trees. Kruskal–Wallis tests were used to compare species richness and cover among the different height zones on the trees. Prior to any parametric statistical testing, variables were inspected to ensure conformity to assumptions. Statistical analyses were carried out using SPSS version 11.0.1 (2001).

Results

Epiphyte diversity

A total of 68 species was recorded on the 24 trees: 28 bryophyte, 39 lichen and one vascular plant species (Table 2). The vascular plant species recorded was a juvenile pteridophyte, not identifiable to genus level. A large number of species were recorded infrequently on the trees studied. The most common species was *Hypnum jutlandicum*,which was recorded on every tree. The five next most commonly recorded species were the lichens *Fuscidea lightfootii*, *Hypotrachyna revoluta* and *Dimerella lutea*, the liverwort *Metzgeria temperata* and the moss *Ulota crispa s.l.*. One of the moss species recorded, *Daltonia splachnoides*, is listed as 'Vulnerable' on the British Red Data List (Church et al. 2001) and is likely to appear on the Irish Red Data List, which is currently in preparation; the discovery of a number of recent records means it may be classified as 'Near Threatened' (D. Holyoak, pers. comm.). Another moss species recorded in this study, *Plagiothecium laetum*, is also likely to appear on the Irish Red Data List (D. Holyoak, pers. comm.): its two records in this study constitute the 3rd and 4th records of this species in Ireland. Two of the lichen species, *Byssoloma subdiscordans* and *Usnea esperantiana*, are listed as 'Near Threatened' in the Conservation Evaluation of British lichens (Woods and Coppins 2003).

A total of 14 species was recorded on the open space edge trees only (seven bryophyte and seven lichen species) and a total of 15 on the forest interior trees only (two bryophyte, twelve lichen and one vascular plant species) (Table 2). However, ten of the species recorded only on the edge trees and 12 of those recorded only on the interior trees were each recorded on only one tree. Of the remaining seven species, six occurred on two trees and one on three. Indicator species analysis revealed no significant indicators for the edge or interior trees.

The mean epiphyte richness per tree was 16.3 (SE = 0.9). There were no significant differences in total epiphyte, bryophyte or lichen richness between the edge and interior trees (Table 3). There were also no significant differences between the edge and interior

Table 2 Epiphyte species recorded in the survey plots

Species	Edge (*n* = 12)	Interior (*n* = 12)
Mosses		
Atrichum undulatum	1	0
Campylopus introflexus	2	0
Campylopus sp.	0	1
Cryphaea heteromalla	1	0
Daltonia splachnoides	1	1
Hypnum andoi	3	1
Hypnum jutlandicum	12	12
Hypnum resupinatum	1	1
Isothecium myosuroides	2	0
Kindbergia praelonga	5	8
Mnium hornum	1	0
Plagiothecium laetum	1	1
Plagiothecium undulatum	5	5
Pseudotaxiphyllum elegans	1	0
Rhytidiadelphus loreus	0	1
Thuidium tamariscinum	0	2
Ulota crispa s.l.	10	9
Ulota phyllantha	2	0
Liverworts		
Calypogeia muelleriana	1	3
Colura calyptrifolia	6	7
Frullania dilatata	5	4
Frullania tamarisci	1	1
Lejeunea cavifolia	1	4
Lophocolea bidentata	7	5
Metzgeria fruticulosa	2	2
Metzgeria furcata	6	6
Metzgeria temperata	11	9
Microlejeunea ulicina	6	6
Radula complanata	1	1
Lichens		
Anisomeridium biforme	0	1
Anisomeridium polypori	0	1
Byssoloma subdiscordans	1	0
Candelariella reflexa	0	1
Cladonia chlorophaea	0	1
Cladonia sp.	1	1
Dimerella lutea	5	11
Dimerella pineti	5	7
Dimerella sp.	0	1
Evernia prunastri	2	1
Lichens continued		
Fellhanera bouteillei	1	3
Fuscidea lightfootii	12	10
Graphis elegans	2	1
Graphis scripta	1	0
Graphis sp.	1	0
Gyalideopsis anastomosans	4	7
Hypogymnia sp.	0	1
Hypogymnia tubulosa	7	5
Hypotrachyna revoluta	10	11
Hypotrachyna sp.	1	1
Lecania cyrtella	0	1

Table 2 continued

Species	Edge (n = 12)	Interior (n = 12)
Lecanora chlarotera	0	1
Lecanora pulicaris	2	4
Lecidella elaeochroma	1	0
Lepraria incana	7	8
Melanelia fuliginosa	0	1
Micarea lignaria	1	1
Micarea peliocarpa	1	5
Micarea prasina	3	5
Micarea sp.	2	6
Parmelia sulcata	3	1
Parmotrema chinense	1	2
Phaeographis smithii	2	1
Physcia adscendens	0	1
Physcia aipolia	1	0
Physcia sp.	2	0
Physcia tenella	5	3
Pseudevernia furfuracea	1	0
Ramalina farinacea	6	3
Ramalina fastigiata	0	1
Ramalina sp.	0	1
Rinodina biloculata	0	1
Trapeliopsis flexuosa	0	3
Usnea esperantiana	1	1
Usnea filipendula	0	1
Usnea flammea	2	0
Usnea sp.	0	2
Xanthoria polycarpa	1	0
Vascular Plants		
Juvenile pteridophyte	0	2

Edge and *Interior* indicate the number of open space edge and forest interior trees each species occurred on respectively

Table 3 The species richness, mean Simpson's reciprocal diversity (1/D) and Whittaker's beta diversity (β_w) of the open space edge and forest interior trees

	Open space edge (n = 12)	Forest interior (n = 12)	p-value
Species richness			
Total epiphyte	15.8 (SE = 1.2)	16.8 (SE = 1.3)	0.297
Bryophyte	7.8 (SE = 0.9)	7.5 (SE = 0.9)	0.695
Lichen	7.9 (SE = 0.7)	9.1 (SE = 0.9)	0.274
Mean diversity (1/D)			
Total epiphyte	1.9 (SE = 0.3)	2.1 (SE = 0.2)	0.389
Bryophyte	1.0 (SE = 0.2)	1.1 (SE = 0.2)	0.428
Lichen	1.1 (SE = 0.1)	1.3 (SE = 0.1)	0.290
Beta diversity (β_w)			
Total epiphyte	3.7 (SE = 0.3)	3.4 (SE = 0.3)	0.259
Bryophyte	3.2 (SE = 0.3)	3.5 (SE = 0.5)	0.622
Lichen	4.9 (SE = 0.5)	4.5 (SE = 0.6)	0.321

trees in the mean total epiphyte, bryophyte or lichen diversity of all plots on each tree, as measured by Simpson's reciprocal, or in the total epiphyte, bryophyte or lichen beta (β_w) diversity on trunks (Table 3).

Vertical distribution

The vertical distribution of epiphytes on the trunks of the trees was examined. Bryophyte richness decreased and lichen richness increased from the tree base to upper zones on both edge and interior trees (Fig. 3). There were significant differences in total epiphyte richness among the four height zones on the edge trees ($H = 9.00$, $p = 0.029$) but not on the interior trees ($H = 0.56$, $p = 0.905$). There were also significant differences in lichen richness among the four height zones on the edge trees ($H = 37.47$, $p < 0.001$) and, despite the lack of any significant differences in total epiphyte richness, there were significant differences in bryophyte and lichen richness among the four height zones on the interior trees (bryophyte: $H = 10.92$, $p = 0.012$; lichen: $H = 28.06$, $p < 0.001$). Further testing revealed a significantly greater total epiphyte and lichen richness in the upper height zone of the edge trees compared to all other zones (Fig. 3). Lichen richness was also significantly greater in the upper zone than in any other zone on the forest interior trees.

As with species richness, bryophyte cover decreased and lichen cover increased from the tree base to the upper zones (Fig. 4). There were significant differences in total epiphyte cover among the four height zones on the edge trees ($H = 15.07$, $p = 0.002$), but not on the interior trees ($H = 4.02$, $p = 0.259$). There were, however, significant differences in bryophyte and lichen cover on both the edge (bryophyte cover: $H = 25.24$, $p < 0.001$; lichen cover: $H = 34.43$, $p < 0.001$) and interior trees (bryophyte cover: $H = 13.03$, $p = 0.005$;

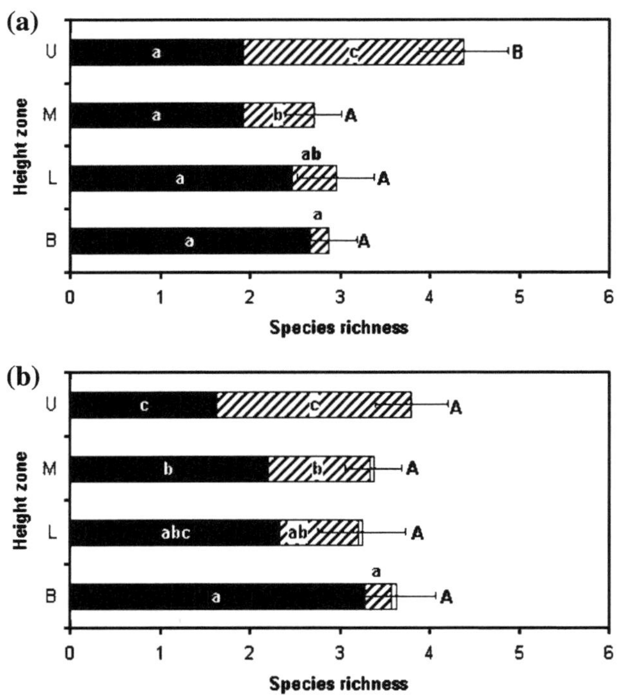

Fig. 3 The mean species richness in the trunk plots in the four height zones (B—tree base, L—lower, M—middle, U—upper) on all (**a**) open space edge and (**b**) forest interior trees. The total epiphyte richness is shown and the relative proportions of bryophyte (shaded), lichen (hatched) and vascular plant species (unshaded) are indicated. Error bars indicate ± standard error. There is no significant difference in total epiphyte (uppercase), bryophyte or lichen (lowercase) richness between two height zones if indicated by the same letter

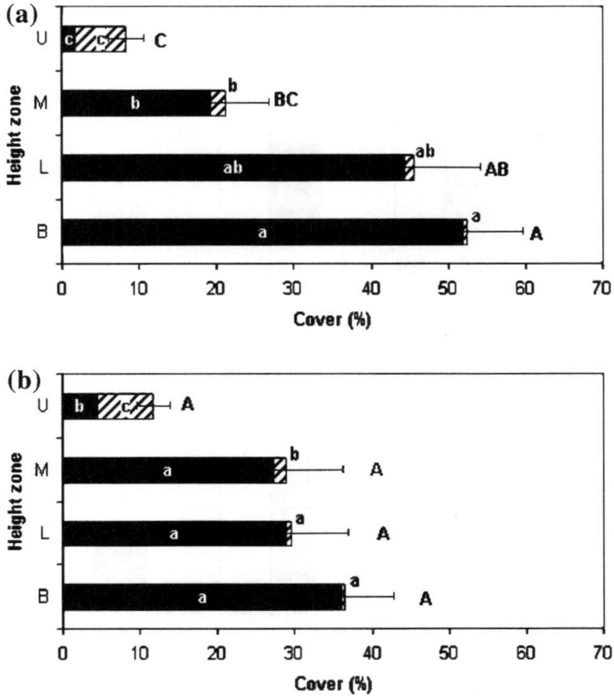

Fig. 4 The mean cover of epiphytes in the trunk plots in the four height zones (B—tree base, L—lower, M—middle, U—upper) on all (**a**) open space edge and (**b**) forest interior trees. The total cover is shown and the relative proportions of bryophyte (shaded) and lichen (hatched) cover are indicated. Error bars indicate ± standard error. There is no significant difference between two height zones for total epiphyte (uppercase), bryophyte or lichen (lowercase) cover if indicated by the same letter

lichen cover: $H = 36.56$, $p < 0.001$). Bryophyte cover was significantly lower and lichen cover significantly higher in the upper height zone than in all other height zones on both the edge and interior trees (Fig. 4). Total epiphyte and bryophyte cover were higher in the tree base and lower height zones on the edge trees compared with the interior trees and the difference was significant for the tree base (paired samples *t*-test: total epiphyte cover, $p = 0.017$; bryophyte cover, $p = 0.020$).

Effects of aspect

The plots on the south side of the trunks of the edge trees supported significantly greater total epiphyte and lichen species richness compared with those on the north side (Wilcoxon signed ranks test: total epiphyte richness, $p = 0.016$; lichen richness, $p = 0.004$) but there was no significant difference in bryophyte richness (Wilcoxon signed ranks test: $p = 0.407$) (Fig. 5). In contrast, there were no significant differences in total epiphyte, bryophyte or lichen species richness between the north and south sides of the trunks of the forest interior trees (Wilcoxon signed ranks test: total epiphyte richness, $p = 0.893$; bryophyte richness. $p = 0.737$; lichen richness, $p = 0.364$). The higher total epiphyte and lichen richness on the south sides of the edge trees was not as a result of an increase in species richness on the south sides of the edge trees relative to the south sides of the forest interior trees (there were

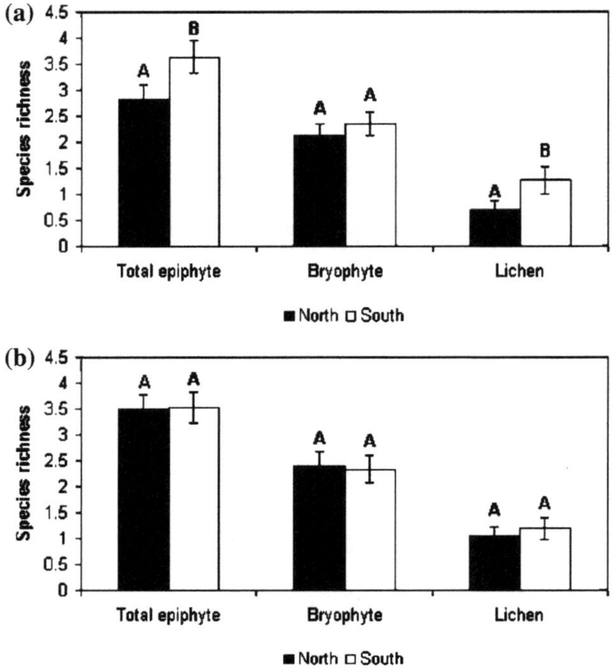

Fig. 5 The mean total epiphyte, bryophyte and lichen species richness in the trunk plots on the north and south sides of the (**a**) open space edge and (**b**) forest interior trees. Error bars indicate ± standard error. There is no significant difference in species richness between the two aspects if indicated by the same letter

no significant differences in total epiphyte, bryophyte or lichen species richness between the south sides of the edge and interior trees (Wilcoxon signed ranks test: total epiphyte richness, $p = 0.998$; bryophyte richness, $p = 0.899$; lichen richness, $p = 0.808$)). The difference was as a result of a significantly lower total epiphyte and lichen richness on the north sides of the edge trees in comparison to (i) the north sides of the interior trees (Wilcoxon signed ranks test: total epiphyte richness $p = 0.018$; bryophyte richness $p = 0.199$; lichen richness $p = 0.008$) and (ii) the south sides of the edge trees.

There was also a significantly greater total epiphyte, bryophyte and lichen cover on the south sides of the trunks of the open space edge trees compared to the north sides (Wilcoxon signed ranks test: total epiphyte cover, $p < 0.001$; bryophyte cover, $p = 0.008$; lichen cover, $p = 0.016$) (Fig. 6). Again, in contrast, there were no significant differences in total epiphyte, bryophyte or lichen cover between the north and south sides of the trunks of the forest interior trees (Wilcoxon signed ranks test: total epiphyte cover, $p = 0.908$; bryophyte cover, $p = 0.473$; lichen cover, $p = 0.344$) (Fig. 6). In contrast to the situation with species richness, the significantly greater cover on the south sides of the edge trees was as a result of an increased cover relative to the south sides of the interior trees; the difference was significant for total epiphyte and bryophyte cover (Wilcoxon signed ranks test: total epiphyte cover, $p = 0.025$; bryophyte cover, $p = 0.015$). There were no significant differences in cover between the north sides of the edge and the north sides of the interior trees (Wilcoxon signed ranks test: total epiphyte cover, $p = 0.373$; bryophyte cover, $p = 0.221$; lichen cover, $p = 0.777$).

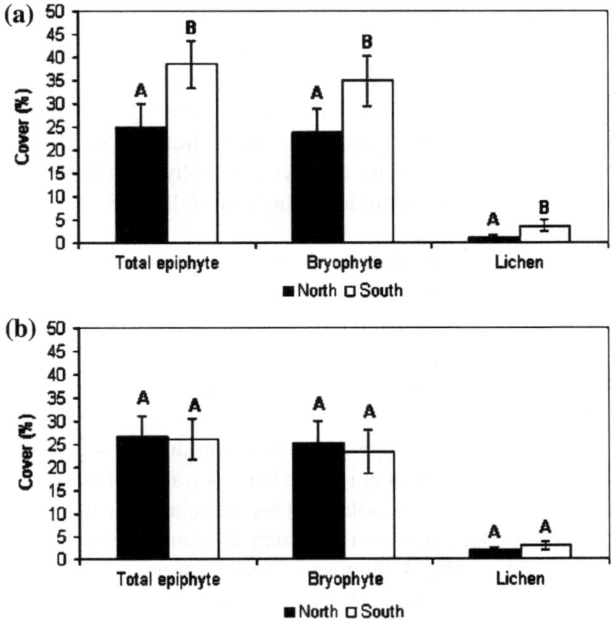

Fig. 6 The mean total epiphyte, bryophyte and lichen cover in the trunk plots on the north and south sides of the (**a**) open space edge and (**b**) forest interior trees. Error bars indicate ± standard error. There is no significant difference in cover between the two aspects if indicated by the same letter

Discussion

Epiphyte diversity

Trees adjacent to open spaces in Sitka spruce plantations can support a reasonably diverse range of epiphyte species, including two relatively rare bryophyte species. One of these rare species, *Daltonia splachnoides*, and the relatively commonly occurring liverwort species, *Colura calyptrifolia*, have a European distribution classed as 'Hyperoceanic Southern-temperate', meaning they are not only more or less restricted in their European range to the Oceanic zone but also markedly western within that zone (Hill and Preston 1998). The spread of *Colura* onto willows growing almost exclusively in or close to conifer plantations has been reported in south and central Wales (Bosanquet 2004). The author suggests three possible explanations for this: that conifer cover protects *Colura* more effectively from harmful frosts than deciduous cover; that conifers act as more efficient spore traps than deciduous trees; or that conifers provide constantly humid conditions. *Colura* has recently been reported growing on conifers at the same Welsh sites (B. Coppins, pers. comm.) Another noteworthy feature is the frequency of the lichen *Physcia tenella*, a species commonly associated with nutrient-rich and -enriched substrata (Purvis et al. 1992).

Owing to the lack of prior studies, it is difficult to put these results into context. However, compared with two studies on epiphytes in British plantations (Bates et al. 1997; Humphrey et al. 2002), and to one from Belgium and France (Vanderpoorten et al. 2004) it appears that plantations in Ireland are relatively rich in epiphytes and particularly in epiphytic bryophytes. The most comparable study, by Humphrey et al. (2002), reported only two bryophyte and two lichen species on substrates other than deadwood (viz. rocks and living trees up to 2 m) in 1 ha plots in four mid-rotation (20–30 years) and four mature (50–80 years) Sitka spruce plantations. In the present study, 23 bryophyte and 7 lichen species were recorded on living trees between 0 m and 1.55 m. The higher epiphyte richness in the present study may be related to the oceanic climate of Ireland. Studies in North America have shown that conifer species lacking epiphytic bryophytes in drier regions quite commonly support rich bryophyte communities in high-rainfall areas of the Pacific Northwest (Glime and Hong 2002).

If open spaces were having an effect on epiphyte diversity, one would expect to see a difference in diversity between open space edge and forest interior trees. Esseen (2006) studied a natural edge in a forest-wetland mosaic in northern Sweden and found reduced abundance of the fruticose lichen *Alectoria sarmentosa* at the edge. However, the sampled edges were open and exposed to prevailing winds and it was concluded that disturbance by wind was the most plausible explanation for the low abundance of the lichen at these sites. Pearson (1969) studied the natural edge between a bog and its surrounding woodland in Minnesota and found that the cover of epiphytic lichens was significantly higher on trees at the edge than in the surrounding woodland. This he related to the lower humidity and higher light levels at the edge. However, no such differences between edge and interior trees were apparent in this study. A number of species were recorded only on either the edge or interior trees, but few of these species were recorded on more than one tree. This means that their discovery on an edge tree rather than an interior tree may have been a matter of chance. The lack of any significant indicator species for either edge or interior trees suggests that the floras of edge and interior trees may not be distinct and that other factors are more important in determining species composition. However, the high number of infrequently occurring species, the small sample size, the variation in the size and configuration of the open spaces studied and environmental variation among sites may also have been important factors in the apparent lack of differentiation between edge and interior trees.

Vertical distribution

Bryophyte richness and cover decreased and lichen richness and cover increased from the tree base to the upper zone on the trunks of the trees. A similar pattern was seen on Sitka spruce growing in an old-growth redwood (*Sequoia sempervirens*) forest in California (Ellyson and Sillett 2003). This pattern of distribution is most likely related to differences in microclimate, as light intensity, wind and evaporation all increase from the base to the top of a forest tree (Barkman 1958). Bryophytes and lichens tend to have differing ecological requirements, with most bryophytes adapted to low light levels and higher humidity (Trynoski and Glime 1982) and lichens preferring lower and more variable humidity (Pearson 1969) and higher light levels (Rose 1974).

On the open space edge trees, total epiphyte richness was significantly higher in the upper plots, mainly as a result of a significantly higher lichen richness. Light intensity is likely to be higher and humidity lower at the edge (Pearson 1969), particularly at this height in the tree. Bryophyte cover was also significantly higher at the tree base on the edge trees than on the interior trees. Live branches were present from near the base of the south sides of the edge trees, resulting in a 'side-canopy' which will have closed the edge to direct sunlight and to air movement (Matlack and Litvaitis 1999). Some lateral penetration of light is likely to have occurred through this side-canopy and this may have produced higher light levels at the base of the edge trees compared to the interior trees. The result may have been optimised light and humidity levels on the edge trees, leading to greater photosynthetic production and therefore increased bryophyte growth, as found by Peck et al. (1995).

Effects of aspect

There was no difference in species richness or cover between the north and south sides of the forest interior trees. This is not surprising since variation with aspect will be less in dense forests (Barkman 1958) and many of the stands studied were dense and unthinned. Peck et al. (1995) also found little difference in liverwort abundance with aspect on Sitka spruce growing on Hotsprings Island, Queen Charlotte Islands, Canada, and related this to the high humidity of the site, which resulted in a relatively uniform distribution of moisture. As previously stated, a number of stands were planted on poorly drained blanket peat soils and others were at high elevations. These factors, combined with the high tree density, are likely to have resulted in general high humidity levels.

In contrast, there were significant differences in species richness and cover between the north and south sides of the open space edge trees. Total epiphyte and lichen richness were lower on the north sides of the edge trees compared to both the south sides of the same trees and the north sides of the forest interior trees. Total epiphyte, bryophyte and lichen cover were significantly higher on the south sides of the edge trees than on the north sides. Total epiphyte and bryophyte cover were also significantly higher on the south sides of the edge trees than on the south sides of the forest interior trees. The presence of live branches from the base of the south sides of the edge trees may have resulted in the north sides of the edge trees being more shaded than the north sides of the forest interior trees, which did not have live branches close to ground level. These low light levels may have excluded all but the most shade tolerant species. The greater bryophyte cover on the south side of trees was mainly as a result of the high cover at the tree base (discussed above). The high lichen cover was mainly in the upper plots, where the higher light levels may have favoured more light demanding species.

Conclusions

This study addresses a major gap in the information on plantation forest biodiversity. Epiphytes are an often-neglected component of studies of plant diversity and this study has shown that Sitka spruce plantations in Ireland can support a reasonably diverse range of epiphytes. However, these plantations were species-poor compared to Irish semi-natural woodlands where as many as 46 lichen species have been recorded on a single oak tree (Fox et al. 2001). In particular they lacked the cyanolichen species of the *Lobarion* community, which is generally accepted to be the major 'climax' community of European post-glacial forests (Rose 1988). It is unlikely that Sitka spruce plantations managed on short rotations will ever develop an epiphytic flora characteristic of semi-natural woodlands owing to the lack of habitat continuity, although it is possible that diversity may increase given time (Humphrey et al. 2002).

Open spaces within Sitka spruce plantations do not appear to have significant effects on epiphyte diversity. This seems to be related to the presence of 'closed' edges at the majority of the open spaces studied, with live branches present from close to ground level. These findings suggest that the inclusion of open spaces within Sitka spruce plantations as a management tool may not benefit epiphyte diversity. However, further research is required.

Many of the open spaces studied also contained broadleaved trees and shrubs. These have been found to be important for epiphytic lichen diversity in managed conifer stands in Oregon (Neitlich and McCune 1997). Study of the epiphytes of these trees and shrubs would give a clearer picture of the overall contribution made by open spaces in forest plantations to epiphyte diversity. Basic research on the epiphytes of Irish semi-natural woodlands is also urgently required to put studies of Irish forestry plantations into context.

Acknowledgements We would like to thank Deirdre Ninaber, Siobhán McNamee and Bastian Egeter for their assistance with fieldwork. We also thank T.H. Blackstock, Dr B. Coppins, H. Fox, Dr D.T. Holyoak, G.P. Rothero and R. Porley for identification of difficult specimens. The comments of three anonymous reviewers are gratefully acknowledged. This work was carried out as part of the BIOFOREST project (http://bioforest.ucc.ie) which was jointly funded by the Environmental Protection Agency (EPA) and the National Council for Forest Research and Development (COFORD) through the National Development Plan.

References

Anon. (2003) State of Europe's forests 2003. MCPFE, Vienna

Barkman JJ (1958) Phytosociology and ecology of cryptogamic epiphytes. Van Gorcum, Assen

Bates JW, Proctor MCF, Preston CD, Hodgetts NG, Perry AR (1997) Occurrence of epiphytic bryophytes in a 'tetrad' transect across southern Britain. I. Geographical trends in abundance and evidence of recent change. J Bryol 19:685–714

Bosanquet SDS (2004) *Colura calyptrifolia* in Wales. Field Bryol 82:3–5

Boudreault C, Bergeron Y, Gauthier S, Drapeau P (2002) Bryophyte and lichen communities in mature to old-growth stands in eastern boreal forests of Canada. Can J For Res 32:1080–1093

Church JM, Hodgetts NG, Preston CD, Stewart NF (2001) British red data books mosses and liverworts. JNCC, Peterborough

Coppins BJ (2002) Checklist of lichens of Great Britain and Ireland. British Lichen Society, London

Cornelissen JHC, ter Steege H (1989) Distribution and ecology of epiphytic bryophytes and lichens in dry evergreen forest of Guyana. J Trop Ecol 5:131–150

Dickinson CH, Thorp TK (1968) Epiphytic lichens on *Corylus avellana* in the Burren, county Clare. Lichenologist 4:66–72

Doyle GJ (1987) *Hypnum jutlandicum* Holmen and Warncke growing as a pendulous epiphyte on *Pinus contorta* in a plantation in the west of Ireland. Lindbergia 12:73–75

Dufrêne M, Legendre P (1997) Species assemblages and indicator species: the need for a flexible assymetrical approach. Ecol Monogr 67:345–366

Ellyson WJT, Sillett SC (2003) Epiphyte communities on Sitka spruce in an old-growth redwood forest. Bryologist 106:197–211

EPA (2004) Ireland's environment 2004. Environmental Protection Agency, Dublin

Esseen P-A (2006) Edge influence on the old-growth forest indicator lichen *Alectoria sarmentosa* in natural ecotones. J Veg Sci 17:185–194

Folan ACM, Mitchell ME (1970) The lichens and lichen parasites of Derryclare wood, Connemara. Proc R Ir Acad 70(B):163–170

Forest Service (2000) Irish national forest standard. Forest Service, Department of the Marine and Natural Resources, Dublin

Fox H, Cullen M, Little DJ, Ryan M, Ciarruiz P, Dwyer R, Boyle GM (2001) Vegetation monitoring and botanical survey of Brackloon wood, Westport, Co. Mayo. Forest Ecosystem Research Group. Department of Environmental Resource Management, University College Dublin, Dublin

French LJ (2005) Ground flora communities in Ireland's plantation forests: their diversity, structure and composition. PhD Thesis, University of Dublin, Trinity College

Glime JM, Hong WS (2002) Bole epiphyte on three conifer species from Queen Charlotte Islands, Canada. Bryologist 105:451–464

Hill MO (1979) The development of flora in even-aged plantations. In: Ford ED, Malcolm DC, Atterson J (eds) The ecology of even aged plantations. Institute for Terrrestrial Ecology, Cambridge, pp 175–192

Hill MO, Preston CD (1998) The geographical relationships of British and Irish bryophytes. J Bryol 20:127–226

Humphrey JW, Davey S, Peace AJ, Ferris R, Harding K (2002) Lichens and bryophyte communities of planted and semi-natural forests in Britain: the influence of site type, stand structure and deadwood. Biol Conserv 107:165–180

Iremonger S, O'Halloran J, Kelly DL, Wilson MW, Smith GF, Gittings T, Giller PS, Mitchell FJG, Oxbrough A, Coote L, French L, O'Donoghue S, McKee A-M, Pithon J, O'Sullivan A, Neville P, O'Donnell V, Cummins V, Kelly TC, Dowding P (2006) Biodiversity in Irish plantation forests. Environmental Protection Agency (EPA) and the National Council for Forest Research and Development (COFORD), Dublin

Joyce PM, O Carroll N (2002) Sitka spruce in Ireland. National Council for Forest Research and Development (COFORD), Dublin

Kelly DL (1981) The native forest vegetation of Killarney south-west Ireland: an ecological account. J Ecol 69:437–472

Kelly DL (2000) Charting diversity in a Killarney oakwood: levels of resolution in floristic recording, and the effects of fencing and felling. In: Rushton BS (eds) Biodiversity: the Irish dimension. Royal Irish Academy, Dublin, pp 76–94

Kirby EN, O'Connell M (1982) Shannawoneen wood, county Galway, Ireland: the woodland and saxicolous communities and the epiphytic flora. J Life Sci R Dubl Soc 4:73–96

Lesica P, McCune B, Cooper SV, Hong WS (1991) Differences in lichen and bryophyte communities between old-growth and managed second-growth forests in the Swan Valley, Montana. Can J Bot 6:283–292

Magurran AE (2003) Measuring biological diversity. Blackwell, Oxford

Matlack GR, Litvaitis JA (1999) Forest edges. In: Malcolm L, Hunter JR (eds) Maintaining biodiversity in forest ecosystems. Cambridge University Press, Cambridge, pp 210–233

McCarthy PM (1980) Vertical zonation of lichens on alder (*Alnus glutinosa* (L.) Gaertn.) near Cork, Ireland. Sci Proc R Dubl Soc 6:397–405

McCarthy PM, Mitchell ME, Schouten MGC (1986) Lichens epiphytic on *Calluna vulgaris* (L) Hull in Ireland. Nova Hedwigia 42:91–98

McCune B, Mefford MJ (1999) Multivariate analysis of ecological data. MjM Software, Glenedon Beach, Oregon

Met Éireann (2006) Rainfall. http://www.meteireann.ie/climate/rainfall.asp. Cited 18 Feb 2006

Mitchell AW, Secoy K, Jackson T (eds) (2002) The global canopy handbook: techniques of access and study in the forest roof. Global Canopy Programme, Oxford

Mitchell ME (1964) Lichens occurring on arbutus at Killarney. Ir Nat J 14:277–278

Mullen K, Fahy O, Gormally M (2003) Ground flora and associated arthropod communities of forest road edges in Connemara, Ireland. Biodivers Cons 12:87–101

Neitlich PN, McCune B (1997) Hotspots of epiphytic lichen diversity in two young managed forests. Conserv Biol 11:172–182

Paton JA (1999) The liverwort flora of the British Isles. Harley Books, Colchester

Pearson LC (1969) Influence of temperature and humidity on distribution of lichens in a Minnesota bog. Ecology 50:740–746

Peck JE, Hong WS, McCune B (1995) Diversity of epiphytic bryophytes on three host tree species, Thermal Meadow, Hotsprings Island, Queen-Charlotte-Islands, Canada. Bryologist 98:123–128

Peterken GF (1996) Natural woodland: Ecology and conservation in northern temperate regions. Cambridge University Press, Cambridge

Phillips EA (1959) Bark bryophyte unions in southern Ireland. Bryologist 62:24–31

Purvis OW, Coppins BJ, Hawksworth DL, James PW, Moore DM (1992) The lichen flora of Great Britain and Ireland. Natural History Publications and The British Lichen Society, London

Ratcliffe PR, Peterken GF (1995) The potential for biodiversity in British upland spruce forests. For Ecol Manage 79:153–160

Rhoades FM (1981) Biomass of epiphytic lichens and bryophytes on *Abies lasiocarpa* on a Mt. Baker lava flow, Washington. Bryologist 84:39–47

Richards PW (1938) The bryophyte communities of a Killarney oakwood. Ann Bryol 11:108–130

Richardson DHS (1987) Lichens as pollution indicators in Ireland. In: Richardson DHS (ed) Biological indicators of pollution. Royal Irish Academy, Dublin

Rose F (1974) The epiphytes of oak. In: Morris MG, Perring FH (eds) The British oak: Its history and natural history. E.W. Classey, Faringdon

Rose F (1988) Phytogeographical and ecological aspects of *Lobarion* communities in Europe. Bot J Linn Soc 96:69–79

Simpson EH (1949) Measurement of diversity. Nature 163:688

Smith AJE (2004) The moss flora of Great Britain and Ireland. Cambridge University Press, Cambridge

Smith GF, Gittings T, Wilson M, French L, Oxbrough A, O'Donoghue S, Pithon J, O'Donnell V, McKee A-M, Iremonger S, O'Halloran J, Kelly DL, Mitchell FJG, Giller PS and Kelly T (2005) Assessment of biodiversity at different stages of the forest cycle. Report to the National Council for Forest Research and Development (COFORD) and the Environmental Protection Agency (EPA). http://www.coford.ie/iopen24/pub/pub/Project_files/312Report.pdf

Sparks TH, Greatorex-Davies JN, Mountford JO, Hall ML, Marrs RH (1996) The effects of shade on the plant communities of rides in plantation woodland and the implications for butterfly conservation. For Ecol Manage 80:197–207

SPSS (2001) SPSS for windows 11.0.1. SPSS Inc

Stace C (1997) New flora of the British Isles. Cambridge University Press, Cambridge

Trynoski SE, Glime JM (1982) Direction and height of bryophytes on four species of northern trees. Bryologist 85:281–300

Vanderpoorten A, Engels P, Sotiaux A (2004) Trends in diversity and abundance of obligate epiphytic bryophytes in a highly managed landscape. Ecography 27:567–576

Whittaker RH (1972) Evolution and measurement of species diversity. Taxon 21:213–215

Woods RG, Coppins BJ (2003) A conservation evaluation of British lichens. British Lichen Society, London

This paper was previously published in Biodiversity and Conservation, Volume 16(14) under doi 10.1007/s10531-007-9203-5

Bird assemblages in pine plantations replacing native ecosystems in NW Patagonia

María Victoria Lantschner · Verónica Rusch · Celina Peyrou

Originally published in the journal Biodiversity and Conservation, Volume 17, No 5, 969–989.
DOI: 10.1007/s10531-007-9243-x

Abstract Forest plantations of exotic conifers represent an important economic activity in NW Patagonia, Argentina. However, there is a remarkable lack of information on the impact of forestry on native biodiversity. We analyzed the effect of *Pinus ponderosa* plantations on bird communities, considering different stand management practices (dense and sparse tree covers), and different landscape contexts where they are planted (*Austrocedrus chilensis* forest and steppe). Ultimately we wished to assess in which way plantations may be designed and managed to improve biodiversity conservation. Bird richness and abundance did not change significantly in the steppe, although community composition did, and was partially replaced by a new community, similar to that of ecotonal forests. In contrast, in the *A. chilensis* forest areas, species richness decreased in dense plantations, but bird community composition remained relatively constant when replacing the native forest with pine plantations. Also, in *A. chilensis* forest, stand management practices aiming at maintaining low tree densities permit the presence of many bird species from the original habitat. In the steppe area in turn, both dense and sparse plantations are unsuitable for most steppe species, thus it is necessary to manage them at higher scales, maintaining the connectivity of the native matrix to prevent the fragmentation of bird populations. We conclude that pine plantations can provide habitat for a substantial number of native bird species, and this feature varies both with management practices and with the landscape context of areas where afforestation occurs.

Keywords *Austrocedrus chilensis* · Bird diversity · Exotic tree plantations · Forest management · Landscape context · Steppe

---

This manuscript is for the special issue: "Biodiversity and Planted Forests—Oxymoron or Opportunity?".

M. V. Lantschner (✉) · V. Rusch · C. Peyrou
Grupo de Ecología Forestal, INTA EEA Bariloche, CC 277, Bariloche 8400, Argentina
e-mail: mvlantschner@bariloche.inta.gov.ar

E.G. Brockerhoff et al. (eds.), *Plantation Forests and Biodiversity: Oxymoron or Opportunity?*
DOI: 10.1007/978-90-481-2807-5_3

Introduction

Biodiversity is an issue of increasing significance for the development and management of plantation forests and for their long-term sustainability (Carnus et al. 2006). Although the primary goal of plantation forestry is the efficient production of timber and pulp, it also offers important opportunities for biodiversity conservation if plantation design and management are appropriate (Lindenmayer 2002).

It is widely thought that exotic plantation forests are less favorable habitats than native forests (Hartley 2002; Carnus et al. 2006). Comparisons of unmanaged forests and plantations have found impoverished flora (Shankar et al. 1998; Humphrey et al. 2002) and fauna (Pomeroy and Dranzoa 1998; Lindenmayer and Munks 2000; Schnell et al. 2003) in the latter. Plantations may be unsuitable for many native species, because of the loss of some of the structural components of native habitats, such as understory vegetation (Yirdaw 2001; Brockerhoff et al. 2003), snags, and old or dead trees (Clout and Gaze 1984; Gjerde and Saetersdal 1997; Humphrey et al. 2002), which are critical for some wildlife. However, the effect of forest plantations on biodiversity depends on the type of plantation and the natural structure of surrounding native forests (Hartley 2002), and plantations can still contribute to biodiversity conservation if they are correctly designed and managed (Hartley 2002; Sayer et al. 2004; Carnus et al. 2006).

In Argentina, plantations of exotic fast growing conifers have been promoted by the state through subsidies in the last decades. In NW Patagonia, plantation forestry is a new activity that replaces traditional sheep production systems, representing an important economic alternative. Consequently, plantation rates show a rapidly increasing tendency (Schlichter and Laclau 1998). Plantations are established mainly in two types of ecosystems: *Austrocedrus chilensis* xeric forests and steppes. In steppes more than 2 million hectares are regarded as potentially useful for pine plantations (SAGPyA 1999).

Austrocedrus chilensis forests belong to the Valdivian Temperate Rainforest ecoregion, and steppe to the Patagonian Steppe ecoregion. Both have high conservation value; they were included in the WWF's "Global 200" conservation strategy (Olson and Dinerstein 1998) because they harbor some of the world's most outstanding and representative biodiversity. Both ecoregions have many endemic species. In the forest area they represent a high proportion (ca. 50% of species), whereas in the steppe area the proportion is smaller (ca. 20% of species) (Vuilleumier 1972). At the same time, they are threatened because of human intervention. The *A. chilensis* forest subregion is the most heavily altered and threatened within the Argentine portion of the Valdivian Temperate Rainforest ecoregion, as it has undergone forest fires, overgrazing and a high pressure for timber extraction (Laclau 1997). These forests currently have a low level of protection. Only 7% of the area is under the Protected Areas system, and mainly within the less restrictive status of national reserves (Vila 2002). In the Patagonian Steppe ecoregion, sheep overgrazing appears to have modified the vegetation and accelerated soil degradation processes (Soriano et al. 1983). About 4% of this ecoregion is under protected areas. Besides, most of these reserves are only nominally so, since little real protection is offered to wildlife (less than 1%) (Walker et al. 2005).

Potential negative environmental impacts of pine plantations have been the focus of great public concern in Patagonia (Rusch and Schlichter 2005). Pine plantations have been defined as "green deserts" and opposition to several afforestation projects has arisen, despite the fact that scientific information regarding their detrimental of beneficial effects is limited. In NW Patagonia there are some studies that suggest that changes in biodiversity in forest plantations depend on certain stand structural characteristics and on the taxa considered (Paritsis 2002; Rusch et al. 2005b; Corley et al. 2006). However, there is scarce information about the impact of pine plantations on biodiversity, particularly with respect to the landscape context.

The aim of this study is to assess the effects of plantation forestry in NW Patagonia, by considering its effects on the avifauna in two different ecosystems (*Austrocedrus chilensis* forest and steppe). We also look at different stand management practices (dense and sparse plantations), in order to provide information aiming at enhancing biodiversity conservation.

Methods

Study area

The study was carried out in NW Patagonia (from 39° 55′ S to 41° 51′ S; and from 71° 03′ W to 71° 33′ W) (Fig. 1). The climate of this area is temperate, dominated by a marked west-to-east decrease in precipitation (from 3,000 to 600 mm in less than 100 km). Vegetation types reflect this climatic pattern, and they are distributed as north–south orientated belts, parallel to the Andes. The study region is located in the forest/steppe ecotone, with vegetation ranging from xeric forests of *A. chilensis* in the west, to a shrubby steppe in the east. In the *A. chilensis* forest area, mean annual rainfall ranges between 900 and 1,200 mm/year (Barros et al. 1983), and the prevailing vegetation is a pure *A. chilensis* arboreal stratum, and an understory of shrubs and trees such as *Aristotelia chilensis, Maytenus boaria* and *Lomatia hirsuta.* In steppe area where pines are planted, mean annual rainfall ranges between 700 and 1,000 (Barros et al. 1983). Vegetation corresponds to that of a cold semi-desert, dominated by bunchgrasses (*Stipa* spp. and *Festuca* spp.) and low shrubs (*Mulinum spinosum* and *Senecio filaginoides*) (Cabrera 1976).

Plantations in the region comprise mainly of three species: ponderosa pine (*Pinus ponderosa*), lodgepole pine (*Pinus contorta*) and douglas fir (*Pseudotsuga menziesii*). We selected three situations (treatments) in both steppe and *A. chilensis* forest habitats (Fig. 1):

- Dense pine plantations: ponderosa pine plantations, where tree density was high, the herbaceous-shrubby cover was less than 15%, and the canopy cover was higher than 80% (N = 8 in steppe area, and N = 10 in *A. chilensis* forest area).
- Sparse pine plantations: ponderosa pine plantations where tree density was low, the herbaceous-shrubby cover was higher than 15% and the canopy cover was less than 80% (N = 9 in steppe area; and N = 6 in *A. chilensis* forest area).
- Native vegetation: continuous areas with the native vegetation, managed in the traditional way (cattle grazing of steppes and light selective logging of *A. chilensis* forests) which were sampled as control, and were located close to plantations (N = 9 in steppe area, and N = 11 in *A. chilensis* forest area).

Independent replicates were called "sites". The size of plantation blocks was between 9 and 200 ha (see Table 1).

Habitat characterization

Vegetation structure and composition were characterized on each site. Herbaceous (0–50 cm height) and shrub (50–200 cm height) cover were estimated through 10 random

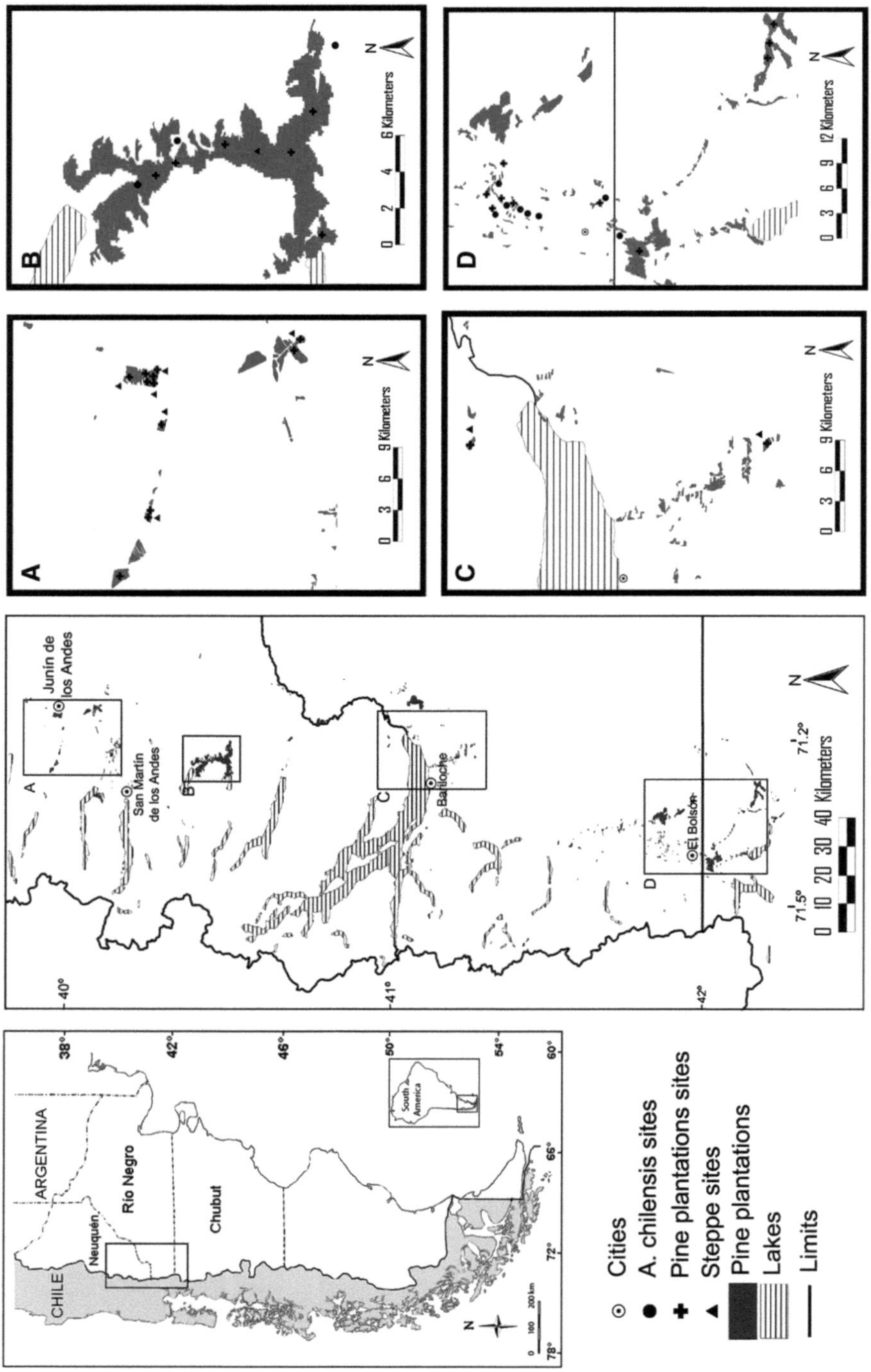

Fig. 1 Map of the study area, North Western Patagonia, Argentina

Table 1 Habitat variable values (mean ± standard error), in the different vegetation types

	Steppe			*A. chilensis* forest		
	NV	SPP	DPP	NV	SPP	DPP
Herbaceous cover (%)	45.5 ± 4.4	24.56 ± 1.9	0.5 ± 0.3	27.2 ± 2.8	24.0 ± 7.0	2.5 ± 0.5
Shrub cover (%)	0.0 ± 0.0	0.0 ± 0.0	0.0 ± 0.0	8.7 ± 2.3	8.2 ± 4.7	0.2 ± 0.2
Arboreal cover (%)	0.0 ± 0.0	50.5 ± 6.3	86.9 ± 1.6	62.5 ± 7.7	66.9 ± 6.6	88.7 ± 1.9
Canopy height (m)	0.0 ± 0.0	9.5 ± 0.6	13.0 ± 1.1	21.6 ± 1.9	11.6 ± 1.1	21.0 ± 1.9
Number of herbaceous-shrubby spp.	31.3 ± 2.45	25.1 ± 1.5	11.6 ± 2.1	44.0 ± 1.5	29.7 ± 3.6	15.4 ± 2.6
Number of arboreal spp.	0.0 ± 0.0	1.1 ± 0.1	1.3 ± 0.2	3.7 ± 0.4	3.7 ± 1.0	2.1 ± 0.4
Mean DBH (cm)	0.0 ± 0.0	18.1 ± 1.8	18.5 ± 1.2	22.8 ± 2.5	20.3 ± 1.6	22.1 ± 1.2
Basal area (m^2/ha)	–	15.1 ± 3.0	29.9 ± 4.1	31.6 ± 4.1	28.8 ± 5.9	42.8 ± 5.5
Precipitation (mm/year)	911 ± 61	922 ± 691	900 ± 77	990 ± 9	1016 ± 40	1000 ± 0
Plantation age (years)	–	12–19	17–22	–	18–23	23–32
Plantation area (ha)	–	106.8 ± 56.5	100.7 ± 74.5	–	51.2 ± 46.7	55.1 ± 53.5

NV, Native vegetation; SPP, Sparse pine plantation; DPP, Dense pine plantation quadrates of 45 cm × 45 cm per plot. All species in these strata present in the plot were recorded in order to estimate richness.

To characterize the arboreal structure, a 500 m^2 circular plot (or 1,000 m^2 in cases where tree density was less than 200 trees per ha) was established in each site. In each plot, diameter at breast height (DBH) of all the trees taller than 5 cm were measured, and canopy height was estimated with clinometer. These data were used to estimate basal area. Arboreal cover for each species was estimated using a densitometer (four readings per point at 10 random points per plot). Although it is very well known that structural elements such as snags and logs may influence birds, we did not include them in the analysis because these forests are not pristine and they almost lack these structural elements. Additionally, mean annual rainfall for each site was obtained from precipitation maps (Barros et al. 1983).

Bird surveys

Bird counts were conducted using fixed 50-m-radius point-counts (Ralph et al. 1993), from December 2001 to March 2002, on clear days, from sunrise to around 10 AM. In each site, seven plots were established, separated by a minimum of 120 m from adjacent points and from vegetation edges. In the cases where plantations were too small to include seven point counts, we established only six or five plots. At each plot, all bird species heard and/or seen were recorded, during a 7 minute-period, once in each site. Birds flying over were not recorded unless they were somehow using the vegetation below them. Taxonomy of the birds follows Narosky and Babarksas (2000). Bird abundance was estimated as the mean number of individuals per point, in each site.

Data analysis

Differences in the habitat variables across the treatments were assessed through one-way ANOVA in the cases where data met normality assumptions (Kolmogorov-Smirnov tests), followed by Waller-Duncan multiple comparisons (as we had an unequal sample size), to determine sources of differences. A Kruskal-Wallis test was performed in the cases where data were not normally distributed.

As sample efforts varied across sites, bird richness in each site was estimated with the Chao 1 estimate, using the program EstimateS 7.5 (Colwell 2005). The Chao 1 estimate is a non-parametric method for estimating total species richness. It is an abundance-based estimator, which uses the number of rare species to estimate the number of missing ones, and allows to standardize the survey effort (Chao 1984). Bird abundance was expressed as the mean number of birds per point, in each site. Differences between avian richness and abundance across vegetation types—both, in the steppe and the *A. chilensis* forest area—were assessed through one-way ANOVA, as data met normality (Kolmogorov-Smirnov tests) and homoscedasticity requirements. Waller-Duncan multiple comparisons for unequal sample sizes were used as post hoc tests. Comparisons of abundance of each species across vegetation types were done using the non-parametric Kruskal-Wallis test, as they were not normally distributed. Spearman rank correlations were performed in order to relate bird species abundance with vegetation variables (Sokal and Rohlf 1981).

As there is a wide variation in the sizes of plantation fragments, and this can affect the ability of birds to explore into the plantation from the surrounding native matrix, Pearson correlations between plantation size and total bird abundance and richness were estimated, to assess if there is a significant relation between these variables. Additionally, Spearman correlations were estimated between plantation size and the abundance of each species in each site, to assess if there is a relation between plantation size and any of the species.

We also assessed whether bird species of conservation importance changed their abundance in sparse and dense pine plantations. The conservation status of each species was obtained from the IUCN Red List of Threatened Species (IUCN 2006). Additionally, we selected the species classed as having high conservation importance by the "SUMIN" index, for Nahuel Huapi National Park, located nearby the study area (Grigera et al. 1996). The SUMIN is an index comprising 12 survival-related variables, which are considered essential for the conservation of bird species. Variables with the greatest influence upon the index value are those related to distribution, space-use plasticity, reproductive potential and trophic amplitude. The index ranges between 2 and 18, and we arbitrarily selected all species with an index of 13 or higher. A Kruskal-Wallis test was performed on all those species with more than 3 detections, to compare the difference in abundance of each species between treatments.

Multivariate methods were used to analyze the distribution of bird species among the different vegetation types and to assess the effect of substitution of the native vegetation by exotic pine plantations on native bird communities. We performed an analysis of similarity (ANOSIM) with the Bray-Curtis similarity index, including all vegetation types together, to determine the existence of differences in the composition of bird communities. ANOSIM is a non-parametric test to establish differences between two or more groups, based on distance measures. Distances are converted to ranks, and then distances between groups are compared with distances within groups. The test uses the statistic R, which can take values between 0 and 1. Large positive R means dissimilarity between groups (Hammer et al. 2001). The significance of the test was determined by permutation of group membership, with 5,000 replicates. The test was performed with PAST 1.46 (Hammer et al. 2001).

The relationship between the composition of the bird species communities and habitat variables across all vegetation types, together, was examined by a Canonical Correlation Analysis (CCA, ter Braak 1986). Ordination axes represent the maximum variability that is attributable to the environmental parameter. Relative effects of individual environmental parameters were then visualized by the relative length of the respective vectors in the ordination space (Kent and Coker 1992). A Monte Carlo permutation test was performed, to test the significance of the relation between species and environmental variables, based on the first ordination axis, and on all canonical axes together (Kent and Coker 1992).

Results

Habitat characterization

In the steppe area, that native vegetation showed significant structural differences from pine plantations (Table 1), particularly to dense plantations, whereas differences from sparse plantations were not so marked. Herbaceous cover and species richness showed significant differences ($F = 58.151$, $P < 0.000$; and $F = 23.475$, $P < 0.000$, respectively): they were highest in native vegetation plots; they decreased in sparse pine plantations, and practically no herbaceous vegetation was recorded in dense pine plantation plots. Mean canopy height, arboreal cover, and basal area were also significantly different across treatments (F = 100.061, $P < 0.000$; and F = 123.951, $P < 0.000$; F = 30.088, $P < 0.000$, respectively): they increased inversely, being higher in dense pine plantations and lower in native vegetation. The dbh values also showed significant differences (F = 78.294, $P < 0.017$), being similar in dense and sparse pine plantation, and null in native vegetation. Mean annual precipitation, and plantations areas were similar in the three vegetation types (Kruskal-Wallis $P > 0.926$; and F = 0.017, $P > 0.983$).

In the *A. chilensis* forest area, in turn, differences in vegetation structure across the different vegetation types were less significant (Table 1), with dense pine plantation plots being the most different. Herbaceous, shrub, and arboreal cover showed significant differences across the treatments (F = 18.348, $P < 0.000$; F = 4.189, $P < 0.028$; and F = 5.746, $P < 0.009$). Herbaceous and shrub cover values were similar in native vegetation and sparse pine plantations, and higher than those of dense pine plantations (Waller-Duncan, $P < 0.050$), whereas arboreal cover was higher in dense pine plantations than in native vegetation and sparse pine plantations (Waller-Duncan, $P < 0.050$). Canopy height also showed significant differences across treatments (F = 7.207, $P < 0.004$), being similar in native vegetation and dense pine plantations, and less high in sparse pine plantations (Waller-Duncan, $P < 0.050$). Herbaceous-shrubby species richness was also significantly different across treatments (F = 41.901, $P < 0.000$), being highest in native vegetation, intermediate in sparse pine plantations and lowest in dense pine plantations (Waller-Duncan, $P < 0.050$). Arboreal species richness, basal area, and dbh, in turn, did not show differences across treatments (F = 3.028, $P > 0.067$; F = 2.106, $P > 0.144$; and F = 0.350, $P > 0.708$). As in the steppe area, mean annual precipitation and plantation areas were similar in the three vegetation types (Kruskal-Wallis, $P > 0.794$; and F = 0.010, $P > 0.990$).

Bird abundance and richness

A total of 41 bird species was recorded in the study area (Table 2). All except one (*Lophortix californica*) were native to the region. Thirty three species were recorded in the steppe area and twenty-six in the *A. chilensis* forest area.

In the steppe area, neither species richness nor bird abundance was significantly different across native vegetation, sparse pine plantations and dense pine plantations (Chao's mean species/site estimate: 9.81, 10.09 and 5.79; and mean individuals/point: 2.61; 2.44 and 1.87, respectively) (ANOVA for richness $F_{2,23} = 1.14$; $P < 0.337$ and ANOVA for abundance $F_{2,23} = 0.61$; $P < 0.553$) (Fig. 2). Species richness and bird abundance did not show a significant correlation (Pearson) with the plantation area, neither in dense nor in sparse plantations (richness: $P > 0.300$; $P > 0.256$, respectively; abundance: $P > 0.099$; $P > 0.565$, respectively).

In *A. chilensis* forest area, species richness did not differ significantly across the different types of vegetations (ANOVA $F_{2,24} = 0.631$; $P < 0.541$) (Fig. 2). (Chao's mean species/site estimate: 8.77, 9.89 and 7.30). In turn, total abundance of birds was significantly different across the different types of vegetation (ANOVA $F_{2,24} = 4.677$; $P < 0.019$) (Fig. 2). Bird assemblages living in native vegetation and sparse plantations (4.51 and 4.19 individuals/point, respectively) had significantly higher abundance than those in dense pine plantations (2.35 individuals/point) (Waller-Duncan $P < 0.050$). As in the steppe area, species richness and bird abundance did not show a significant correlation with the plantation area, neither in dense nor in sparse plantations (richness: $P > 0.388$; $P > 0.454$, respectively; abundance: $P > 0.526$; $P > 0.900$, respectively).

Table 2 Mean number of bird species per point, in the different types of habitats

Species	Abbreviation	SUMIN[a]	Steppe			*A. chilensis* forest		
			NV	SPP	DPP	NV	SPP	DPP
Agriornis sp. (Shrike-Tyrant)	Agr sp	–	0.019	0.000	0.000	0.000	0.000	0.000
Anairetes parulus (Tufted Tit-Tyrant)	Ana par	10	0.082	0.138	0.099	0.183	0.433	0.031
Aphrastura spinicauda (Thorn-tailed Rayadito)	Aph spi	12	0.000	0.143	0.021	0.487	0.390	0.173
Asthenes pyrrholeuca (Lesser Canastero)	Ast pyr	8	0.206	0.019	0.025	0.013	0.298	0.014
Buteo polyosoma (Red-backed Hawk)	But pol	11	0.000	0.016	0.000	0.000	0.000	0.000
Caprimulgus longirostris (Band-winged Nightjar)	Cap lon	9	0.000	0.016	0.000	0.000	0.000	0.000
Carduelis barbata (Black-chinned Siskin)	Car bar	9	0.175	0.368	0.265	0.247	0.157	0.017
Cinclodes sp. (Cinclodes)	Cin sp.	–	0.000	0.016	0.000	0.000	0.000	0.000
Cistothorus platensis (Grass Wren)	Cis pla	6	0.019	0.000	0.000	0.000	0.000	0.000
Colaptes pitius (Chilean Flicker)	Col pit	11	0.000	0.000	0.000	0.087	0.000	0.060
Colorhamphus parvirostris (Patagonian Tyrant)	Col par	13	0.000	0.000	0.000	0.013	0.000	0.000
Columba araucana (Chilean Pigeon)	Col ara	18	0.000	0.000	0.000	0.109	0.000	0.000
Columba picazuro (Picazuro Pigeon)	Col pic	–	0.016	0.251	0.154	0.013	0.000	0.020
Coragyps atratus (Black vulture)	Cor atr	9	0.000	0.016	0.000	0.000	0.000	0.000
Curaeus curaeus (Austral Blackbird)	Cur cur	8	0.016	0.032	0.208	0.009	0.067	0.000
Diuca diuca (Common Diuca-Finch)	Diu diu	7	0.474	0.019	0.057	0.000	0.462	0.000
Elaenia albiceps (White-crested Elaenia)	Ela alb	6	0.241	0.733	0.583	2.472	1.203	1.220
Falco sparverius (American Kestrel)	Fal spa	4	0.037	0.019	0.000	0.000	0.000	0.000
Geositta cunicularia (Common Miner)	Geo cun	6	0.056	0.000	0.000	0.000	0.000	0.000
Leptasthenura aegithaloides (Plain-mantled Tit-Spinetail)	Lep aeg	8	0.000	0.000	0.000	0.018	0.024	0.040
Lophortix californica (Californian Quail)	Lop cal	–	0.111	0.000	0.000	0.000	0.000	0.000
Milvago chimango (Chimango Caracara)	Mil chi	6	0.016	0.000	0.000	0.031	0.000	0.000
Muscisaxicola sp. (Ground-Tyrant)	Cus sp	–	0.016	0.000	0.000	0.000	0.000	0.000
Phrygilus fruticeti (Mourning Sierra-Finch)	Phr fru	7	0.079	0.000	0.000	0.000	0.000	0.000

Table 2 continued

Species	Abbreviation	SUMIN[a]	Steppe			*A. chilensis* forest		
			NV	SPP	DPP	NV	SPP	DPP
Phrygilus gayi (Gray-hooded Sierra-Finch)	Phr gay	7	0.071	0.000	0.000	0.018	0.000	0.071
Phrygilus patagonicus (Patagonian Sierra-Finch)	Phr pat	13	0.000	0.053	0.158	0.009	0.117	0.396
Polyborus plancus (Crested Caracara)	Pol pla	6	0.000	0.016	0.036	0.000	0.000	0.000
Pteroptochos tarnii (Black-throated Huet-Huet)	Pte tar	11	0.000	0.000	0.000	0.000	0.033	0.000
Pygarrhichas albogularis (White-throated Treerunner)	Pyg alb	13	0.000	0.000	0.000	0.028	0.000	0.000
Scelorchilus rubecula (Chucao Tapaculo)	Sce rub	10	0.019	0.000	0.000	0.279	0.105	0.063
Scytalopus magellanicus (Andean Tapaculo)	Scy mag	12	0.000	0.000	0.000	0.000	0.000	0.014
Sicalis lebruni (Patagonian Yellow-Finch)	Sic leb	–	0.032	0.000	0.000	0.000	0.000	0.000
Strix rufipes (Rufous-legged Owl)	Str ruf	13	0.000	0.000	0.061	0.015	0.000	0.014
Sturnella loyca (Long-tailed Meadowlark)	Stu loy	7	0.280	0.000	0.000	0.000	0.000	0.000
Tachycineta leucopyga (Chilean Swallow)	Tac leu	6	0.103	0.016	0.036	0.031	0.213	0.000
Troglodytes aedon (House Wren)	Tro aed	3	0.079	0.183	0.018	0.285	0.371	0.088
Turdus falcklandii (Austral Thrush)	Tur fal	6	0.257	0.272	0.122	0.096	0.156	0.105
Tyto alba (Barn Owl)	Tyt alb	6	0.000	0.000	0.000	0.000	0.000	0.000
Vanellus chilensis (Southern Lapwing)	Van chi	6	0.066	0.000	0.000	0.000	0.000	0.000
Zenaida auriculata (Eared Dove)	Zen aur	5	0.019	0.000	0.025	0.013	0.024	0.000
Zonotrichia capensis (Rufous-collared Sparrow)	Zon cap	2	0.124	0.114	0.000	0.052	0.142	0.020

NV, Native vegetation; SPP, Sparse pine plantation; DPP, Dense pine plantation in the steppe and the *A. chilensis* forest area

[a] SUMIN index (Grigera et al. 1996), it indicates the conservation importance of the bird species, based in survival-related variables (higher numbers indicate higher conservation importance)

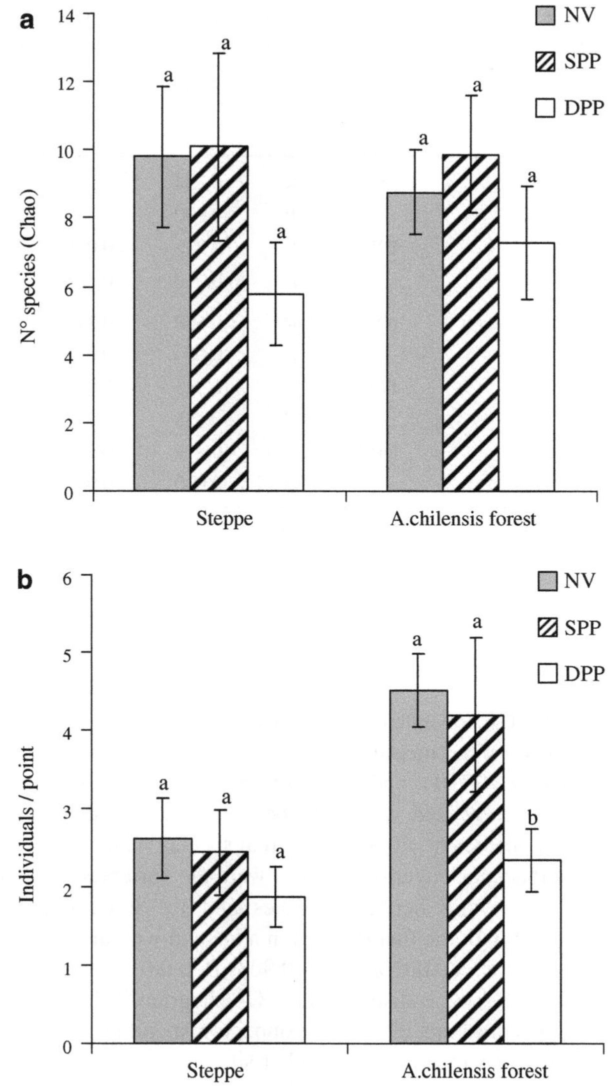

Fig. 2 Bird community parameters in the different types of vegetation (NV: native vegetation, SPP: sparse pine plantation, DPP: dense pine plantation) in steppe and *A. chilensis* forest habitats. (**a**) Richness (Chao estimate) and (**b**) abundance (individuals/point) of birds (means ± standard error)

Community analysis

The ANOSIM showed that, in the steppe area (Table 3), bird communities of sparse and dense pine plantations were similar, whereas those of native vegetation differed. Although species richness in native vegetation and pine plantations was similar, differences between bird communities can mainly be accounted for by differences in species composition. Thirteen bird species present in the native vegetation were exclusive to this habitat (52% of the total number of species), and twelve (48%) were shared with pine plantations.

In *A. chilensis* forest areas, species composition in native vegetation sites, and sparse and dense pine plantations, was similar, most species (93%) being present in these vegetation types. The ANOSIM (Table 3) showed differences between native vegetation and dense plantations, but these were due to lower species richness and abundance (i.e., only

Table 3 Analysis of similarity for bird communities (ANOSIM) in native vegetation (NV), sparse pine plantation (SPP) and dense pine plantation (DPP) for both steppe and *A. chilensis* forest areas

			Steppe			*A. chilensis* forest	
			NV	SPP	DPP	NV	SPP
Steppe	SPP	R	0.354				
		P	0.000				
	DPP	R	0.235	0.028			
		P	0.010	0.333			
A. chilensis forest	NV	R	0.688	0.343	0.510		
		P	0.000	0.003	0.000		
	SPP	R	0.223	0.040	0.221	0.207	
		P	0.033	0.297	0.054	0.055	
	DPP	R	0.519	0.267	0.322	0.261	0.289
		P	0.000	0.005	0.003	0.003	0.018

R (which can take values between 0 and 1) reflects the degree of separation of the assemblages based on their species composition; *P* is significant at the α-level of 0.05

58% of the total species found in the *A. chilensis* forest area was also observed in dense plantations).

The CCA revealed that 16.3% of the total variance in species dispersion can be explained by the measured environmental variables. Canopy height ($r = -0.792$), arboreal cover ($r = -0.791$), and the number of arboreal species ($r = -0.637$) were the strongest variables correlated with the first axis. On the other hand, the richness of herbaceous-shrubby species ($r = 0.633$), arboreal species richness ($r = 0.450$), shrub cover ($r = 0.311$), and herbaceous cover ($r = 0.307$) were the variables most correlated with the second axis. The relationship between species and the environmental variables was significantly correlated with the first ordination axis, and with all canonical axes together (Monte Carlo $F = 4.164$; $P < 0.002$ and $F = 0.908$; $P < 0.014$, respectively).

Figure 3a and b displays the CCA diagram. When analyzing all the sites together, the diagram shows that there is a group, corresponding to steppe native vegetation sites, which is clearly separated from the other sites, with species such as *Vanellus chilensis, Milvago chimango, Sturnella loyca, Phrygilus gayi, Melanodera xanthogramma* and *Geositta cunicularia*. This group is associated with a high herbaceous cover, and presented a group of bird species characteristic of open areas. The rest of the sites were distributed close together, and presented a bird community composed mainly of species typical of native forests, such as *Phrygilus patagonicus, Aphrastura spinicauda* and *Elaenia albiceps*. Sparse and dense pine plantation sites in steppe areas tended to be grouped, and were associated with a low number of arboreal species, low shrub cover, and low canopy height. Dense pine plantation in *A. chilensis* forests areas were also distributed close together, and showed an association with high arboreal cover. *A. chilensis* forests were associated with a high number of arboreal species, shrub cover, and canopy height; whereas sites corresponding to sparse pine plantation in *A. chilensis* forest areas were distributed between pine plantations in steppe areas and native *A. chilensis* forest vegetation sites.

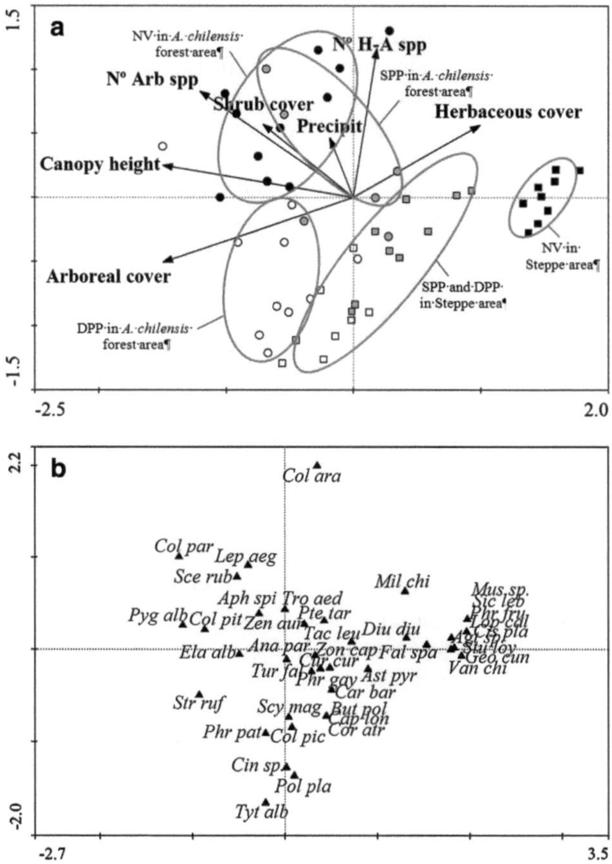

Fig. 3 Ordination plots from canonical correspondence analysis of the species/environment data in the studied sites. (**a**) Environmental variables (rows) and Sites (■ Steppe native vegetation, ▣ Sparse pine plantation plantation on steppes, □ Dense pine plantation on steppes, • *A. chilensis* forest native vegetation, ● Sparse pine plantation on *A. chilensis* forest, ○ Dense pine plantation on *A. chilensis* forest). The arrows are plotted pointing in the direction of maximum change of the environmental variable across the diagram, and the length of the arrow is proportional to the magnitude of change in that direction. (**b**) Species, indicated with the three first letters of the genus and the species (see Table 2)

Relationships between bird species and habitat structure

In steppe areas, four species were significantly more abundant in native vegetation: *Sturnella loyca*, *Asthenes pyrrholeuca*, *Phrygilus gayi* and *Diuca diuca* (Kruskal-Wallis $X^2 = 8.519$, $P < 0.014$; $X^2 = 9{,}428$, $P < 0.009$; $X^2 = 6.132$, $P < 0.047$ and $X^2 = 6.713$; $P < 0.035$, respectively). The first three species were positively correlated with herbaceous cover ($P < 0.032$; $P < 0.014$ and $P < 0.009$, respectively) and negatively correlated with arboreal cover ($P < 0.011$; $P < 0.008$ and $P < 0.034$, respectively), and basal area ($P < 0.011$; $P < 0.024$ and $P < 0.034$, respectively). *Diuca diuca* was negatively correlated with arboreal cover ($P < 0.034$) and basal area ($P < 0.030$). On the other hand, *A. spinicauda* was significantly more abundant in sparse pine plantations (Kruskal-Wallis $X^2 = 7.581$; $P < 0.023$), *Columba picazuro* was significantly more abundant in both sparse and dense pine plantations than in the other habitats (Kruskal-Wallis $X^2 = 6.655$; $P < 0.036$) and showed a positive association with arboreal cover ($P < 0.039$); and *Tyto alba* was most abundant in dense pine plantations (Kruskal-Wallis $X^2 = 7.313$; $P < 0.026$), showing a negative association with herbaceous cover ($P < 0.020$). Only one species showed a significant correlation with plantation size: *A. spinicauda*, which was negatively correlated ($P < 0.047$).

In *A. chilensis* forest areas, the abundance of four species differed significantly between the different types of vegetation: *Elaenia albiceps* which was most abundant in native vegetation (Kruskal-Wallis $X^2 = 9.254$; $P < 0.010$); *Anairetes parulus*, which was most abundant in native vegetation and sparse pine plantations (Kruskal-Wallis $X^2 = 6.315$; $P < 0.043$), and showed a positive association with herbaceous cover ($P < 0.021$); *Diuca diuca* which was most abundant in sparse pine plantations, and was negatively associated with canopy height (Kruskal-Wallis $X^2 = 11.328$; $P < 0.003$); and *Tachycineta leucopyga* which was most abundant in sparse pine plantations (Kruskal-Wallis $X^2 = 6.374$; $P < 0.041$), showing a positive association with herbaceous cover ($P < 0.036$) and a negative association with arboreal cover ($P < 0.002$) and canopy height ($P < 0.026$). No species were correlated with plantation size.

None of the species recorded in the study area was listed as threatened in the IUCN Red List (IUCN 2006); they were all at low risk. Also, one species recorded in the steppe area and five in the *A. chilensis* forest habitat, presented a SUMIN index (Grigera et al. 1996) equal to or higher than 13 (Table 2). Among these, *Phrygilus patagonicus* appeared in both, steppe and *A. chilensis* forest areas. In the steppe area there were no significant differences in abundance between habitat types (Kruskal-Wallis $X^2 = 3.130$; $P < 0.209$), whereas in the *A. chilensis* forest area, it was not abundant enough to determine significant differences across treatments. *Columba araucana* was only recorded in the *A. chilensis* forest although differences between habitats were not significant, it did not present significant differences across treatments either. *Colorhamphus parvirostris, P. albogularis*, and *Strix rufipes* were also recorded in the *A. chilensis* forest, but because of their low abundance, we could not determine significant differences in their densities across the different vegetation types.

Discussion

Changes in bird abundance and richness due to the replacement of native vegetation with exotic pine plantations differed depending on which type of vegetation was replaced. In the steppe areas, there were no changes in species richness and abundance, whereas in the *A. chilensis* forests, bird richness decreased in dense pine plantations.

Several studies in different regions of the world have found that conifer plantations support fewer bird species and lower total density of birds than native vegetation (Driscoll 1977; Carlson 1986; Leberton and Pont 1987; Mitra and Sheldon 1993; Estades 1994; Gjerde and Saetersdal 1997; Pomeroy and Dranzoa 1998; Marsden et al. 2001; Paritsis 2002; Lindenmayer et al. 2003; Zurita et al. 2006). However, others have noted that the avifauna in plantation forests, may be as diverse and abundant as in the natural vegetation they replace (Clout and Gaze 1984; Estades and Temple 1999; Vergara and Simonetti 2004; Lantschner 2005; Gonzalez-Gomez et al. 2006). Our results are in line with the view that changes in bird richness and abundance depend strongly on site and regional characteristics.

A recurring dilemma for land managers worldwide, however, is the trade-off between managing for maintenance of total species diversity and the need to pay special attention to the species belonging to the original native systems (Petit and Petit 2003). When native vegetation was replaced with pine plantations in the steppe areas we studied, species composition changed substantially although numbers of species and individuals changed little overall. In the *A. chilensis* forest area, in turn, some changes in species richness occurred but species composition remained quite similar. These findings suggest that replacing native vegetation with exotic conifer plantations does not always lead to changes in overall bird species richness or bird abundance, but that changes in the species composition of the bird community may occur and must also be considered.

Stand-management practices influence the presence of several bird species, and sparse plantations, particularly in the *A. chilensis* forest area, tend to have less impact on the native bird communities than dense pine plantations. Sparse plantations have different origins; some are sparse simply because trees are still small, while others have undergone thinning, or were planted at low densities. In all cases, however, the fact that they have low canopy cover, leads to increased light availability, and consequently, the development of a higher shrub and herbaceous cover (Miller 2001).

Native vegetation inside pine plantations may partly account for the composition of the bird communities observed in the study. Past studies in South American temperate forests have concluded that the presence of native vegetation is one of the most important factors determining the use of plantations by native birds (Estades 1994; Estades and Temple 1999; Vergara and Simonetti 2004; Lantschner and Rusch 2007). Our results confirm observations by Estades and Temple (1999) for Chilean temperate forests that the abundance of many bird species, (e.g., *L. aegithaloides, S. rubecula, C. parvirostris*) is positively associated with the amount of native vegetation in the understory, particularly in the *A. chilensis* area. Understory vegetation may provide escape cover against predators, safe nesting sites, and food resources for birds. Its importance for some native bird species was also documented by several studies in the Chilean temperate forests, in sites associated with other disturbances, like agriculture, livestock grazing, and logging (Willson et al. 1994; Sieving et al. 1996, 2000; Reid et al. 2004; Díaz et al. 2004, 2005; Willson 2004; Castellón and Sieving 2006).

The landscape context also affects the composition of bird communities. Small changes resulted when plantations replaced native forest having a similar vegetation structure, whereas changes were more marked when plantations replaced steppe habitats. Consequently, when plantations were established in forest areas, the surrounding matrix affected the composition of the avifauna in plantations, and hence most forest birds were as likely to be found and as abundant in sparse plantations as they were in native vegetation. In dense plantations, richness decreased, but the assemblage found was a subset of the assemblage found in native vegetation. In contrast, when plantations were established in steppe areas, structural changes of the vegetation were so important that most steppe bird species did not find suitable habitat inside the plantation, particularly those that feed and nest in open grasslands (e.g. *S. loyca, P. gayi, A. pyrrholeuca, M. chimango*) (Christie et al. 2004). Therefore the steppe bird community was partially replaced with a new community made up of generalist birds and species typical of the forest ecotonal areas located near the study area, such as those that use trees to feed, nest, or take refuge (i.e., *A. spinicauda, T. alba, C. picazuro*) (Christie et al. 2004). The fact that steppe bird communities were more affected than *A. chilensis* communities can be also explained by the fact that bird species in Patagonian temperate forests have broad niches and wider distributions across habitats than steppe birds (Vuilleumier 1972, 1985) because these forests are isolated and have evolved as 'islands' (Vuilleumier 1985). This ability of forest birds to adapt to different types of habitats might enable them to adopt pine plantations more easily as new forest habitats, which was suggested by Estades and Temple (1999) for pine plantations in Chile's temperate regions.

According to these results, in the *A. chilensis* forest area, the habitat connectivity for most of the forest bird species may be maintained in sparse pine plantations but may be affected in dense pine plantation. Almost all bird species that were present in the native vegetation were also actively using sparse pine plantations, whereas dense plantations supported fewer species. In steppe areas, in turn, most steppe bird species were absent in both sparse and dense pine plantations. Large plantations could therefore act as barriers for many species and fragment their habitat (Fahrig 2001).

In NW Patagonia, plantation forests still occupy relatively small areas within native vegetation at present and exist as small scattered patches. Thus their impact on the connectivity of native bird populations may be less serious than in other parts of the world where plantations cover more extensive areas (Clout and Gaze 1984; Estades and Temple 1999; Lindenmayer et al. 2003). Nevertheless, strong subsidies on exotic forest plantations can increase planted areas substantially. Planning for biodiversity conservation in managed areas based on local information is, then, urgently needed in the region to anticipate severe negative impacts.

Changes at species level

In the steppe area, some species were clearly affected by the replacement of native vegetation with pine plantations where their abundance was greatly reduced. This was mainly the case with those species that require open areas, such as *Sturnella loyca* and *Phrygilus gayi*, which feed on the soil in open grasslands (Christie et al. 2004); *Asthenes pyrrholeuca*, which forages and hides within low bushes in open areas; and *Milvago chimango*, which usually looks for carrion in open lands, roads, and forest openings (Christie et al. 2004).

On the other hand, plantations benefited some other bird species, particularly those typical of ecotonal forest areas, which use trees to feed or take refugee. *Aphrastura spinicauda* was more abundant in sparse plantations, probably because it feeds on insects on the trees, but it was absent in dense plantations, which indicates that these are unsuitable for this species probably because they are too dense. On the other hand, *Tyto alba*, which is a nocturnal raptor owl that hides in holes in trees or caves during the day, appeared only in dense plantations. Finally *Columba picazuro* was more abundant in sparse and dense plantations than in native vegetation. This is a common species in other regions that has expanded its distribution to the south in the last decades (Narosky and Babarksas 2000), and has been reported as common in agricultural areas and artificial forests (Christie et al. 2004).

In the *A. chilensis* forest area only one species, *Elaenia albiceps*, was significantly more abundant in native vegetation than in pine plantations. Although this species was abundant in plantations, it was twice as abundant in native vegetation. These results are in line with other studies carried out in Chile (Estades and Temple 1999). The explanation for this pattern could be the greater heterogeneity of the native forest vegetation which provides a greater diversity and abundance of foraging resources for this species. In addition, three species were more abundant in sparse plantations than in native vegetation and dense plantations: *Diuca diuca*, *Anairetes parulus* and *Tachycineta leucopyga*. These species are typical of ecotonal forests, and they may benefit from sparse pine plantations because they prefer open or disturbed forests areas where they look for insects in the soil, shrubs or air (Christie et al. 2004).

Five of the species recorded in the *A. chilensis* forest area (*E. albiceps, S. rubecula, P. tarnii, A. spinicauda* and *S. rufipes*), were considered focal species for the South American temperate rainforests, due to their habitat needs, range sizes, and/or importance in the food chain (Vila 2002; Rusch et al. 2005a). All these species were present in pine plantations, which implies that these are not unsuitable for those birds. However, the foliage insectivorous *E. albiceps*, as seen above, was negatively affected by the replacement of native forest with pine plantations. In addition, the understory insectivorous *S. rubecula* and the foliage insectivorous *A. spinicauda* also tend to decrease in pine plantations, particularly in dense plantations, although the differences were not significant. As pine plantations appear to be less suitable than native forests for some keystone species, detailed studies of population dynamics should be conducted to improve our understanding of the conservation value of these anthropogenic homogenous ecosystems. On the other hand, the nocturnal raptor owl *S. rufipes* and the understory bird *P. tarnii* did not show any negative tendency in pine plantations. Both species have been seen nesting in pine plantations (Vergara and Simonetti 2003).

As the bird surveys were carried out without taking into account differences in detectability of species between habitat types, species abundance values should be interpreted with caution. The abundance numbers of some species was probably underestimated. However, we do not think that there are important differences in abundance values across the different species and habitats as we only recorded the bird species heard or seen inside the first 50 m radius of each plot.

Management and Conservation Implications

The traditional view of conservation reserves is of large, untouched areas. However, few landscapes provide the opportunity to preserve large tracts of land, and conserving biodiversity within the matrix of multiple-use lands becomes essential (Lindenmayer and Franklin 1997). Our results show that the type of management applied to pine plantations influences their suitability as habitat for birds, and so appropriate changes in design and management regimes of pine plantations can contribute to biodiversity conservation.

At the stand-scale, the maintenance of some forest structural elements is likely to permit the conservation of forest birds in planted forests. In this sense, the presence of native understory vegetation is of great importance, thus ideally the management of pine plantations should enhance the native understory vegetation to provide additional conservation benefits (Estades and Temple 1999). Based on our results, one of the most important ways to promote the presence of native vegetation in the understory is to plant at low densities or by early thinning (Zobrist and Hinckley 2005). Additionally, management practices such as herbicide application, removal of the stumps and roots of native trees, and other soil disturbances may reduce habitat quality for these birds (Vergara and Simonetti 2003). On the other hand, retention of biological legacies such as long-lived trees, snags and downed wood within plantation stands gives plantations a structure more similar to natural stands (Clout and Gaze 1984; Gjerde and Saetersdal 1997). In our study, different canopy heights and the number of tree species also proved to be important in bird community composition.

At the landscape scale, the most important factors to consider are plantation size, shape, and location (Díaz et al. 1998), extent to which a landscape has been and will be planted, the similarity of plantation structure to natural vegetation (Gjerde and Saetersdal 1997), and what habitats are being converted into plantation (Hartley 2002). Our results show that pine plantations, particularly in the steppe area, are unsuitable for some local bird species. This implies that it is important to consider landscape design alternatives for maintaining or enhancing diversity in planted landscapes, and avoiding the fragmentation of bird populations. In this regard, special areas of high diversity and those areas that provide habitat for threatened, rare, or endangered species should be identified and specifically managed. In addition, it would be desirable to consider the connectivity of the remaining natural habitats. Unlike to other studies, which found that the number of bird species was higher in small patches of plantations (Curry 1991; Lindenmayer et al. 2002), our results showed that plantation size was not an important factor of the ability of birds to use the plantations. However, it would be necessary to carry out further studies covering a wider range of sizes to provide more certainty about this.

The impact of plantation forestry on biodiversity also depends on the degree to which the landscape is natural versus degraded. While the conversion of natural ecosystems to plantation forests will rarely be desirable from a biodiversity point of view, planted forests often replace other land uses (Carnus et al. 2006). Thus, an objective assessment of the potential or actual impacts of planted forests on biodiversity requires appropriate reference points. In our study area, the native vegetation sites were not pristine prior to the establishment of plantation forests (i.e., cattle grazing occurred in the steppe area and light selective logging in *A. chilensis* forests). Therefore it is necessary to consider that bird communities in the replaced habitats were already affected by other disturbances before the replacement with forest plantations, and pristine systems may have shown stronger effects.

However, our study covered a relatively short-term, and long-term records may be necessary to better understand the effects of land management on the biodiversity of these environments. Additionally, further studies should include other taxa, such as mammals or invertebrates, which have requirements that differ from those of birds, and which may perceive the impact of pine plantations in different ways.

Conclusions

The results of this study suggest that pine plantations can provide habitat for a substantial number of native bird species, and that this varies with the landscape context. Plantations established in a forest matrix generate less impact on bird communities than those in a steppe matrix. Thus, in the *A. chilensis* forest areas, stand management practices aiming at maintaining low tree densities enhance the retention of many bird species, as they enable the persistence of some critical structural elements of native vegetation. In steppe areas, in turn, both dense and sparse plantations are unsuitable for many species. In those areas it is necessary to manage plantations with consideration of higher scales, maintaining the connectivity of the native vegetation remnants to minimize the fragmentation of bird populations.

Landscapes comprising mosaics of native vegetation and forest plantations are more desirable from a conservation perspective than other land uses that are more structurally simplified, such as agriculture (Moore and Allen 1999), or intensive livestock grazing (Lantschner 2005). Thus, when analyzing the impact of plantation forestry on biodiversity, the ecological context of planted forest development, as well as the social and economic context shaping land-use changes must be considered (Carnus et al. 2006). The definition of management objectives linked to sustainability, considering endangered and functional keystone species, and the integrated analysis of different spatial scales are important to find a balance between intensive land use and biodiversity conservation.

Acknowledgments We thank the landowners who provided access to their lands where fieldwork was conducted. We also thank M. Sarasola and staff from APN (Administración de Parques Nacionales) for assistance during the fieldwork, P. Willems during statistical analysis, and J. Corley and four anonymous referees for the comments on this manuscript. This study was funded by SAGPyA (Secretaría de Agricultura, Ganadería, Pesca y Alimentos) through the project PIA 01/00, INTA (Instituto Nacional de Tecnología Agropecuaria, "Manejo Sustentable de Plantaciones" Project) and Turner Foundation.

References

Barros VR, Cordon VH, Moyano CL, Méndez RJ, Forquera JC, Pizzio O (1983) Cartas de Precipitación de la zona oeste de las Provincias de Río Negro y Neuquén. Informe de la Facultad de Ciencias Agrarias, Universidad Nacional del Comahue, Cinco Saltos, Neuquén, Argentina

ter Braak CJF (1986) Canonical correspondence analysis: a new eigenvector technique for multivariate direct gradient analysis. Ecology 67:1167–1179

Brockerhoff E, Ecroyd CE, Leckie AC, Kimberley MO (2003) Diversity and succession of adventive and indigenous vascular understory plants in *Pinus radiata* plantation forests in New Zealand. For Ecol Manage 185(3):307–326

Cabrera A (1976) Regiones fitogeográficas argentinas. Enciclopedia Argentina de Agricultura y Jardinería. ACME, Buenos Aires, Argentina

Carlson A (1986) A comparison of birds inhabiting pine plantation and indigenous forest patches in a tropical mountain area. Biol Conserv 35:195–204

Carnus JM, Parrotta J, Brockerhoff EG, Arbez M, Jactel H, Kremer A, Lamb D, ÓHara K, Walters B (2006) Planted forests and biodiversity. J For 104(2):65–77

Castellón TD, Sieving KE (2006) An experimental test of matrix permeability and corridor use by an endemic understory bird. Conserv Biol 20:135–145

Chao A (1984) Non-parametric estimation of the number of classes in a population. Scand J Stat 11:265–270

Christie MI, Ramilo EJ, Bettinelli MD (2004) Aves del noroeste patagónico: Atlas y Guía. L.O.L.A, Buenos Aires, Argentina

Clout MN, Gaze PD (1984) Effects of plantation forestry on birds in New Zealand. J Appl Ecol 21:795–815

Colwell RK (2005) Estimates: statistical estimation of species richness and shared species from samples. Version 7.5, <purl.oclc.org/estimate>

Corley J, Sackmann P, Rusch V, Bettinelli J, Paritsis J (2006) Effects of pine silviculture on the ant assemblages (Hymenoptera: Formicidae) of the Patagonian steppe. For Ecol Manage 222:162–166

Curry GN (1991) The influence of proximity to plantation edge on diversity and abundance of bird species in an exotic pine plantation in north-eastern New South Wales. Wildl Res 18:299–314

Díaz AI, Armesto JJ, Reid S, Sieving KE, Willson MF (2005) Linking forest structure and composition: avian diversity in successional forests of Chiloé Island, Chile. Biol Conserv 123:91–101

Díaz IA, Armesto JJ, Willson MF (2004) Mating success of the endemic Des Murs' Wiretail (*Silviorthorhynchus desmursii*) in Chilean fragmented rainforests. Austral Ecol 31(1):13–21

Díaz M, Carbonell R, Santos T, Tellería JL (1998) Breeding bird communities in pine plantations of the Spanish plateaux: biogeography, landscape and vegetation effects. J Appl Ecol 35:562–574

Driscoll PV (1977) Comparison of bird counts from pine forests and indigenous vegetation. Aust Wildl Res 4:281–288

Estades CF (1994) Impacto de la sustitución del bosque natural por plantaciones de *Pinus radiata* sobre una comunidad de aves en la Octava Región de Chile. Boletín Chileno de Ornitología 1:8–14

Estades CF, Temple S (1999) Deciduous-forest bird communities in a fragmented landscape dominated by exotic pine plantations. Ecol Appl 9:573–585

Fahrig L (2001) How much habitat is enough? Biol Conserv 100:65–74

Gjerde I, Saetersdal M (1997) Effects on avian diversity of introduced spruce *Picea* spp. plantations in the native pine *Pinus sylvestris* forests of western Norway. Biol Conserv 79:241–250

González-Gómez PL, Estades CF, Simonetti JA (2006) Strengthed insectivory in a temperate fragmented forest. Oecologia 148:137–143

Grigera D, Úbeda C, Reca A (1996) Estado de conservación de las aves del Parque y Reserva Nacional Nahuel Huapi. Hornero 14:1–13

Hammer O, Harper DAT, Ryan PD (2001) PAST: paleontological statistics software package for education and data analysis. Paleontol Electron 4(1):1–9
Hartley MJ (2002) Rationale and methods for conserving biodiversity in plantation forests. For Ecol Manage 155:81–95
Humphrey JW, Davey S, Peace AJ, Ferris R, Harding K (2002) Lichens and bryophyte communities of planted and semi-natural forests in Britain: the influence of site type, stand structure and dedwood. Biol Conserv 107(2):165–180
IUCN (2006) 2006 IUCN Red List of Threatened Species. http://www.iucnredlist.org. Downloaded on 11 August 2006
Kent M, Coker P (1992) Vegetation description and analysis: a practical approach. Belhaven Press, London
Laclau P (1997) Los ecosistemas forestales y el hombre en el sur de chile y argentina. Fundación Vida Silvestre Argentina. Boletín Técnico N° 34
Lantschner MV (2005) Estructura de las comunidades de aves de los fondos de valle y laderas bajas de la cuenca del río Foyel (Prov. Río Negro), en relación a los disturbios antrópicos. Undergraduate dissertation, Universidad Nacional del Comahue, Bariloche, Argentina, p 116
Lantschner MV, Rusch V (2007) Impacto de diferentes disturbios antrópicos sobre las comunidades de aves de bosques y matorrales de *Nothofagus antarctica* en el NO Patagónico. Ecología Austral 17:99–112
Leberton P, Pont B (1987) Avifaune et altérations forestières. Acta OEcologica 8:227–235
Lindenmayer DB (2002) Plantation design and biodiversity conservation. Rural Industries Research and Development Corporation, Canberra, Australia
Lindenmayer DB, Cunningham RB, Donnelly CF, Nix H, Lindenmayer BD (2002) Forest fragmentation on bird assemblages in a novel landscape context. Ecol Monogr 72(1):1–18
Lindenmayer DB, Franklin JF (1997) Managing stand structure as part of ecologically sustainable forest management in Australian mountain ash forests. Conserv Biol 11:1053–1068
Lindenmayer DB, McIntyre S, Fischer J (2003) Birds in eucalypt and pine forests: landscape alteration and its implications for research models of faunal habitat use. Biol Conserv 110(1):45–53
Lindenmayer DB, Munks S (2000) Principles and recommendations. In: Munks S, McArthur C (eds) Plantation design and fauna conservation. Fauna Issues and Plantation Design Workshop, Horbat, Australia
Marsden SJ, Whiffin M, Galetti M (2001) Bird diversity and abundance in forest fragments and *Eucalyptus* plantation around an Atlantic forest reserve, Brazil. Biodivers Conserv 10:737–751
Miller JH (2001) Pine plantation communities: how do we begin to manage for plant diversity? Paper presented in the 54th Annual Southern Weed Science Society Meeting, Biloxi, Mississippi, 22–24 January 2001
Mitra SS, Sheldon FH (1993) Use of an exotic tree plantation by Bornean lowland forest birds. The Auk 110(3):529–540
Moore SE, Allen HL (1999) Plantation forestry. In: Hunter ML Jr (ed) Maintaining biodiversity in forest ecosystems. Cambridge University Press, New York
Narosky T, Babarksas M (2000) Aves de la Patagonia, Guía para su reconocimiento. Vázquez Mazzini Editores, Buenos Aires
Olson DM, Dinerstein E (1998) The Global 200: a representation approach to conserving the earth's most biologically valuable ecoregions. Conserv Biol 12(3):402–515
Paritsis J (2002) Efectos de las plantaciones de coníferas exóticas sobre el ensamble de plantas vasculares, coleópteros epigeos y aves del bosque de *Nothofagus dombeyi*. Undergraduate dissertation, Universidad Nacional del Comahue, Bariloche, Argentina, p 93
Petit LJ, Petit DR (2003) Evaluating the importance of human-modified lands for neotropical bird conservation. Conserv Biol 17(3):687–694
Pomeroy D, Dranzoa C (1998) Do tropical plantations of exotic trees Uganda and Kenya have conservation value for birds? Bird Popul 4:23–36
Ralph CJ, Geupel GR, Pyle P, Martin TE, De Sante DF (1993) Handbook of field methods for monitoring landbirds. Pacific Southwest Research Station, Albany, CA
Reid S, Díaz IA, Armesto JJ, Willson MF (2004) Importance of native bamboo for understory birds in Chilean temperate forest. The Auk 121(2):515–525
Rusch V, Sarasola M, Schlichter T (2005a) Indicadores de biodiversidad en bosques *Nothofagus*. IDIA 8: 8–14
Rusch V, Schlichter T, Corley JC (2005b) Integration of spatial scales for the conservation of biodiversity in forest plantations. Paper presented at the "IUFRO International Conference: Biodiversity and Conservation Biology in Plantation Forests". Bordeaux, France, 26–29 April 2005

Rusch V, Schlichter T (2005) El empleo de principios, criterios e indicadores. ¿Quiénes se benefician con su uso? Paper presented at the "Congreso Forestal Argentino y Latinoamericano". Corrientes, Argentina, 6–9 May, 2005

SAGPyA (Secretaría de Agricultura, Ganadería, Pesca y Alimentos) (1999) Argentina, oportunidades de inversión en bosques cultivados. Buenos Aries, Argentina

Sayer J, Chokkalingam U, Poulson J (2004) The restoration of forest biodiversity and ecological values. For Ecol Manage 201:3–11

Schlichter T, Laclau P (1998) Ecotono estepa-bosque y plantaciones en la Patagonia norte. Ecología Austral 8:285–296

Schnell MR, Pik AJ, Dangerfield JM (2003) Ant community succession within eucalypt plantations on used pasture and implications for taxonomic sufficiency in biomonitoring. Austral Ecol 28(5):553–565

Shankar U, Lama SD, Bawa KS (1998) Ecosystem reconstruction through "taungya" plantations following commercial logging of a dry, mixed deciduous forest in Darjeeling Himalaya. For Ecol Manage 102(2):131–142

Sieving KE, Willson MF, De Santo TL (1996) Habitat barriers to movement of understory birds in south-temperate rainforest. Auk 113(4):944–949

Sieving KE, Willson MF, De Santo TL (2000) Defining corridors for endemic birds in fragmented south-temperate rainforest. Conserv Biol 14(4):1120–1132

Sokal RR, Rohlf FJ (1981) Biometry. Freeman, San Francisco, USA

Soriano A, Volkheimer W, Walter H, Box EO, Marcolín AA, Vaierini JA, Movia CP, Leon RJ, Gallardo JM, Rumboli M, Canevari M, Canevari P, Vasina WG (1983) Deserts and semi-deserts of Patagonia. In: West NE (ed) Ecosystems of the world. Temperate deserts and semi-deserts. Elsevier, Amsterdam

Vergara PM, Simonetti JA (2003) Forest fragmentation and Rhinocryptid nest predation in central Chile. Acta Oecologica 24:285–288

Vergara PM, Simonetti JA (2004) Avian responses to fragmentation of the Maulino forest in central Chile. Oryx 38(4):383–388

Vila AR (ed) (2002) Visión de la biodiversidad para la eco-región de los bosques templados valdivianos. CD-Rom, Fundación Vida Silvestre Argentina, Buenos Aires, Argentina

Vuilleumier F (1972) Brid species diversity in Patagonia (Temperate South-America). Am Nat 106:266–304

Vuilleumier F (1985) Forest birds of Patagonia. Ornithol Monogr 36:255–304

Walker S, Novaro A, Funes M, Baldi R, Chehebar C, Ramilo E, Ayesa J, Bran D, Vila A (2005) Rewilding Patagonia. Wild Earth, Fall Winter 2004–2005:36–41

Willson MF (2004) Loss of habitat connectivity hinders pair formation and juvenile dispersal of Chucao tapaculos in Chilean rainforest. The Condor 106:166–171

Willson MF, De Santo T, Sabag C, Armesto JJ (1994) Avian communities of fragmented south-temperated rainforests in Chile. Conserv Biol 8(4):508–520

Yirdaw E (2001) Diversity of naturally regenerated native woody species in forest plantations in the Ethiopian highlands. New For 22:159–177

Zobrist KW, Hinckley TM (2005) A literature review of management practices to support increased biodiversity in intensively managed Douglas-fir plantations. Final Technical Report to the National Commission on Science for Sustainable Forestry (NCSSF), USA

Zurita GA, Rey N, Varela DM, Villagra M, Bellocq MI (2006) Conversion of Atlantic Forest into native and exotic tree plantations: effects on bird communities from the local and regional perspectives. For Ecol Manage 235:164–173

Identifying practical indicators of biodiversity for stand-level management of plantation forests

George F. Smith · Tom Gittings · Mark Wilson · Laura French · Anne Oxbrough · Saoirse O'Donoghue · John O'Halloran · Daniel L. Kelly · Fraser J. G. Mitchell · Tom Kelly · Susan Iremonger · Anne-Marie McKee · Paul Giller

Originally published in the journal Biodiversity and Conservation, Volume 17, No 5, 991–1015.
DOI: 10.1007/s10531-007-9274-3

Abstract Identification of valid indicators of biodiversity is a critical need for sustainable forest management. We developed compositional, structural and functional indicators of biodiversity for five taxonomic groups—bryophytes, vascular plants, spiders, hoverflies and birds—using data from 44 Sitka spruce (*Picea sitchensis*) and ash (*Fraxinus excelsior*) plantation forests in Ireland. The best structural biodiversity indicator was stand stage, defined using a multivariate classification of forest structure variables. However, biodiversity trends over the forest cycle and between tree species differ among the taxonomic groups studied. Canopy cover was the main structural indicator and affected other structural variables such as cover of lower vegetation layers. Other structural indicators included deadwood and distances to forest edge and to broadleaved woodland. Functional indicators included stand age, site environmental characteristics and management practices. Compositional indicators were limited to more easily identifiable plant and bird species. Our results suggest that the biodiversity of any one of the species groups we surveyed cannot act as a surrogate for all of the other species groups. However, certain subgroups, such as forest bryophytes and saproxylic hoverflies, may be able to act as surrogates for each other. The indicators we have identified should be used together to identify stands of potentially high biodiversity or to evaluate the biodiversity effects of silvicultural management practices. They are readily assessed by non-specialists, ecologically meaningful and applicable over a broad area with similar climate conditions and silvicultural systems. The approach we have used to develop biodiversity indicators,

G. F. Smith · L. French · S. O'Donoghue · D. L. Kelly · F. J. G. Mitchell · S. Iremonger · A.-M. McKee
BIOFOREST Project, Department of Botany, Trinity College Dublin, Dublin, Ireland

T. Gittings · M. Wilson · A. Oxbrough · J. O'Halloran · T. Kelly · P. Giller
BIOFOREST Project, Department of Zoology, Ecology and Plant Science, University College Cork, Cork, Ireland

Present Address:
G. F. Smith (✉)
Atkins, 150–155 Airside Business Park, Swords, Co. Dublin, Ireland
e-mail: george.f.smith@atkinsglobal.com

E.G. Brockerhoff et al. (eds.), *Plantation Forests and Biodiversity: Oxymoron or Opportunity?*
DOI: 10.1007/978-90-481-2807-5_4

including stand structural types, is widely relevant and can enhance sustainable forest management of plantations.

Keywords Biodiversity · Forest management · Indicator · Plantation · Species richness · Stand structure · Sustainable forest management

Abbreviations

CWD	Coarse woody debris
Dbh	Diameter at breast height (1.3 m)
GPS	Geographical positioning system
IndVal	Indicator value
NMS	Non-metric multidimensional scaling
PCA	Principal components analysis
Se	Standard error
SR	Species richness

Introduction

Comprehensive biodiversity inventories of natural forests are virtually impossible to undertake because of the time and effort involved (Lawton et al. 1998). Similarly, the resources necessary to complete biodiversity inventories of plantation forests are not usually available, despite the often simplified nature of plantation ecosystems. Therefore, biodiversity assessment and management in plantation forests must rely on the use of biodiversity indicators (Lindenmayer 1999; Noss 1999; Lindenmayer et al. 2000; Larsson 2001). Despite the clear need, however, most indicators that have been published or exist in the 'grey' literature are the product of conventional wisdom and lack scientific validation (Noss 1999; Lindenmayer et al. 2006).

In order for indicators to be practical for sustainable forest management, it is important that they are repeatable, cost-effective, ecologically meaningful and easy to assess, particularly by forest managers or other non-ecologists (Ferris and Humphrey 1999). Indicators can be used by forest managers to assess the effect of site management on biodiversity or to identify sites that potentially are of high biodiversity value, in order to comply with national forest standards (e.g. Forest Service 2000b, c; Forestry Commission 2004) or the requirements of forestry grant schemes (e.g. Forest Service 2000a, 2006). In sites where few indicators are present, management can be reviewed and improved. Forest stands identified as being of potentially high biodiversity can be surveyed and assessed more thoroughly, and management for biodiversity can be prioritised in forest planning and operations. Forest biodiversity indicators can be developed at the regional or landscape scales for use in forest planning, but stand-scale indicators may be the most practical, as most management operations are carried out at this level (Similä et al. 2006).

At the level of the forest stand, compositional indicators can be particular species or species groups (Noss 1990). The universality and applicability of surrogacy relationships among species groups—where the diversity of one group reflects diversity in another, unrelated group—are the focus of much recent conservation biology research, with mixed results (e.g. Howard et al. 1998; Vessby et al. 2002; Sætersdal et al. 2003; Anand et al. 2005;

Oertli et al. 2005; Williams et al. 2006; Similä et al. 2006). Important elements of forest structure that may serve as structural indicators include tree size, vertical foliage distribution, horizontal canopy distribution and density and abundance of deadwood (Noss 1990; Spies 1998). Functional indicators can include processes such as productivity, nutrient cycling rates, disturbance regime and management practices (Noss 1990).

Aspects of stand structure have the potential to be particularly useful biodiversity indicators, especially as structure is the product of site environment and management and directly affects biodiversity and ecosystem function (Spies 1998). Comparison of managed forests and old, natural forests has found that managed forests often lack old-growth features, such as large trees, vertical heterogeneity, diverse tree species assemblages and large-diameter dead wood, that may be important for promoting biodiversity (Halpern and Spies 1995; Hodge and Peterken 1998; Humphrey 2005). Accordingly, many studies of forest biodiversity have paid special attention to stand structure (e.g. Pitkänen 1997; Humphrey et al. 1999, 2002; Ferris et al. 2000). However, quantification of stand structure can be difficult due to its multivariate nature (McElhinny et al. 2005).

In this paper, we develop potential indicators for biodiversity of five groups of plants and animals in plantation forests. These indicators can be used by non-specialists as tools to assess the effectiveness of current management practices in maintaining forest biodiversity and/or to identify stands or forests of potentially high biodiversity value. We pay particular attention to stand structure by developing a forest stand structure classification and assessing how changes in stand structure are reflected by changes in biodiversity.

Methods

Species groups

This study was part of a larger research programme on biodiversity in commercial forestry plantations in Ireland (O'Halloran et al. 2004; Smith et al. 2005; Iremonger et al. 2007). We were not able to survey all taxonomic groups present in plantation forests, and thus we focused our efforts on five groups: bryophytes, vascular plants, spiders, hoverflies (Diptera: Syrphidae) and birds. These groups vary in mobility and the scales at which forest environment and management are likely to affect their diversity. The ecology and taxonomy of these groups are well-known. Forest understorey vegetation provides food and structural diversity that can be exploited by dependent fauna. Vascular plants in particular are a well-known group in Ireland and have been used as surrogates for total biodiversity in other countries (Ferris and Humphrey 1999; Niemi and McDonald 2004). Bryophytes are an important component of native forest flora, and in oceanic regions attain levels of diversity comparable with higher plants (Kelly 1981, 2005). Spiders represent an intermediate trophic level, and because of their relatively small ranges, they are responsive to changes at the stand and smaller scales (Niemela et al. 1996). Hoverflies are quite mobile and are therefore more sensitive to conditions at larger scales than spiders. They are a diverse group in terms of trophic and habitat requirements and have been used as indicators of disturbance or habitat quality (Sommaggio 1999). Birds range over wider areas than members of any of the other taxa, and are therefore affected by environmental variation at the plantation and landscape scales (Pithon et al. 2005). Species assemblages present in the sites surveyed are analysed in more detail in related work (French et al. in press; Oxbrough et al. 2005; Smith et al. 2005; Wilson et al. 2006).

Study design

The study sites were in 44 plantation forests distributed across the Republic of Ireland (Fig. 1). In 12 sites, ash (*Fraxinus excelsior*) stands were sampled, and in 20 sites, Sitka spruce (*Picea sitchensis*) stands were sampled. In the remaining 12 sites, both ash and Sitka spruce stands were present in a non-intimate mix. These two species were chosen as the most commonly planted native broadleaf and the most commonly planted exotic conifer in Ireland. Each forest was in its first rotation and was at least 4 ha in area. The forests ranged in age from 5 yr to 81 yr at the time of surveying. The majority of the study sites are owned by Coillte Teoranta, the semi-state forestry company, and only a few of the youngest forests were privately owned. Sites were surveyed in 2001 and 2002.

We employed a chronosequence approach where we sampled different sites at different stages of maturity. We selected age classes that would represent the major structural changes that take place over the course of a commercial rotation:

- 5 years: prior to canopy closure (4 pure spruce, 4 pure ash and 4 spruce/ash mix sites)
- 8–15 years: canopy closure phase (4 pure spruce, 4 pure ash and 4 spruce/ash mix sites)
- 20–30 years: mid-rotation, beginning of thinning operations (4 pure spruce sites)

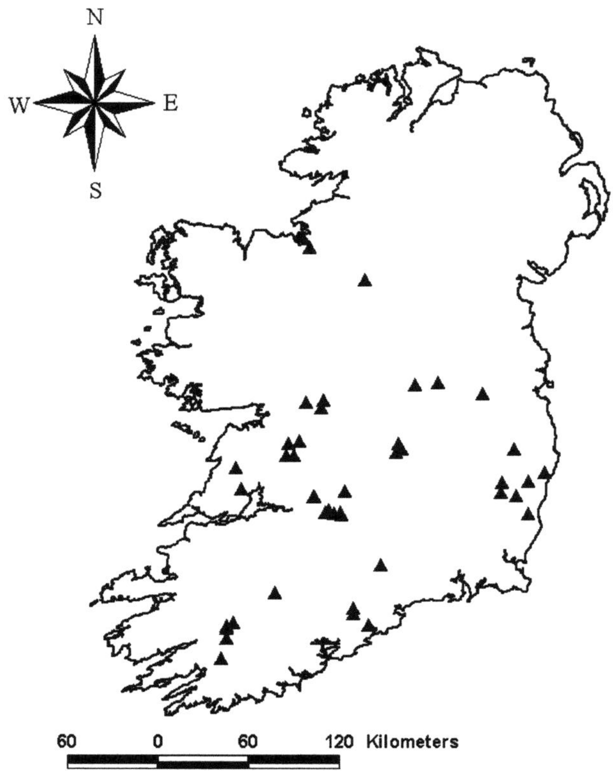

Fig. 1 Location of the 44 forests surveyed

- 35–50 years: approaching commercial maturity of Sitka spruce (8 pure spruce and 4 spruce/ash mix sites)
- 50–81 years: approaching commercial maturity of ash (4 pure ash sites)

To reduce confounding variation among age classes due solely to site environmental factors, study sites were clustered geographically, with each cluster including sites across the range of age × species combinations. It was not possible to match pure ash sites, as few such sites existed that met our forest size site selection criterion.

Because of logistical difficulties largely caused by an outbreak of foot-and-mouth disease in 2001, and also because of loss of invertebrate traps to disturbance, it was not possible to sample every site for all taxonomic groups. We indicate sample size or degrees of freedom for all statistical tests performed.

Field survey

Bryophytes and vascular plants were surveyed in three representative 100 m^2 plots at least 50 m apart at each site. Percent cover of each species was estimated to the nearest 5%. Forest structure was also assessed in these plots. Top height of the dominant trees and average spacing between trees was measured. Diameter at breast height (dbh, measured at 1.3 m) was measured for all trees in the plot, or for a random subsample of 10 trees in dense stands 15 yrs old or less. The percentage cover of the forest canopy was estimated by eye by two and usually three researchers jointly to reduce variation in estimates. Height, canopy cover, spacing and mean dbh for each plot were then averaged to produce means for each site. Volume of coarse woody debris (CWD) > 7 cm diameter was measured in each plot.

Spiders were sampled using pitfall traps (Curtis 1980) arranged in 16 m^2 plots established at least 50 m apart. Where possible, spider plots were adjacent to vegetation plots. Five pitfall traps were established in each plot, and five plots were established in each monoculture site. In mix sites, five plots were sampled in the spruce component and two in the smaller ash component. Pitfall traps were run for nine weeks and emptied and changed every third week. Cover of litter, bare soil and vegetation in three layers (<10 cm, >10–50 cm and >50–200 cm) were estimated in each plot using the Braun-Blanquet scale (Mueller-Dombois and Ellenberg 1974).

Hoverflies were sampled using two Malaise traps (Southwood 2000) in each monoculture site and the spruce component of mix sites, and one Malaise trap in the ash component of mix sites. Traps were located in canopy gaps rather than under a closed canopy to increase their effectiveness. Traps were located at least 100 m from each other and were run for a minimum of six weeks. The presence of wet microhabitat features, such as streams and flushes, within 100 m of each trap was recorded. Frequency of standing and fallen CWD in the vicinity of each trap was recorded in four 10 × 100 m transects radiating from the trap towards the four cardinal compass points.

Birds were surveyed using 4–9 point counts (Bibby et al. 1992) per site, depending on the size and structural variation of the site. Points were located at least 100 m from each other, and their location was marked using a GPS. Point counts were conducted for 10 minutes, during which birds more than 50 m away were recorded and the positions of birds within 50 m were estimated. The distance of each sampling point to the nearest forest edge was determined using ArcView GIS. Canopy height (m) and the cover (nearest 5%) of three vegetation layers—canopy, shrub (woody plants 0.5–2 m tall, excluding young planted trees) and field/ground layer (non-woody vascular plants and bryophytes)—were recorded within 30 m of each point.

Soil samples were collected in each 100 m^2 vegetation plot; subsamples were taken from the four corners and then bulked. The pH of field-moist soils was measured using a pH meter with a glass electrode on a soil:distilled water (1:2) suspension. The soils were then air-dried and sieved prior to further chemical and physical analyses. Available P was extracted using Morgan's reagent and quantified by a colorimetric method using a spectrophotometer (Allen et al. 1986). Further environmental data, such as elevation, were collected at the sampling unit or site level, as appropriate. Distance from the site to the nearest old woodland and the area of old woodland within 1 km were determined using 1:10,560 Ordnance Survey maps published from 1900 to 1915. Management information was obtained from the Coillte inventory and forest managers.

For further details on survey methodology, see French et al. (in press) for vegetation, Oxbrough et al. (2005) for spiders, Wilson et al. (2006) for birds and Smith et al. (2005) for all taxonomic groups, environmental and management data and overall study design. Nomenclature follows Smith (2004) for mosses, Stace (1997) for vascular plants and Beaman (1994) for birds.

Data analysis

Stand structural types

Preliminary analyses of stand structural variables, such as canopy cover and tree size, showed high variability within a given age class. Forest age is only one of many factors that affect stand structure. Other factors include environmental parameters, such as climate and soil fertility, and management factors, such as thinning regime. Although stand age per se can influence biodiversity, particularly through the operation of dispersal and colonisation mechanisms, changes in stand structure in plantation forestry may have a stronger affect on biodiversity through modification of the below-canopy environment. Accordingly, many studies of forest biodiversity focus on stand structure rather than stand age (e.g. Pitkänen 1997; Humphrey et al. 1999, 2002; Ferris et al. 2000).

To improve our investigations of biodiversity and structural changes over the forest cycle, we developed a small number of stand structural types to summarise the structural characteristics of our study sites. Separate analyses of forest structure were conducted for each tree species (ash or Sitka spruce) using data from the vegetation plots. PCA ordination using covariance matrices was conducted on site means of canopy cover, tree height, dbh and spacing. Percent variation explained by individual axes was calculated by dividing the eigenvalue of each axis by the sum of all eigenvalues. Sites were assigned to structural stages using Ward's hierarchical clustering (Legendre and Legendre 1998). All variables were transformed to a 0–1 scale by ranging (Sneath and Sokal 1973) prior to analysis to place them on equivalent scales. Ordinations were performed using PC-Ord (McCune and Mefford 1997).

Biodiversity measures

Species richness of each of the taxonomic groups was calculated at the sample unit and site levels. We focus on species richness as our primary measure of biodiversity, as this is the most basic and universal method (Gaston 1996; Magurran 2004). However, total species richness does not indicate whether the species involved are of conservation significance and will also underestimate the conservation value of important but naturally species-poor habitats. To address this issue, we have analysed species richness of various subgroups: species characteristic of forest in Ireland (calculated separately for bryophytes, vascular plants, spiders, hoverflies and birds); hoverflies dependent on deadwood (saproxylic species), wet substrates, ground debris or semi-natural habitats (anthropophobic species: Speight and Castella 2001); and ground-nesting and cavity-nesting birds. Species characteristic of native woodland in Ireland will be referred to as 'forest species', although they may not be typical forest species elsewhere. Plantation forests may have an important role in providing habitat for forest species in regions where semi-natural forests are rare, as is the case in Ireland where less than 1% of the island is occupied by semi-natural forest (Cross 1998).

Species assemblages were identified using non-metric multidimensional scaling ordination (NMS) and flexible-beta clustering (Legendre and Legendre 1998). The results of these analyses will be briefly referred to, but space precludes a complete presentation of the analysis and results. For further details, see French et al. (in press) for vegetation, Oxbrough et al. (2005) for spiders, Wilson et al. (2006) for birds and Smith et al. (2005) for all taxonomic groups. Indicator species of vegetation cluster groups were identified using indicator species analysis (Dufrêne and Legendre 1997). The method assesses the constancy and fidelity of species to defined assemblages and produces an indicator value score (IndVal) ranging from 0 to 100 which can be validated using Monte Carlo tests. Ordinations and indicator species analysis were performed using PC-Ord (McCune and Mefford 1997).

Indicators

Changes in species richness over forest structural stages were compared for the five taxonomic groups and the five forest species subgroups. Relationships between potential indicators and species richness were analysed using ANOVA/t-tests for categorical variables and correlation (Pearson's r) for continuous variables. Prior to analysis, variables were inspected for conformity to assumptions of parametric statistics and transformed where necessary. In some cases, transformation was inadequate, and Kruskal–Wallis tests or Spearman's rank correlation as appropriate were used instead (Sokal and Rohlf 1995). Correlations among predictive variables were also investigated. As distance to semi-natural woodland and area of woodland within 1 km were negatively correlated with forest age, partial correlations between species richness and woodland variables were performed to control for the effects of forest age. Differences in species richness between ash and spruce stands were tested using nested ANOVAs, with stand structural stage as the nested factor, to partition variation due to structural stage and not tree species. These analyses were performed using SPSS (2001).

Results

Stand structural types

Sitka spruce

The PCA ordination (Fig. 2) shows that the sites form a continuum of changing stand structure over the plantation cycle. Axis 1 explains 73% of the variance in the structural

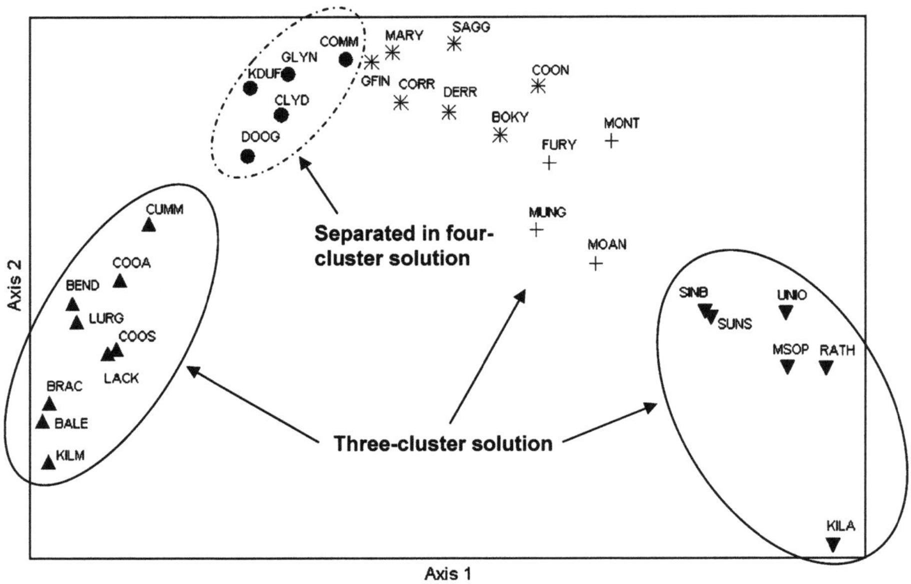

Fig. 2 PCA ordination and Ward's cluster analysis of mean canopy cover, tree height, dbh and spacing in 31 Sitka spruce forests. Sites are identified by four-letter codes. At the three-cluster stage, the two groups enclosed by solid rings are separate from the middle group of sites. At the four-cluster stage, the group enclosed by a dashed ring is divided from the middle group. Symbols indicate the stand types based on the five-cluster solution: ▲ = Stage I, ● = Stage II, ✳ = Stage III, + = Stage IV and ▼ = Stage V. Axis 1 of the ordination explained 73.0% of the variance in the data ($\lambda_1 = 7.087$) and axis 2 explained 21.0% of the variance ($\lambda_2 = 2.044$). Total variance of the dataset was 9.715

variables and mainly represents increasing height and dbh from left to right (Fig. 2, Table 1). Axis 2 explains 21% of the variation in the data and is most closely associated with increasing canopy cover (Fig. 2, Table 1). Spacing contributes similarly to both axes, with spacing increasing on Axis 1 and decreasing on Axis 2. Several structural variables were highly correlated with each other and with stand age (Table 2). The lowest correlations were generally with canopy cover, which has a hump-shaped relationship with tree height, dbh and age.

Defining five stand stages appeared to provide the best compromise between parsimony and adequate description. Means of structural variables at each stage are shown in Table 3. In the four-cluster solution, there was considerable variation in canopy cover and tree height in amalgamated group III/IV. There was substantial overlap in the age ranges of the oldest three structural stages (Table 3).

Table 1 Eigenvectors of the first two axes of the Sitka spruce and ash stand structure PCA ordinations showing the relative contributions of canopy cover, tree height, dbh and spacing

	Height	Dbh	Spacing	Canopy cover
Sitka spruce				
Axis 1	0.707	0.590	0.330	0.209
Axis 2	−0.024	−0.057	−0.419	0.906
Ash				
Axis 1	0.593	0.551	–	0.587
Axis 2	−0.368	−0.462	–	0.807

Table 2 Pearson correlations among Sitka spruce forest structural variables and stand age

	Height	Dbh	Spacing	Age
Canopy cover	0.38*	0.33	−0.11	0.33
Height		0.92***	0.69***	0.95***
Dbh			0.74***	0.92***
Spacing				0.61***

Statistically significant correlations are indicated: $*P \leq 0.05$, $**P \leq 0.01$, $***P \leq 0.001$. N = 31 sites

Table 3 Mean (and range of site means in brackets) canopy cover (%), tree height (m), dbh (cm) and spacing (m) in Sitka spruce stand structural stages defined by a Ward's hierarchical clustering analysis. Also shown is mean (and range) stand age (yr)

Stand stage	Canopy cover (%)	Height (m)	Dbh (cm)	Spacing (m)	Age (yr)
I	29.6 (11.7–43.3)	2.5 (1.4–3.8)	3.7 (1.6–7.0)	1.6 (1.0–2.0)	4.6 (3–10)
II	80.3 (60.0–93.3)	5.9 (4.3–7.3)	12.4 (10.4–16.5)	1.9 (1.5–2.0)	10.7 (9–13)
III	86.9 (78.3–95.0)	12.7 (9.8–15.7)	19.3 (14.7–24.3)	1.7 (1.4–2.0)	26.3 (14–39)
IV	70.8 (63.3–80.0)	18.8 (16.8–20)	22.4 (21.0–24.8)	2.3 (2.0–2.8)	36.1 (25–43)
V	54.7 (40.0–60.0)	21.1 (18.3–23.0)	39.0 (31.6–44.8)	3.9 (3.0–6.0)	42.6 (37–47)

Ash

When the four structural variables were analysed for the ash sites, the resulting clusters joined sites that varied widely in tree size and separated others largely on the basis of spacing. Analyses were then performed using only canopy cover, tree height and dbh. This simplified classification was better at forming coherent groupings of larger-tree sites and also clusters of smaller-tree sites.

The PCA ordination of the ash sites does not show as simple a structural pattern as was found for the Sitka spruce sites (Fig. 3). Axis 1 explains 86.9% of the variation in the three structural variables and is positively associated with all structural variables (Table 1). Axis 2 explains 11.4% of the variation in the data; it is positively associated with canopy cover and negatively associated with tree size. As with Sitka spruce stands, the strongest correlations were among tree height, dbh and stand age (Table 4). As these three variables increase, canopy cover increases asymptotically.

In the six-cluster solution and in the ordination diagram (Fig. 3), KILA was separated as an outlier. It was a 45 yr old stand that did not fit well into the stand types with similar canopy cover or tree size (Table 5). The six-cluster solution was therefore accepted as defining the stand stages, and KILA was excluded from the stand type classification. Mean stand ages in ash Stages I and II and Stages IV and V were quite similar (Table 5).

Indicators

Table 6 summarises the significant relationships between species richness of taxonomic groups and subgroups and structural, compositional and functional variables. See Smith et al. (2005) for further details.

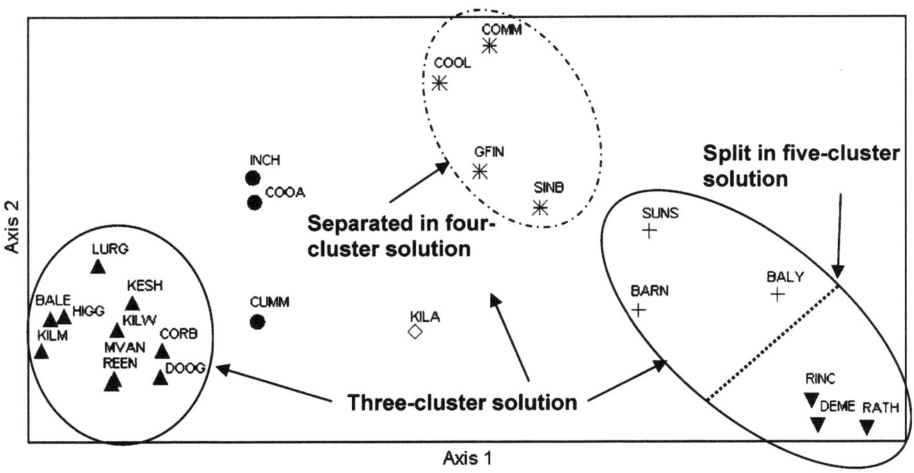

Fig. 3 PCA ordination and Ward's cluster analysis of mean canopy cover, tree height and dbh in 24 ash forests. Sites are identified by four-letter codes. At the three-cluster stage, the two groups enclosed by solid rings are separate from the middle group of sites. At the four-cluster stage, the group enclosed by a dashed ring is divided from the middle group. At the five-cluster stage, RINC, DEME and RATH were separate from the others in their group. Symbols indicate the six-cluster solution: ▲ = Stage I, ● = Stage II, ✳ = Stage III, + = Stage IV, ▼ = Stage V and ◇ = not assigned. Axis 1 of the ordination explained 86.9% of the variance in the data ($\lambda_1 = 6.028$) and axis 2 explained 11.4% of the variance ($\lambda_2 = 0.793$). Total variance of the dataset was 6.939

Table 4 Pearson correlations among ash forest structural variables and stand age

	Height	Dbh	Spacing	Age
Canopy cover	0.76***	0.72***	0.26	0.65***
Height		0.95***	0.54*	0.90***
Dbh			0.64**	0.93***
Spacing				0.63**

Statistically significant correlations are indicated: *$P \leq 0.05$, **$P \leq 0.01$, ***$P \leq 0.001$. N = 24 sites

Table 5 Mean (and range of site means in brackets) canopy cover (%), tree height (m), dbh (cm) and spacing (m) in ash stand structural stages defined by a Ward's hierarchical cluster analysis. KILA was structurally different than the remainder of the sites and was assigned to its own group in the six-cluster solution

Stand stage	Canopy cover (%)	Height (m)	Dbh (cm)	Spacing (m)	Age (yr)
I	12.2 (5.0–21.7)	3.1 (1.3–5.0)	3.8 (0.9–9.1)	2.1 (1.5–3.5)	6.7 (3–11)
II	57.8 (45.0–80.0)	4.4 (3.0–6.0)	6.3 (4.8–8.9)	1.8 (1.5–2.0)	6.6 (5–10)
III	77.1 (70.0–88.3)	9.0 (6.8–11.5)	10.0 (7.8–13.85)	2.3 (1.8–3.0)	18.3 (8–37)
IV	75.6 (66.7–81.7)	18.8 (16.3–22.0)	17.3 (15.8–19.7)	2.4 (1.5–3.8)	52.7 (44–62)
V	72.2 (70–73.3)	21.6 (18.5–25.0)	29.1 (27.6–30.9)	4.0 (3.0–6.0)	63.3 (47–81)
KILA	43.3	10.5	12.9	2.5	45

Note that mean spacing was not used in the cluster analysis. Also shown is mean (and range) stand age (yr)

Table 6 Summary of relationships between structural, compositional and functional variables and species richness of taxonomic groups and subgroups

	Bryophytes		Vascular plants		Spiders			Hoverflies		Birds		
	All	Forest Spp	All	Forest Spp	All	Open Spp	Forest Spp	All	Saproxylic Spp	All	Generalist	Open Spp
Canopy cover	+[a]	+++[b]	– – –	+++[b]		–						
Shrub cover											++	
Field layer cover						+	– – –					
Ground layer cover						–	+					
Conifer litter cover							++					
Coarse woody debris	++	+++							++			
Distance to forest edge										– – –	– –	– –[c]
Distance to native woodland		0		– – –								
Area of native woodland within 1 km		0		+++								
Sitka spruce[d]	+++	0			+++		+++	0		0		
Ash[e]		0	+++	+				0	++	0		
Age		+++		+++			++			–[f]		– –[f]
Wet microhabitats								++[g]	+++			
Elevation										–[h]		
Available P			+[i]									

Strength of relationships are indicated as follows: + $P < 0.05$, ++ $P < 0.01$ and +++ $P \leq 0.001$. Negative relationships are indicated similarly. Notable lack of relationship is indicated by 0. See text and Smith et al. (2005) for further details

[a] But bryophyte species richness low at high values of canopy cover in spruce stands

[b] In ash forests; in spruce forests species richness highest at intermediate levels of canopy cover

[c] In Old (mainly Stage IV–V) forests

[d] Relative to ash forests; only positive relationships shown

[e] Relative to Sitka spruce forests; only positive relationships shown

[f] In Intermediate (Stage II–III) sites

[g] Anthropophobic species and species associated with wet substrates and ground debris. See Table 8

[h] Except ground-nesters

[i] In Stage V Sitka spruce forests

Structural

Species richness varied among structural stages, but there were differences among taxonomic groups in trends across the forest cycle. For example, vascular plant and to a lesser extent spider species richness trends mirrored changes in Sitka spruce canopy cover, whereas bryophyte species richness increased and hoverfly species richness decreased as spruce forests matured (Fig. 4a). Similar patterns were observed in ash forests, most notably an increase in bryophyte species richness, a decrease in hoverfly species richness and a decline in vascular plant species richness corresponding with an increase in canopy cover (Fig. 4b). There were also differences in resolution among different species groups in biodiversity trends among structural stages. The trends of hoverflies and birds were more

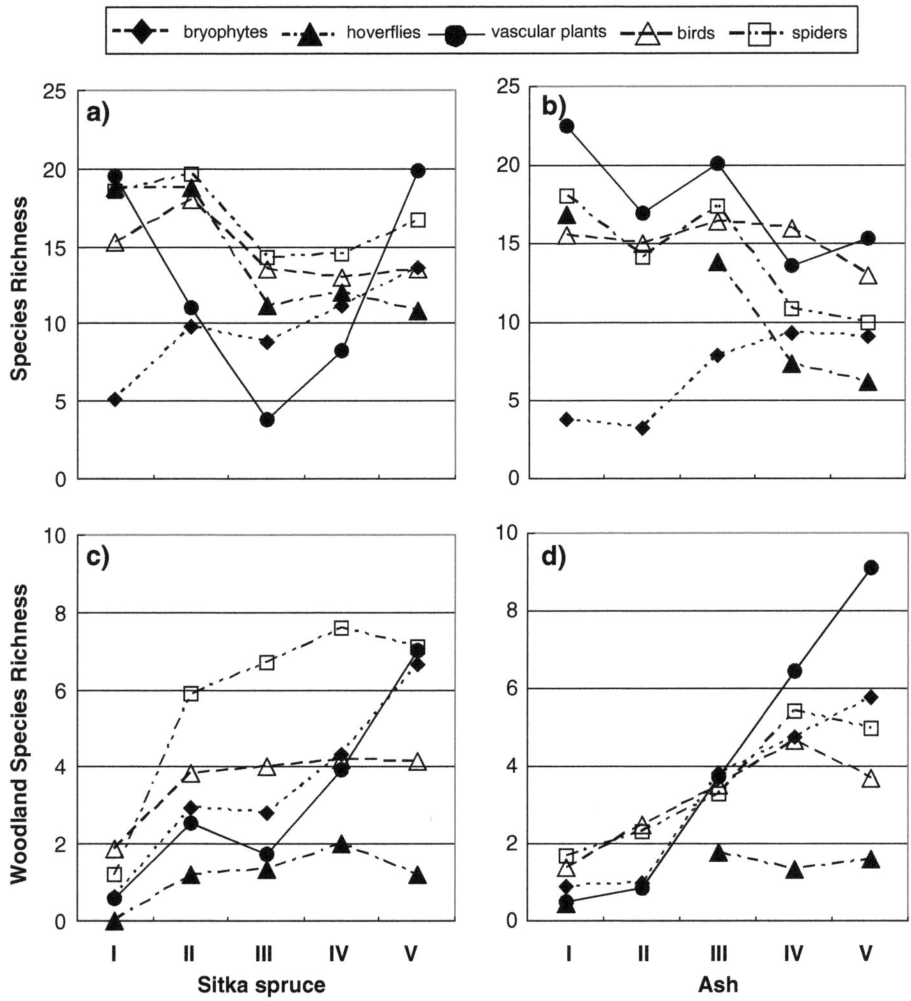

Fig. 4 Trends in species richness over the forest cycle for five taxonomic groups: (**a**) total species richness in Sitka spruce stands, (**b**) total species richness in ash stands, (**c**) woodland species richness in Sitka spruce stands and (**d**) woodland species richness in ash stands. For birds, data from spruce/ash mix sites were combined with data from monoculture sites for both Sitka spruce and ash categories

coarse than that of vascular plants (Fig. 4a; Wilson et al. 2006). Richness of forest species in all taxonomic groups increased with forest maturity in Sitka spruce and ash forests, but again there were differences in scale of response (Fig. 4c, d).

Although birds are shown in Fig. 4 for comparison with other species groups, preliminary analyses found that birds were not as sensitive to changes in forest structure or tree species recorded at the stand scale. Therefore, a separate stand type classification was performed using the structural data from the bird point counts of ash and Sitka spruce forests combined. This resulted in three bird habitat subgroups, Younger, Intermediate and Older (Wilson et al. 2006), corresponding to structural Stages I–II, Stages II–III and Stages III–V, respectively. Indicators for bird diversity were developed in the context of this simplified structural classification.

The importance of canopy cover, particularly in spruce forests, is emphasised by the negative relationship between vascular plant species richness and canopy cover in forests beyond the initial, pre-thicket structural stage (Fig. 5). Canopy cover in turn influenced the amount of vegetation cover in the lower strata. Canopy cover in more mature (i.e. Stages II–V) Sitka spruce vegetation plots was negatively correlated with cover of other structural layers, such as graminoids (Spearman rho $= -0.44$, $P = 0.0003$) and forbs (Spearman rho $= -0.38$, $P = 0.002$) and positively correlated with cover of conifer litter (Spearman rho $= 0.80$, $P \leq 0.0001$). Cover in several structural layers was associated with species richness of plant and animal groups or subgroups. In Older ash and spruce forests, the species richness of generalist birds was positively associated with shrub cover ($r = 0.61$, $n = 19$, $P = 0.006$). Similarly, species richness of spider assemblages typical of more open ash and spruce forests was positively correlated with cover of field layer vegetation (<50 cm tall) (young spruce/ash group: $r = 0.45$, $n = 20$, $P = 0.05$; young ash group: $r = 0.40$, $n = 34$, $P = 0.02$; open spruce group: $r = 0.26$, $n = 44$, $P = 0.09$). In contrast, forest spiders in the open spruce assemblage were negatively correlated with field layer

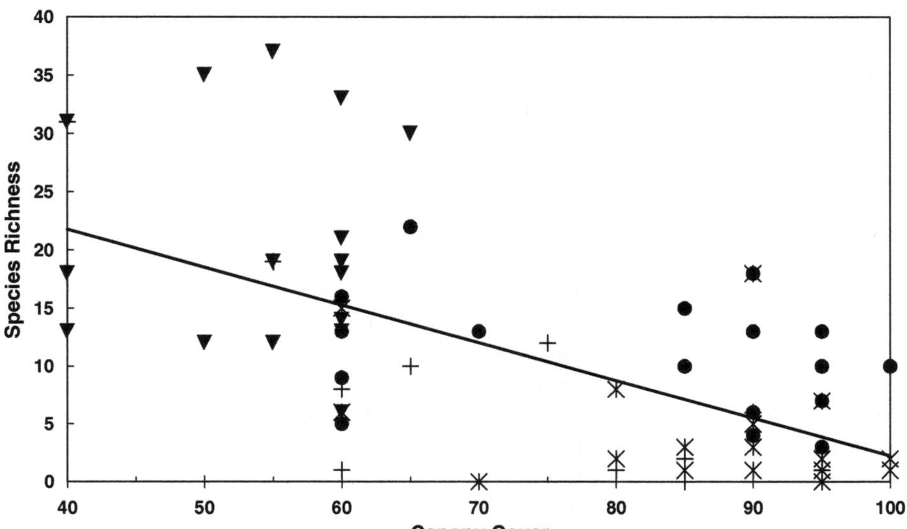

Fig. 5 Relationship between vascular plant species richness and canopy cover (%) in 67 100 m^2 plots in Sitka spruce forests (excluding Stage I). Symbols indicate structural stage: ● = Stage II, ✱ = Stage III, + = Stage IV and ▼ = Stage V. Predictions of a linear regression model are shown by a line ($r^2 = 0.392$, $P \leq 0.0001$)

cover ($r = -0.48$, $n = 44$, $P \leq 0.001$) and positively correlated with conifer litter cover ($r = 0.46$, $n = 44$, $P = 0.002$).

Bryophyte species richness was positively associated with volume of CWD in Stage II–V ash and spruce plots ($r = 0.27$, $n = 105$, $P = 0.003$, 1-tailed). When restricted to spruce forests only, the relationship was stronger ($r = 0.42$, $n = 68$, $P = 0.0002$, 1-tailed). Forest bryophytes were also positively associated with CWD volume in Stage II–V ash and spruce plots ($r = 0.35$, $n = 105$, $P = 0.0001$, 1-tailed). The species richness of saproxylic hoverflies was positively correlated with frequency of standing CWD ($r = 0.62$, $n = 20$, $P = 0.002$, 1-tailed) and fallen CWD ($r = 0.57$, $n = 20$, $P = 0.004$, 1-tailed) in spruce forests older than 20 years.

Structural features at the landscape scale were also associated with species richness of some taxonomic groups. In Older ash and spruce forests, distance to the forest edge was negatively correlated with species richness of birds ($r = -0.72$, $n = 19$, $P = 0.001$). Species richness of forest vascular plants in Stage II–V ash and spruce forests was negatively correlated with distance to native woodland ($r = -0.75$, $n = 30$, $P \leq 0.001$) and positively correlated with area of native woodland within 1 km ($r = 0.75$, $n = 30$, $P \leq 0.001$) (both are partial correlations controlling for variation due to forest age).

Compositional

Differences in species richness between Sitka spruce and ash stands, when variation due to structure is removed from the analysis, varied by taxonomic group. Vegetation plots in Sitka spruce stands supported a significantly higher number of bryophyte species (9.3 ± 0.6 se) than in ash stands (5.9 ± 0.5 se) ($F_{1,152} = 18.9$, $P < 0.001$). In contrast, ash stands supported significantly more vascular plant species (19.0 ± 0.9 se) than Sitka spruce stands (13.0 ± 1.0 se) ($F_{1,152} = 17.97$, $P < 0.001$). When both plant groups are combined, differences in species richness are not significant ($F_{1,152} = 1.95$, $P = 0.17$). There were also no differences in mean hoverfly species richness between ash (12.5 ± 1.2) and Sitka spruce (11.5 ± 0.7) ($F_{1,55} = 0.05$, $P = 0.82$). Species richness of saproxylic hoverflies, however, was significantly higher in ash compared to Sitka spruce: 2.1 ± 0.3 se in ash and 1.6 ± 0.3 se in Sitka spruce ($F_{1,55} = 11.0$, $P = 0.002$). Spider species richness was higher in Sitka spruce stands (16.4 ± 0.4 se) than in ash (14.4 ± 0.8 se) ($F_{1,170} = 13.2$, $P < 0.001$). There were no significant differences in bird species richness between tree species in any of the three bird habitat subgroups.

Over the three bird habitat subgroups, the strongest positive correlations between bird species richness and the abundance of particular bird species were with abundances of Dunnock (*Prunella modularis*) ($r = 0.55$, $P < 0.001$), Wren (*Troglodytes troglodytes*) ($r = 0.49$, $P < 0.001$) and Blackbird (*Turdus merula*) ($r = 0.47$, $P = 0.001$). Abundances of Goldcrest (*Regulus regulus*) were negatively correlated with bird species richness ($r = -0.30$, $P = 0.048$). When considering Older forests only, ten bird species were significantly correlated with total bird species richness (Table 7a).

Five plant species assemblages were identified in the more mature Sitka spruce stands (Smith et al. 2005), two of which supported significantly higher total plant species richness than the others. Four significant indicator species with indicator values greater than 25 were found for cluster A, representing a more open, vascular-plant rich community, and four for cluster B, a community type particularly rich in bryophyte species (Table 7b). Indicator plant species were not identified for ash forests, as there were no significant differences in total plant species richness among ash plant communities. There was a

Table 7 Compositional indicators of biodiversity. (a) Significant correlations of bird species abundance with bird species richness in Older forests ($n = 19$); (b) Significant indicator values > 25 (IndVal) of plant species for species-rich (SR) vegetation communities in Stage II–V Sitka spruce stands derived by flexible-beta clustering (Legendre and Legendre 1998)

Species	Scientific name	r	*P*
a)			
Wren	*Troglodytes troglodytes*	0.78	0.0001
Dunnock	*Prunella modularis*	0.63	0.0039
Blackbird	*Turdus merula*	0.58	0.0087
Pheasant	*Phasianus colchicus*	0.57	0.0111
Robin	*Erithacus rubecula*	0.53	0.0204
Treecreeper	*Certhia familiaris*	0.52	0.0225
Stonechat	*Saxicola torquata*	0.52	0.0232
Greenfinch	*Carduelis chloris*	0.51	0.0272
Great Tit	*Parus major*	0.50	0.0282
Blue Tit	*Parus caeruleus*	0.47	0.0409
	Species	IndVal	*P*
b)			
Sitka spruce[a]	*Rubus fruticosus*	93	0.001
Cluster A	*Dryopteris dilatata*	73	0.001
SR = 32.7 ± 2.9	*Agrostis capillaris*	72	0.006
n = 15	*Thuidium tamariscinum*	56	0.006
Sitka spruce	*Hypnum jutlandicum*	65	0.002
Cluster B	*Dicranum scoparium*	64	0.006
SR = 27.2 ± 2.9	*Kindbergia praelonga*	54	0.017
n = 14	*Plagiothecium undulatum*	25	0.084

[a] Mean species richness in the other clusters: C—19.4 ± 1.6 ($n = 14$), D—14.0 ± 1.3 (n = 15), E—4.7 ± 1.1 (n = 7). See Smith et al. (2005) for further details

significant difference in forest species richness between two groups of communities, but this simply contrasted more mature forests with a characteristic woodland flora (Stages III–V) with younger forests that supported a grassy understorey (Stages II–III).

Functional

There were fewer clear functional indicators of biodiversity than structural and compositional indicators. Species richness of forest vascular plants increased with forest age ($r^2 = 0.53$, $n = 43$, $P < 0.0001$). Forest bryophyte species richness also increased with forest age, but the rate of increase declined in older forests. When forest age was log-transformed, a significant linear relationship was observed ($r^2 = 0.74$, $n = 43$, $P < 0.0001$). Species richness of forest spiders increased with forest age ($r^2 = 0.22$, $n = 31$, $P = 0.008$), but was lower in the oldest ash forests.

Species richness of some hoverfly groups was higher in Stage III–V sites where wet microhabitats (e.g. streams or flushes) occurred than in sites without these features (Table 8). In addition, in spruce sites more than 20 years old, saproxylic hoverfly species

Table 8 Species richness (±se) per Malaise trap of three hoverfly species groups in wet and dry Stage III-V Sitka spruce stands. Sample size (n) is the number of traps.

	Wet sites (n = 17)	Dry sites (n = 8)	t	*P*
Anthropophobic species	1.9 ± 0.2	0.9 ± 0.2	3.15	0.005
Wet substrate species	4.2 ± 0.2	0.8 ± 0.2	13.3	<0.001
Ground debris species	3.5 ± 0.3	1.9 ± 0.2	4.5	<0.001

were more abundant in wet sites (2.5 ± 0.1 se) than in dry sites (1.1 ± 0.4 se) ($t_{19} = 4.0$, $P < 0.001$), reflecting the greater amounts of standing and fallen CWD in wet sites.

Total bird species richness in Older forests was negatively correlated with site elevation ($r = -0.50$, $P = 0.031$, $n = 19$). However, when the nesting habits of bird species were considered, species richness of ground-nesters was positively correlated with elevation ($r = 0.61$, $P = 0.006$, $n = 19$) whereas species richness of cavity-nesters was negatively correlated with elevation ($r = -0.61$, $P = 0.006$, $n = 19$). Available P was positively correlated with vascular plant species richness in Stage V Sitka spruce forests ($r = 0.71$, $P = 0.022$, $n = 18$).

Discussion

We have found that biodiversity of bryophytes, vascular plants, spiders, hoverflies and birds vary in Sitka spruce and ash plantation forests across the forest cycle (Fig. 4). Patterns of variation differ among species groups and subgroups with respect to forest structure and tree species. Several structural, compositional and functional variables correlated with species richness in one or more groups (Table 6) have the potential to be used as biodiversity indicators in forest management at the stand scale.

Surrogacy

Given the species groups we have surveyed, an important question is whether the biodiversity in these groups can act as surrogates for other groups. The variation in responses among the species groups to changes in forest structure and canopy species, and the different sets of indicators identified for them, suggest that no one group can act as a surrogate for biodiversity of all other groups. Other studies have come to similar conclusions (Prendergast et al. 1993; Lawton et al. 1998; Jonsson and Jonsell 1999; Vessby et al. 2002; Oertli et al. 2005; Similä et al. 2006). However, some studies have found that at least some groups of vascular plants can serve as surrogates for other taxa in forest ecosystems (Pharo et al. 1999; Negi and Gadgil 2002; Sætersdal et al. 2003; Kati et al. 2004). In our study, patterns of total species richness in vascular plants, spiders and, to a lesser extent, birds, show similar trends across the forest structural cycle (Fig. 4a, b). Groups with similar ecological requirements are more likely to act as adequate surrogates for each other, such as typical forest species (Fig. 4c, d) or bryophytes that benefit from deadwood habitats and saproxylic hoverflies (Table 6). Similarly, Similä et al. (2006) found that saproxylic beetles and polypore fungi have the potential to act as surrogates for each other in Finnish boreal forests.

For the above reasons, it is necessary to cover a range of different taxonomic groups to make an adequate assessment of the biodiversity of a particular site. Although we have attempted to do this, it is likely that inclusion of additional taxonomic groups in our study might have revealed additional patterns of variation in biodiversity. For example, forests are an important habitat for some species of bats, all of which are of conservation value in Ireland (Hayden and Harrington 2001). As the features that are important for bats, such as crevices or hollows in trees suitable for roosting, are probably not as important for the taxa we surveyed, patterns of bat diversity would probably be different than those in this study.

Evaluation of indicators

The indicators we have identified are summarised in Table 9. Previous authors have identified characteristics that indicators should ideally possess if they are to be effective management tools (Noss 1990; Ferris and Humphrey 1999; Lindenmayer 1999). Preferably, indicators should: (1) show clear links to particular aspects of biodiversity, (2) be sensitive to changes in those features, (3) be applicable over a broad geographical area, (4) be easy and cost-efficient to measure, and (5) be ecologically meaningful. Our study design and data analysis ensure that the indicators we developed meet the first four criteria. The last criterion is addressed in the interpretation of our results below.

Structural

The structural indicators we have presented are linked to ecological processes that affect the biodiversity of our surveyed groups. When considering biodiversity over the forest cycle, the clearest indicator for the majority of taxonomic groups is stand structural stage, demonstrating the value of a multivariate structural classification. Other studies have successfully used similar multivariate classifications (e.g. Pitkänen 1997; Leppäniemi et al. 1998), whereas others have used more subjective classifications (e.g. Humphrey et al. 1999; Ferris et al. 2000) or have used age as a surrogate (e.g. Currie and Balmford 1982; Brockerhoff et al. 2003; Eycott et al. 2006). In preliminary analyses of biodiversity patterns, we found that our original age categorisation of stand development did not adequately account for structural variation due to such variables as site fertility and thinning regime (Smith et al. 2005). Trends in biodiversity were usually better predicted by structural type rather than age class. The influence of thinning and stand age on structure and diversity are discussed in more detail below.

There are differences in resolution among different species groups in species richness at different structural stages. For example, species richness of vascular plants exhibited marked differences among stages in Sitka spruce and ash, whereas species richness of birds and hoverflies was less variable (Fig. 4). A fundamental distinction in forest structure to which virtually all taxonomic groups responded was between the pre-thicket forests of Stage I and structural stages post-canopy closure (Stages II–V). Some species groups, such as vascular plants and spiders, showed a unimodal response to stand structural stage, with high species richness in Stage I, low species richness in intermediate stages and increased richness in later stages. Other groups, such as typical forest plant, invertebrate and bird species, increased through the course of the structural cycle. These patterns have also been identified during succession in natural Douglas-fir (*Pseudotsuga menziesii*) forests (Spies 1998) and over the

Table 9 Summary of stand-scale structural, compositional and functional biodiversity indicators and the taxonomic groups to which they apply.

Indicator	Taxonomic group
Structural	
Structural stage	All[a]
Canopy openness	Vascular plants, spiders, birds
Shrub cover	Birds
Vegetation 11–50 cm tall cover	Spiders
Conifer litter	Forest spiders
CWD	Bryophytes, saproxylic hoverflies
Proximity to forest edge	Birds
Proximity to native woodland	Forest vascular plants
Compositional	
Ash[b]	Vascular plants, saproxylic hoverflies
Sitka spruce[c]	Bryophytes, spiders
Bird species (Table 7a)[d]	Birds
Rubus fruticosus, *Dryopteris dilatata*, *Agrostis capillaris*, *Thuidium tamariscinum*[d]	Vascular plants
Hypnum jutlandicum, *Dicranum scoparium*, *Kindbergia praelonga*, *Plagiothecium undulatum*[d]	Bryophytes
Functional	
Stand age	Forest bryophytes, forest vascular plants, forest spiders
Wet microhabitats	Hoverflies
Lower elevation	Birds
Higher elevation	Ground-nesting birds
Available P	Vascular plants
Thinning frequency	All

[a] Relationships between biodiversity and stand structural stage vary by taxonomic group and subgroup, particularly forest species. See text for details

[b] Relative to Sitka spruce

[c] Relative to ash

[d] As these are all common species, they should be used as targets to be achieved by management rather than indicators of high biodiversity stands

silvicultural cycle in plantation forests in Britain (Hill 1979; Ferris et al. 2000; Eycott et al. 2006) and New Zealand (Brockerhoff et al. 2003).

Canopy openness was a key biodiversity indicator for vegetation, particularly vascular plants. Several other indicators of biodiversity for plants, spiders and birds were associated with this key factor, such as cover of shrubs, graminoids, conifer litter and all vegetation 11–50 cm tall. Stands with a more open canopy support greater abundance and diversity of understorey vegetation (Hill 1979; Ferris et al. 2000; Eycott et al. 2006), which have the potential to increase diversity of invertebrates (Day et al. 1993; Fahy and Gormally 1998;

Humphrey et al. 1999; Oxbrough et al. 2005) and birds (Currie and Balmford 1982; Bibby et al. 1989; Duffy et al. 1997; O'Halloran et al. 1999).

Deadwood volumes increased over the forest cycle, as has been noted in previous studies (Spies et al. 1988; Spies 1998; Humphrey and Peace 2003). Biodiversity of bryophytes (total and forest) and saproxylic hoverflies were positively associated with volume of deadwood. Several other studies have found that deadwood is correlated with diversity of bryophytes and lichens (e.g. Engelmark and Hytteborn 1999; Humphrey et al. 2002), fungi (e.g. Humphrey et al. 2000) and invertebrates (e.g. Berg et al. 1994; Similä et al. 2006).

The plantation-scale structural indicators—distance to forest edge for birds and distance to and area of native woodland near plantations for forest plants—reflect the availability of additional habitats in the immediate landscape to act as supplementary habitat or population sources. Forest edges may provide habitat for bird species that prefer open or shrubby habitats, and the presence of broadleaf scrub at plantation forest edges would improve the habitat value of the edge for birds (Currie and Balmford 1982; Bibby et al. 1989; Duffy et al. 1997; Iremonger et al. 2006). Native woodlands in close proximity to plantation forests can act as seed sources for forest plants, which are often dispersal-limited (Ehrlén and Eriksson 2000; Verheyen et al. 2003; Devlaeminck et al. 2005). The same relationship was not observed for forest bryophytes, most likely because bryophyte spores disperse more easily over longer distances.

Compositional

For an indicator to be easy and cost-effective to measure, it must be capable of ready assessment by non-specialists (Ferris and Humphrey 1999), such as forest inventory staff or individual landowners. Because of the low species diversity of birds in Ireland and the lack of forest specialists (Wilson et al. 2006), the number of indicator bird species is low (Table 7). In fact most of the indicator species are common birds in Ireland and will be familiar to most non-specialists (Coombes et al. 2002). The vascular plant and bryophyte species we have listed are readily identifiable with some practice; they are also common species in Ireland. We have not selected any invertebrate species as compositional biodiversity indicators, as they require more time-consuming and expensive sampling methods and are not easily identifiable by non-specialists.

Tree species, a potential compositional indicator, produced contrasting results among species groups. Bryophyte and spider species richness were higher in Sitka spruce stands, vascular plant and saproxylic hoverfly species richness were higher in ash stands and no differences in total hoverfly or bird species richness were observed between tree species (Table 6). These results can be explained by the biology of the different taxonomic groups. A greater diversity of vascular plants is facilitated by the lighter ash canopy and also by the more fertile conditions in which the ash stands occurred. Bryophyte species richness is encouraged by the high humidity and lower competition from larger plants in spruce stands. Species richness in spiders is strongly influenced by vegetation structure (Greenstone 1984; Dennis et al. 1998; McNett and Rypstra 2000), and it is likely that the reduction in cover of larger plants in spruce stands facilitated development of ground and lower field layer vegetation that favours ground-dwelling spiders. Our results, however, cannot be extrapolated to other broadleaved or coniferous tree species. For example, the number of plant-feeding invertebrate species—which form part of the diet of spiders and birds—associated with ash in Britain is relatively low compared to other native broadleaved trees (Key 1995). Therefore, comparisons between Sitka spruce and plantations of other native broadleaved tree species might produce greater contrasts in biodiversity than those we found. Oak (*Quercus* spp.) and birch (*Betula* spp.) would be particularly interesting to study in this context as they support high numbers of plant-feeding invertebrate species (Jones 1959; Atkinson 1992; Key 1995). While some invertebrate species may specialise on other conifer species, differences in forest structure may have a more important potential effect on biodiversity. In particular, pines (*Pinus* spp.) and larches (*Larix* spp.) tend to allow greater light penetration through the canopy and therefore allow greater development of vascular ground flora during the middle part of the forest cycle (Hill 1979; Ferris et al. 2000; French et al. in press).

Functional

Stand age was an adequate positive biodiversity indicator only for forest plants and forest spiders and a negative indicator for birds in intermediate-aged stands, especially species characteristic of open habitats. The increase of forest plant and spider species richness is the result of dispersal and colonisation of a suitable habitat, successional processes seen in natural forests and also in plantation forests. Previous research in Britain (Hill and Jones, 1978; Ferris et al. 2000) and New Zealand (Brockerhoff et al. 2003) has also found that mature plantations of native and exotic species can acquire floras characteristic of native forests. In our study, the flora of mature spruce plantations somewhat resembles that of native acid oak woodland and the flora of mature ash plantations is similar to that in semi-natural ash-hazel woodland (French et al. in press). Dispersal is an important mechanism in the succession of forest vascular plants, as demonstrated by the relationship with proximity to old woodland. For spiders, whose young can disperse long distances by ballooning, a more important factor may be the development of suitable habitat in the form of high cover of conifer litter or bryophytes (Table 9; Oxbrough et al. 2005). For birds, dispersal is likely to be much less important than the development of suitable or unsuitable forest structures with age. Tree size and density were strongly correlated with stand age (Tables 2 and 4), and it is these factors that are probably responsible for the negative relationship between open-habitat bird species richness and age. In contrast, canopy cover was less well correlated with stand age, and species groups strongly influenced by canopy cover will as a result have weaker relationships with stand age.

One functional process that is partly responsible for the lower correlation between stand age and canopy cover is thinning. Thinning operations decrease canopy cover, at least temporarily, promote larger diameter trees in the longer term and increase deadwood volume. It was not possible to analyse the relationships between thinning and biodiversity explicitly because of the difficulty in obtaining stand-specific information on thinning regime. However, such information as we were able to obtain suggests that Stage II and III Sitka spruce stands were mainly unthinned and that Stage V spruce stands had been subjected to at least three thinning operations. Therefore, thinning can be considered an indicator of biodiversity in Sitka spruce forests (Table 9). In contrast, Brockerhoff et al. (2003) in New Zealand found that more heavily thinned stands supported a greater proportion of exotic species than more dense stands; however, maximum canopy closure in their pine stands was 80%, lower than the average in our Stage III stands. As thinning is a management practice, it is the easiest of our functional indicators to change and may be the most efficient method of influencing stand biodiversity. We recommend that spruce plantations be thinned early and at regular intervals so as to prevent complete canopy closure (Smith et al. 2005).

We identified few other functional indicators, possibly in part as a result of our strategy of clustering sites to reduce environmental variation. The presence of wet microhabitats, such as flushes and temporary ponds or streams, in plantation forests increased hoverfly biodiversity by meeting the habitat requirements for the larval stages of a particular set of species. Wet microhabitats may also increase stand biodiversity of other plant or animal species groups. Wet spruce sites also supported higher amounts of deadwood, due to greater frequency of windthrow, and thereby greater diversity of saproxylic hoverflies (Table 8). Site elevation as an indicator of bird diversity appears to reflect the differences between upland and lowland plantations and surrounding landscapes. Higher available P in Stage V Sitka spruce stands indicates greater richness of vascular plants, perhaps because greater soil fertility permits coexistence of a wider range of species. Eycott et al. (2006) also found that more fertile sites supported higher vascular plant species richness in pine plantations in lowland England, and Similä et al. (2006) reached the same conclusion in managed boreal forests in Finland. Our sites were of low overall fertility, however, and the true relationship between fertility and vascular plant diversity may be unimodal rather than linear (Grime 1979; Pausas and Austin 2001).

Using indicators

The biodiversity indicators we identify (Table 9) can be used to assess the effect of site management practices on biodiversity and to identify sites that are potentially of high biodiversity value. Used as the former, they are targets to be achieved by management. Presence of few or no indicators in forest stands suggests that management methods should be improved; the indicators also provide information on the changes required. Forest stands identified as being of potentially high biodiversity can be surveyed and assessed more thoroughly, and management for biodiversity can be prioritised in forest planning and operations. Indicators cannot substitute for thorough ecological surveys, particularly when sites of major biodiversity importance may be involved, but they can be employed as a first step in management assessment or in identifying sites of biodiversity value. These indicators cannot identify sites where rare species are present, a failure general to the indicator approach (Niemi and McDonald 2004).

Our indicators should be considered preliminary until they are verified using independent data (Noss 1990). In addition, the context in which they have been identified, i.e. first rotation Sitka spruce and ash stands managed under a clearfelling system, must be taken into consideration. These indicators should generally be employed at the stand level, rather than at the level of the whole plantation or landscape. They should be used in conjunction: in general, it is misleading to label a stand as having high biodiversity (or not) on the basis of just one or two indicators. We recommend the presence of at least four indicators from two or more groups (compositional, structural and functional) as a general guideline for designating sites or stands as potentially having high biodiversity. We are not aware of any similar recommendations regarding the number of indicators required for a favourable biodiversity assessment.

Conclusions

We have developed a set of provisional, stand-scale indicators of biodiversity for Sitka spruce and ash plantation forests for use in Ireland; they are likely to be applicable over a wider area with similar climates, such as northern Britain. The approach we have used to develop indicators has a wider potential for application. The indicators can be employed by non-specialists as a first step in identifying potentially high biodiversity stands or assessing the biodiversity implications of management interventions. Structural and functional indicators are particularly useful, as their assessment is often relatively simple. Functional indicators can represent management interventions, such as thinning, that can be changed if required; structural indicators provide targets for management to reach. Stand structure strongly affects species richness. A multivariate classification of stand structure into a small number of stages can be used to summarise biodiversity changes over the forest cycle. As species groups vary in their response to changes in stand structure, caution is required when using one group as a surrogate for the biodiversity of another.

Acknowledgements We thank Pat Neville and Aileen O'Sullivan for their help in all aspects of our work and Conor Clenaghan, John Cross, Jonathan Humphrey and Tor-Björn Larsson for advice on the design of this project. We are grateful to Coillte and private landowners for access to their forest plantations and information on site management. We thank Jacqueline Bolli, Maire Buckley, Noirín Burke, John Cleary, Sinéad Cummins, Gerry Farrell and Richard Jack for assistance in the field and with sample sorting. Two anonymous reviewers provided helpful comments on an earlier draft. This study is a contribution from the BIOFOREST Project, funded by the National Development Plan through the Council on Forest Research and Development (COFORD) and the Environmental Protection Agency (EPA).

References

Allen SE, Grimshaw HM, Rowland AP (1986) Chemical analysis. In: Moore PD, Chapman SB (eds) Methods in plant ecology. Blackwell, Oxford, pp 285–344

Anand M, Laurence S, Rayfield B (2005) Diversity relationships among taxonomic groups in recovering and restored forests. Conserv Biol 19:955–962

Atkinson MD (1992) Biological flora of the British isles: *Betula pendula* Roth (*B. verrucosa* ehrh.) and *B. pubescens* Ehrh. J Ecol 80:837–870

Beaman M (1994) Palearctic birds. Harrier, Stonyhurst

Berg Å, Ehnström B, Gustafsson L et al (1994) Threatened plant, animal, and fungus species in swedish forests: distribution and habitat associations. Conserv Biol 8:718–731

Bibby C, Burgess ND, Hill DA (1992) Bird census techniques. Academic Press, London

Bibby CJ, Bain CG, Burges DJ (1989) Bird communities of highland birchwoods. Bird Study 36:123–133

Brockerhoff EG, Ecroyd CE, Leckie AC et al (2003) Diversity and succession of adventive and indigenous vascular understorey plants in *Pinus radiata* plantation forests in New Zealand. For Ecol Manage 185:307–326

Coombes RH, Crowe O, Lysaght L et al (2002) Countryside bird survey report 1998–2000. Bird Watch Ireland, Dublin

Cross JR (1998) An outline and map of the potential natural vegetation of Ireland. Appl Veg Sci 1:241–252

Currie FA, Balmford R (1982) The value to birdlife of retaining small conifer stands beyond normal felling age within forests. Q J For 76:153–160

Curtis D (1980) Pitfalls in spider community studies. J Arach 8:271–280

Day KR, Marshall S, Heaney C (1993) Associations between forest type and invertebrates: ground beetle community patterns in a natural oakwood and juxtaposed conifer plantations. Forestry 66:37–50

Dennis P, Young M, Gordon I (1998) Distribution and abundance of small insects and arachnids in relation to structural heterogeneity of grazed, indigenous grasslands. Ecol Entomol 23:253–264

Devlaeminck R, Bossuyt B, Hermy M (2005) Seed dispersal from a forest into adjacent cropland. Agric Ecosys Environ 107:57–64

Duffy BL, O'Halloran J, Kelly TCK et al (1997) The breeding bird communities of Balrath wood, Co. Meath, 1996: a preliminary investigation. Ir For 54:2–8

Dufrêne M, Legendre P (1997) Species assemblages and indicator species: the need for a flexible asymmetric approach. Ecol Monogr 67:345–366

Ehrlén J, Eriksson O (2000) Dispersal limitation and patch occupancy in forest herbs. Ecology 81:1667–1674

Engelmark O, Hytteborn H (1999) Coniferous forests. Acta Phytogeogr Suec 84:55–74
Eycott AE, Watkinson AR, Dolman PM (2006) Ecological patterns of plant diversity in a plantation forest managed by clearfelling. J Appl Ecol 43:1160–1171
Fahy O, Gormally M (1998) A comparison of plant and beetle communities in an Irish oak woodland with a nearby by conifer plantation and clearfelled site. For Ecol Manage 110:263–273
Ferris R, Humphrey JW (1999) A review of potential biodiversity indicators for application in British forests. Forestry 72:313–328
Ferris R, Peace AJ, Humphrey JW, et al (2000) Relationships between vegetation, site type and stand structure in coniferous plantations in Britain. For Ecol Manage 136:35–51
Forest Service (2000a) Afforestation grant and premium schemes. Forest Service, Johnstown Castle
Forest Service (2000b) Code of best forest practice– Ireland. Forest Service, Johnstown Castle
Forest Service (2000c) Forest biodiversity guidelines. Forest Service, Johnstown Castle
Forest Service (2006) Forest environment protection (afforestation) scheme. Forest Service, Johnstown Castle
Forestry Commission (2004) The UK forestry standard. Forestry Commission, Edinburgh
French LJ, Smith GF, Kelly DL et al (in press) Ground flora communities in temperate oceanic plantation forests and the influence of silvicultural, geographic and edaphic factors. For Ecol Manage
Gaston KJ (1996) Species richness: measure and measurement. In: Gaston KJ (ed) Biodiversity: a biology of numbers and difference. Blackwell Science, Oxford, pp 396
Greenstone M (1984) Determinants of web spider species diversity: vegetation structural diversity vs. prey availability. Oecologia 62:299–304
Grime JP (1979) Plant strategies and vegetation processes. Wiley & Sons, Chichester
Halpern CB, Spies TA (1995) Plant species diversity in natural and managed forests of the Pacific northwest. Ecol Appl 5:913–934
Hayden T, Harrington R (2001) Exploring Irish mammals. Town House, Dublin
Hill MO (1979) The development of flora in even-aged plantations. In: Ford ED, Malcolm DC, Atterson J (eds) The ecology of even aged plantations. Institute for Terrestrial Ecology, Cambridge, pp 175–192
Hill MO, Jones EW (1978) Vegetation changes resulting from afforestation of rough grazing in Caeo Forest, South Wales. J Ecol 66:433–456
Hodge SJ, Peterken GF (1998) Deadwood in British forests: Priorities and a strategy. Forestry 71:99–112
Howard PC, Viskanic P, Davenport TRB et al (1998) Complementarity and the use of indicator groups for reserve selection in Uganda. Nature 394:472–475
Humphrey JW (2005) Benefits to biodiversity from developing old-growth conditions in British upland spruce plantations: a review and recommendations. Forestry 78:33–53
Humphrey JW, Davey S, Peace AJ et al (2002) Lichens and bryophyte communities of planted and semi-natural forests in Britain: the influence of site type, stand structure and deadwood. Biol Conserv 107:165–180
Humphrey JW, Hawes C, Peace AJ et al (1999) Relationships between insect diversity and habitat characteristics in plantation forests. For Ecol Manage 113:11–21
Humphrey JW, Newton AC, Peace AJ et al (2000) The importance of conifer plantations in northern Britain as a habitat for native fungi. Biol Conserv 96:241–252
Humphrey JW, Peace AJ (2003) Deadwood. In: Humphrey J, Ferris R, Quine C (eds) Biodiversity in Britain's planted forests. Forestry Commission, Edinburgh, pp 41–49
Iremonger S, Gittings T, Smith GF et al (2006) Investigation of experimental methods to enhance biodiversity in plantation forests. Report for COFORD and EPA, Dublin
Iremonger S, O'Halloran J, Kelly DL et al (2007) Biodiversity in Irish plantation forests. Environmental Protection Agency, Dublin
Jones EW (1959) Biological flora of the British isles: *Quercus* L. J Ecol 47:169–222
Jonsson BG, Jonsell M (1999) Exploring potential biodiversity indicators in boreal forests. Biodivers Cons 8:1417–1433
Kati V, Devillers P, Dufrêne M, et al (2004) Testing the value of six taxonomic groups as biodiversity indicators at a local scale. Conserv Biol 18:667–675
Kelly DL (1981) The native forest vegetation of Killarney, south-west Ireland: an ecological account. J Ecol 69:437–472
Kelly DL (2005) Woodland on the western fringe: Irish oakwood diversity and the challenges of conservation. Bot J Scot 57:21–40
Key RS (1995) Invertebrate conservation and new woodland in Britain. In: Ferris-Kaan R (ed) The ecology of woodland creation. John Wiley & Sons, Chichester
Larsson T-B (2001) Biodiversity evaluation tools for European forests. Ecological bulletins, vol 50. Blackwell, Oxford

Lawton JH, Bignell DE, Bolton B et al (1998) Biodiversity inventories, indicator taxa and effects of habitat modification in tropical forest. Nature 391:72–76
Legendre P, Legendre L (1998) Numerical ecology. Developments in environmental modelling, vol 20. 2nd English edn. Elsevier, Amsterdam
Leppäniemi P, Hallikainen V, Mikkola K et al (1998) Forest structure classes in central Finnish Lapland. Scand J For Res 13:442–450
Lindenmayer DB (1999) Future directions for biodiversity conservation in managed forests: indicator species, impact studies and monitoring programs. For Ecol Manage 115:277–287
Lindenmayer DB, Franklin JF, Fischer J (2006) General management principles and a checklist of strategies to guide forest biodiversity conservation. Biol Conserv 131:433–445
Lindenmayer DB, Margules CR, Botkin DB (2000) Indicators of biodiversity for ecologically sustainable forest management. Conserv Biol 14:941–950
Magurran AE (2004) Measuring biological diversity. Blackwell Publishing, Oxford
McCune B, Mefford MJ (1997) PC-ORD for Windows, version 4.25. MjM Software, Glenedon Beach, Oregon
McElhinny C, Gibbons P, Brack C et al (2005) Forest and woodland stand structural complexity: its definition and measurement. For Ecol Manage 218:1–24
McNett B, Rypstra A (2000) Habitat selection in a large orb-weaving spider: vegetational complexity determines site selection and distribution. Ecol Entomol 25:423–432
Mueller-Dombois D, Ellenberg H (1974) Aims and methods of vegetation ecology. Wiley, New York
Negi HR, Gadgil M (2002) Cross-taxon surrogacy of biodiversity in the Indian Garhwal Himalaya. Biol Conserv 105:143–155
Niemela J, Haila Y, Punttila P (1996) The importance of small-scale heterogeneity in boreal forest-floor invertebrates across the succession gradient. Ecography 19:352–368
Niemi GJ, McDonald ME (2004) Application of ecological indicators. Annu Rev Ecol Evol Syst 35:89–111
Noss RF (1990) Indicators for monitoring biodiversity: a hierarchical approach. Conserv Biol 4:355–364
Noss RF (1999) Assessing and monitoring forest biodiversity: a suggested framework and indicators. For Ecol Manage 115:135–146
O'Halloran J, Gittings T, Smith GF et al (2004) Biodiversity of plantation forests in Ireland: Bioforest project. In: MacLennan L (ed) Opportunities for Biodiversity Enhancement in Plantation Forests. COFORD, Dublin, Cork, pp 29–38
O'Halloran J, Walsh PM, Giller PS et al (1999) An assessment of avian biodiversity and opportunities for enhancement in Ireland's forests: preliminary results. Ir For 55:2–14
Oertli S, Müller A, Steiner D et al (2005) Cross-taxon congruence of species diversity and community similarity among three insect taxa in a mosaic landscape. Biol Conserv 126:195–205
Oxbrough AG, Gittings T, O'Halloran J et al (2005) Structural indicators of spider communities across the forest plantation cycle. For Ecol Manage 212:171–183
Pausas JG, Austin MP (2001) Patterns of plant species richness in relation to different environments: an appraisal. J Vegetat Sci 12:153–166
Pharo EJ, Beattie AJ, Binns D (1999) Vascular plant diversity as a surrogate for bryophyte and lichen diversity. Conserv Biol 13:282–292
Pithon J, O'Halloran J, Moles R (2005) The influence of coniferous afforestation on lowland farmland bird communities in Ireland: different seasons and landscape contexts. Landscape Urban Plann 71:91–103
Pitkänen S (1997) Correlation between stand structure and ground vegetation: an analytical approach. Plant Ecol 131:109–126
Prendergast JR, Quinn RM, Lawton JH et al (1993) Rare species, the coincidence of diversity hotspots and conservation strategies. Nature 365:335–337
Sætersdal M, Gjerde I, Blom HH et al (2003) Vascular plants as a surrogate species group in complementary site selection for bryophytes, macrolichens, spiders, carabids, staphylinids, snails, and wood living polypore fungi in a northern forest. Biol Conserv 115:21–31
Similä M, Kouki J, Mönkkönen M et al (2006) Co-variation and indicators of species diversity: can richness of forest-dwelling species be predicted in northern boreal forests? Ecol Indicators 6:686–700
Smith AJE (2004) The moss flora of Britain and Ireland. 2nd edn. Cambridge University Press, Cambridge
Smith G, Gittings T, Wilson M et al (2005) Assessment of biodiversity at different stages of the forest cycle. Report for COFORD and EPA, Dublin http://www.coford.ie/iopen24/pub/pub/Project_files/312Report.pdf
Sneath PHA, Sokal RR (1973) Numerical taxonomy. W.H. Freeman, San Francisco
Sokal RR, Rohlf FJ (1995) Biometry. 3rd edn. W.H. Freeman and Co., New York
Sommagio D (1999) Syrphidae: can they be used as environmental bioindicators? Agric Ecosys Environ 74:343–356

Southwood TRE (2000) Ecological methods with particular reference to the study of insect populations. Blackwell Science, Oxford
Speight MCD, Castella E (2001) An approach to interpretation of lists of insects using digitised biological information about the species. J Insect Conserv 5:131–139
Spies TA (1998) Forest structure: a key to the ecosystem. Northwest Sci 72:34–39
Spies TA, Franklin JF, Thomas TB (1988) Coarse woody debris in Douglas-fir forests of western Oregon and Washington. Ecology 69:1689–1702
SPSS (2001) SPSS for Windows, version 11.0.1. SPSS, Chicago
Stace C (1997) New flora of the British Isles. 2nd edn. Cambridge University Press, Cambridge
Verheyen K, Guntenspergen GR, Biesbrouck B et al (2003) An integrated analysis of the effects of past land use on forest herb colonization at the landscape scale. J Ecol 91:731–742
Vessby K, Söderström B, Glimskär A et al (2002) Species-richness correlations of six different taxa in Swedish seminatural grasslands. Conserv Biol 16:430–439
Williams P, Faith D, Manne L et al (2006) Complementarity analysis: Mapping the performance of surrogates for biodiversity. Biol Conserv 128:253–264
Wilson MW, Pithon J, Gittings T et al (2006) Effects of growth stage and tree species composition on breeding bird assemblages of plantation forests. Bird Study 53:225–236

Options for biodiversity conservation in managed forest landscapes of multiple ownerships in Oregon and Washington, USA

Nobuya Suzuki · Deanna H. Olson

Originally published in the journal Biodiversity and Conservation, Volume 17, No 5, 1017–1039.
DOI: 10.1007/s10531-007-9301-4

Abstract We examine existing and developing approaches to balance biodiversity conservation and timber production with the changing conservation roles of federal and nonfederal forest land ownerships in the US Pacific Northwest. At landscape scales, implementation of the reserve-matrix approach of the federal Northwest Forest Plan in 1994 was followed by proposals of alternative designs to better integrate disturbance regimes or to conserve biodiversity in landscapes of predominantly young forests through active management without reserves. At stand scales, landowners can improve habitat heterogeneity through a host of conventional and alternative silvicultural techniques. There are no state rules that explicitly require biodiversity conservation on nonfederal lands in the region. However, state forest practices rules require retention of structural legacies to enhance habitat complexity and establishment of riparian management areas to conserve aquatic ecosystems. Habitat Conservation Plans (HCPs) under the US Endangered Species Act provide regulatory incentives for nonfederal landowners to protect threatened and endangered species. A state-wide programmatic HCP has recently emerged as a multi-species conservation approach on nonfederal lands. Among voluntary incentives, the Forest Stewardship Council certification comprehensively addresses fundamental elements of biodiversity conservation; however, its tough conservation requirements may limit its coverage to relatively small land areas. Future changes in landscape management strategies on federal lands may occur without coordination with nonfederal landowners because of the differences in regulatory and voluntary incentives between ownerships. This raises

This paper was previously published in Biodiversity and Conservation, Volume 16(13) under doi 10.1007/s10531-007-9198-y

N. Suzuki (✉)
Department of Zoology, Oregon State University, Corvallis, OR 97331, USA
e-mail: nobi.suzuki@oregonstate.edu

D. H. Olson
Pacific Northwest Research Station, USDA Forest Service, Corvallis, OR, USA

N. Suzuki
Pacific Northwest Research Station, 3200 SW Jefferson Way, Corvallis, OR 97331, USA

concerns when potentially reduced protections on federal lands are proposed, and the capacity of the remaining landscape to compensate has been degraded.

Keywords Forest certification · Forest practices rules · Habitat Conservation Plans · Matrix · Nonfederal lands · Northwest Forest Plan · Regulatory incentives · Reserves · Voluntary incentives · United States

Abbreviations

CA	California
DBH	Diameter at Breast Height
ESA	US Endangered Species Act
FSC	Forest Stewardship Council
HCP	Habitat Conservation Plan
NWFP	Northwest Forest Plan
OR	Oregon
SFI	Sustainable Forestry Initiative
US	United States
WA	Washington

Introduction

Biodiversity conservation is a conundrum for a forested landscape, such as the temperate coniferous forests of western Oregon (OR) and Washington (WA), US, because lands are owned and managed by a mixture of federal and nonfederal ownerships that differ in goals and objectives, laws and regulations, and management practices and land use patterns. The situation becomes more complex when conservation measures of one landowner are implemented contingent upon land management decisions occurring on adjacent lands. As measures become further nested within plans and policies across adjacent landowners, the situation can become untenable if these subsequently become altered over time. We examine how biodiversity conservation is addressed in the multiple ownerships of the US Pacific Northwest, at landscape and stand scales for federal and nonfederal ownerships, and synthesize regulatory and voluntary incentives to conserve species and biodiversity for nonfederal ownerships.

Central to understanding current approaches to conserve forest biodiversity in this region is a review of relevant events over the past decade. Public concerns over conservation of rare native species associated with old-growth forests brought forest management of economically valuable conifers, largely comprised of Douglas-fir (*Pseudotsuga menziesii*), to a standstill across 9.8 million ha of federal land in the late 1980s (USDA and USDI 1993). The decision to list the northern spotted owl (*Strix occidentalis*), marbled murrelet (*Brachyramphus marmoratus*), and anadromous salmonid fishes (*Oncorhynchus* species) under the US Endangered Species Act of 1973 (ESA) in the early 1990s raised further concerns for listing 1,098 other species that were potentially associated with late-successional or old-growth forest conditions in the region (Thomas et al. 1993). Restrictions on forest management activities also were imposed on nonfederal lands where ESA-listed species and their habitats were found (Epstein 1997). As a solution to the region's forest management crisis, the Northwest Forest Plan was developed in 1993 (USDA and USDI 1993, 1994) to balance social, economic and ecological values of federal forests (Fig. 1) and to ease burden of species conservation on nonfederal lands (Thomas et al. 2006).

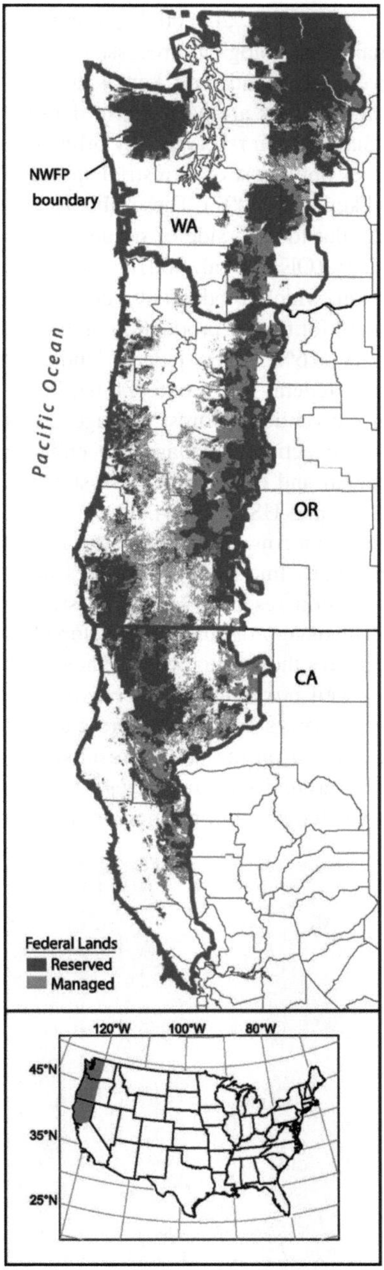

Fig. 1 Forest landscape area of the US Pacific Northwest with the range of the northern spotted owl delineated as the federal Northwest Forest Plan (NWFP) boundary. Federal forest lands in reserved and managed (i.e., planted: matrix, adaptive management area) land use allocations extend from northwestern Washington (WA), through Oregon (OR) into northwestern California (CA)

Under the Northwest Forest Plan (NWFP), a network of large reserves was established on federal lands as a coarse filter approach to conserve habitats of the northern spotted owl, marbled murrelet, and other species that were associated with late-successional and old-growth ecosystems (Fig. 1, Thomas et al. 2006). Aquatic conservation strategies were developed to restore and maintain ecological processes of aquatic and riparian habitat by establishing riparian reserves and identifying key watersheds as targets of restoration (Reeves et al. 2006). Stream networks are particularly dense in the forested landscape of western OR and WA, such that riparian reserves may offer protection not only for aquatic- and riparian-dependent species but a host of terrestrial species as well (Olson et al. 2007b; Rundio and Olson 2007; Rykken et al. 2007). These ribbons of protected federal forest may extend to ~150 m perpendicular to each side of streams, dependening on the presence of fish, stream size and hydrology (Olson et al. 2007b); hence, riparian reserves may be the key foundation element upon which biodiversity conservation rests (Figs. 2 and 3). Meanwhile, forest management for timber production on federal lands became restricted to the lands classified as matrix (16% of the federal lands; 1,609,000 ha) and Adaptive Management Areas (6% of the federal lands; 616,000 ha; Fig. 1 [USDA and USDI 1994]). Instead of simply applying conventional forest management techniques for timber production, these lands have been actively managed to enhance biodiversity by retaining components of late-successional and old-growth ecosystems, including large green trees, snags, and down wood (USDA and USDI 1994).

To further address species' concerns relative to forest management activities on matrix and Adaptive Management Areas, fine filter approaches also were applied, such as the deployment of 40-ha conservation reserves around existing spotted owl nests (Thomas et al. 2006). The survey-and-manage program was developed as an additional fine filter for ~400 rare or little-known species thought to be associated with late-successional and old-growth forests that were not well protected by reserves (Molina et al. 2006b). Under the survey-and-manage program, surveys for species presence were conducted before trees were harvested from stands. Based on the survey results, harvest plans were potentially modified or protection buffers around known species sites were established. The survey-and-manage program also collected new information on rare or little-known species (Olson

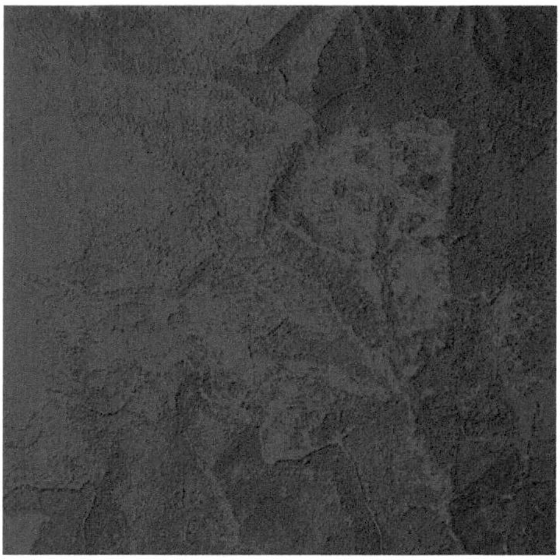

Fig. 2 A study site of the US Bureau of Land Management's density management and riparian buffer study in western Oregon (Cissel et al. 2006) showing a mix of silvicultural approaches offering benefits to native forest-dependent species, including: aggregated retention harvest or leave islands of three circular sizes (0.1, 0.2, 0.4 ha), riparian reserves of differing widths (6–70 m on each side of streams), and thinned forest where the pre-harvest managed stand of ~600 trees per ha (tph) was reduced to a range of densities, ~100–300 tph. Species' responses to these treatments suggest this was a relatively benign disturbance (Wessell 2005; Olson and Rugger 2007; Rundio and Olson 2007)

Fig. 3 Patchwork of forest types created by diverse land ownerships and forest management practices in Oregon and Washington, US, result in a landscape mosaic, where a variety of conservation measures for biodiversity are embedded. Dense stream networks in this region result in riparian reserves occurring in most managed stands, as shown by the ribbon of green trees in the central harvested patch. This specific landscape is an area of ''checkerboard'' ownership near the H.J. Andrews Experimental Forest in the western Cascade Range, Oregon. Photograph by Al Levno, July 2005, courtesy of the USDA Forest Service, Pacific Northwest Research Station and the Oregon State University Forest Science Data Bank

et al. 2007a) and developed management strategies for species persistence on federal lands (Molina et al. 2006b; Thomas et al. 2006).

The NWFP was intended to relieve the burden of species management from nonfederal lands. This has been a successful enterprise, with interesting consequences. Nonfederal lands in the region of the plan comprise 13.2 million ha of multiple ownerships, including states, native American tribes, private industry, and small woodland owners (USDA and USDI 1993). Nonfederal forests in the region are predominantly plantations and often are managed at shorter rotations, such as 40–60 years rotations (e.g., Curtis 1997; Curtis et al. 1998), in comparison to federal matrix lands (80 years rotations, USDA and USDI 1994). The trend of private landowners to favor short rotations and grow smaller and more uniform trees has been intensified since the implementation of the NWFP. This is mainly due to the closure of mills that could process large logs previously supplied by federal lands and to the decline in the export market of large logs to Asia in the mid-1990 (Barbour et al. 2006). Furthermore, 85% of the region's timber harvest has occurred on private lands between 2000 and 2005 (Bormann et al. 2006). Consequently, over the last 10 plus years, two distinct habitat patterns have developed across the US northwest forest landscape between ownerships: older forests dominate on federal lands and young forests dominate on nonfederal lands (Molina et al. 2006a).

Important biodiversity conservation issues arise from this bifurcation of forest land pattern (Fig. 3). Timber production priorities dominate management of most nonfederal forest lands in the region, while a more balanced approach for timber harvest and species conservation occurs on federal lands, as per the design of the NWFP. On nonfederal lands, there are generally fewer protective measures for biodiversity or habitats such as riparian areas relative to federal lands, except ESA-listed species receive some protection. Meanwhile, significantly, large proportions (>40%) of high quality nesting habitat for species such as the marbled murrelet and the northern spotted owl (Bormann et al. 2006) and the majority of best spawning and rearing habitats of coho salmon (*Oncorhynchus kisutch*) remain on nonfederal lands without the level of protection assured by the NWFP (Barbour et al. 2006). This adds to the conundrum for how to best provide for species by the differing yet interdependent approaches offered by federal and nonfederal lands across a mosaic of landowners.

Additionally, it needs to be pointed out that the NWFP is not a static entity. While adaptive management is a foundation element of the NWFP, its basic acceptance within the region has been tenuous, at best. There have been continuous debates and lawsuits to eliminate or modify the NWFP over the first 10 years of its implementation (Thomas et al. 2006). In particular, the survey-and-manage program has fallen under scrutiny. In 2004, this program was terminated in response to a settlement agreement for a lawsuit brought by the timber industry (Molina et al. 2006b). However, a court ruling reinstated the survey-and-manage program in 2006, in response to a counter lawsuit brought by environmental groups, citing inconsistencies and deficiencies in the analyses for termination (Molina et al. 2006b). Another current discussion has been whether to adopt an alternative forest management plan on Oregon Bureau of Land Management (BLM) lands (~1 million ha) which could potentially eliminate large reserves, survey-and-manage, and special status species considerations. This consideration arose in response to a 2003 settlement agreement to another lawsuit, contending that Oregon BLM lands were to be available for sustainable timber production under the OR and CA Revested Railroad Lands Act of 1937 (O&C Act of 1937, Public Land Foundation 2005). Elimination of large reserves and rare species provisions would increase the area of actively managed matrix lands. Hence the role and capacity of these federal lands for biodiversity conservation would be significantly altered. Changes in the NWFP to lower current conservation standards might potentially have ramifications of compensation elsewhere; as biodiversity conservation measures are reduced in one area, adjacent landowners or other stakeholders may be compelled to heighten their measures (Molina et. al 2006a; Stritthold et al. 2006).

While the NWFP's implementation and its consequences over the last 10 years provide a necessary back-drop for understanding our current situation in the US Pacific Northwest, the science and policies of biodiversity conservation are not wholly represented by this narrow focus. The following sections present and evaluate additional approaches that have been conceived or are being implemented in the region. First, we present recently developed or already existing landscape- and stand-scale approaches for forest management plans, with an example of one design that integrates elements from both scales. Second, we review a number of conservation policies and incentives that are particularly relevant to nonfederal landowners.

Forest management approaches at landscape and stand scales

Management approaches at landscape scales

The NWFP has been the most comprehensive approach for conservation of biodiversity at the regional landscape scale in the US, characterized by its reserve-matrix approach to balance conservation of biodiversity and commodity production (Spies and Turner 1999). An alternative landscape management design was proposed to better integrate natural disturbance regimes into this reserve-matrix design (Cissel et al. 1998, 1999). In their alternative, Cissel et al. (1998, 1999) modified the reserve-matrix approach by assigning matrix lands to three categories of historical fire regimes, based on fire frequency and intensity. These matrix lands were then managed according to patterns of the three fire regimes, where rotation lengths matched fire frequencies and levels of harvest (number of retained trees) matched fire intensities. The alternative design was predicted to yield less timber volume but produce more late-successional habitats with large patch sizes, higher canopy heterogeneity, and greater landscape connectivity. Thus, the landscape managed by this alternative approach could potentially contribute more to conservation of biodiversity than the original matrix-reserve approach (Cissel et al. 1998, 1999).

In dry fire-prone landscapes of the US northwest, where fire suppression has altered forest structure, large areas of forested reserves may be lost to fires over time, and a reserve-matrix approach to conservation may not be effective (Spies et al. 2006). Instead, active management of the entire landscape could restore forest stand structure to a natural state, while commodity production could support the cost of stand restoration (Spies et al. 2006). Management objectives for biodiversity conservation in fire-prone landscapes could be to restore open late-successional forests that are resistant to stand-replacement fires and to create habitat islands of dense layered forests within the fire resistant forests for rare species associated with dense forests, such as the northern spotted owl (Spies et al. 2006). For example, a combination of thinning from below and fuel treatments based on vegetation patterns and historic fire regimes would be one way to achieve desired forest stand conditions in a fire-prone landscape (Spies et al. 2006).

Conservation approaches without reserves also are inevitable in landscapes of predominantly young forests heavily focused on timber production, such as most nonfederal forest lands in the US northwest region. In such landscapes, one approach for biodiversity conservation would be to use active management to provide a landscape with full representation of forest stands in different structural stages of development. An example of such a landscape management plan, referred to as "structure-based management," was developed in 2000 by the OR Department of Forestry for a landscape of predominantly young even-aged forests primarily intended for commodity production (Bordelon et al. 2000; ODF 2001). The core strategy of their structure-based management was to actively manage and maintain shifting mosaics of five structural stages of forest stands across the landscape in pre-determined proportions (Bordelon et al. 2000). Various densities of thinning were applied to create stand conditions that met target allocations and stand configurations (Bordelon et al. 2000).

Management approaches at stand scales

To conserve biodiversity at forest stand scales, recent innovative silvicultural approaches incorporate processes of natural stand development and patterns of natural disturbance that are responsible for habitat heterogeneity in natural forest stands (Hunter 1993; Franklin et al. 2002). Variable retention harvest has been proposed as a means to quickly restore function, structure, and composition of late-successional forests at stand scales by retaining key structural legacies of original stands to which various biota have strong associations, including large live trees, snags, down wood, undisturbed layers of forest floor, and understory plant communities (Franklin et al. 1997; Palik et al. 2003). Dispersed retention of dominant or co-dominant trees may provide well-distributed sources of soil energy, future snags and down wood, habitat for late-successional species as well as mitigate microclimate or hydrological processes evenly throughout a stand (Hansen et al. 1995; Franklin et al. 1997). Aggregated retention, also called patch reserves or leave islands, may be used to provide lifeboats for low-mobility species from removal of their entire habitat during stand harvest operations. Retaining leave islands of old trees, snags, down wood, or deciduous trees in conifer stands would provide habitat for some low-mobility species, such as lichens, vascular plants, arthropods, mollusks, and amphibians (Neitlich and McCune 1997; Duncan 1999; Wessell 2005).

Another approach to promote heterogeneity in managed forest stands is to create irregular distributions and densities of trees in a stand through either planting at irregular spacing or thinning at variable densities (McComb et al. 1993, Carey and Curtis 1996, Hayes et al. 1997). Thinning with varying target densities among stands could potentially be used to promote horizontal heterogeneity among stands across the landscape (Hayes et al. 1997). To enhance horizontal habitat heterogeneity within stands, sections within a forest stand would be thinned to two or more densities using a series of variable density thinning operations (Carey and Curtis 1996). Overtime, differences in tree growth among these stand sections induced by variable density thinning would increase overall vertical heterogeneity of the thinned stand (Carey and Curtis 1996). Thinning heavily to low tree densities could be used to accelerate the creation of large diameter trees and potentially be used to recruit large snags and down wood through an artificial means (Carey and Curtis 1996; Hayes et al. 1997). Depending on thinning intensities and locations of stands in a landscape, windthrow may also create snags, down wood and additional patchiness in thinned stands (Carey and Curtis 1996). Meanwhile, shade-tolerant conifers, such as western hemlock and western redcedar (*Thuja plicata*), and hardwood, such as bigleaf maple (*Acer Macrophyllum*), can be planted under canopy gaps created by heavy thinning to further enhance vertical layering of thinned stands (Carey and Curtis 1996; Cissel et al. 2006). With a series of carefully planned thinning operations, rotation age of stands between 40 and 80 years could be extended to 70–240 years (Carey and Curtis 1996; Curtis 1997; Franklin et al. 1997) to provide for species associated with late-successional forests without diminishing potential of stands for timber volume production. Alternatively, a combination of thinning operations and artificial planting or natural regeneration of seedlings would be used to convert even-aged stands to structurally diverse uneven-aged stands to enhance biodiversity (McComb et al. 1993; Cissel et al. 2006).

The Applegate Watershed Design

Management designs that are being developed for the federal lands of the Applegate Watershed integrate many of the landscape- and stand-scale themes above. This watershed occurs in a fire-prone landscape primarily in southern OR (Fig. 4), and is within the larger Klamath-Siskiyou ecoregion that has been identified for its unique diversity of species and habitats (DellaSala et al. 1999). The Applegate Watershed includes nonfederal lands that are predominantly managed as commercial plantation forests, and federal lands administered by both the US Forest Service and BLM. Federally managed forests in this area are designated as an "Adaptive Management Area," a land use allocation where programmed timber harvest is allowed (federally managed lands: Fig. 4a). Much of this watershed has been logged and replanted, and current activities include planning for regeneration harvests, fuels treatments (i.e., thinning) near human communities in areas designated as Wildland-Urban Interface (Fig. 4b), and management of rare species and special habitats.

Addressing multiple species and habitat concerns has resulted in a landscape mosaic of federal and nonfederal planted forest lands interspersed with federal reserves to preserve biodiversity and ecological functions of old forests (Fig. 4a). The larger blocks of federal reserved lands in the south and southeast portion of the watershed were considered by Strittholt and DellaSala (2001) to contribute significantly to the conservation of the region's biodiversity by preserving a multitude of unique habitats and species. North of these large reserves, throughout the major portion of the watershed in the federal Adaptive Management Area is a network of linear riparian reserves along streams to protect fish and

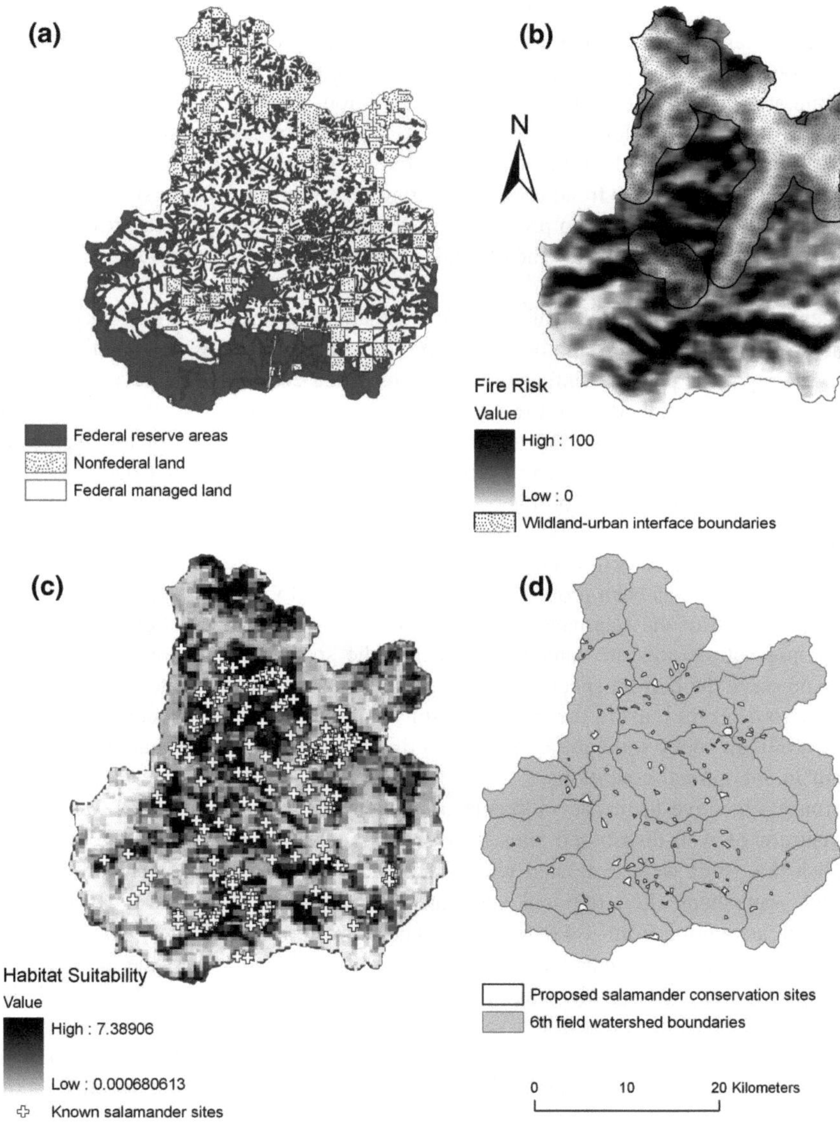

Fig. 4 Integrated approaches of joint biodiversity and planted forest management occurs in the Applegate Watershed of southern Oregon, US, depicting: a) federal and nonfederal land ownerships and federal land-use allocations, managed or reserved lands, including large block reserves to the south, linear riparian reserves and botanical and owl set-aside reserves north of these blocks; b) fire risk and Wildland-Urban Interface delineating communities at risk of fire, reflecting areas designated for fuels treatments such as forest thinning projects; c) habitat suitability and all known sites for the endemic Siskiyou Mountains salamander (*Plethodon stormi*) a species of concern in the area; and d) Siskiyou Mountains salamander sites selected as high priority for species management, to maintain well-distributed populations in the watershed

aquatic habitats. Riparian reserves are buffers at least as wide as 300 feet (91 m) or as two site-potential tree heights on each side if the stream has fish and at least as wide as 150 feet (46 m) or as one site-potential tree height on each side if the stream does not have fish (USDA and USDI 1994). After closer assessment during planning of forest management projects, these widths can be adjusted contingent upon site conditions. Forest management within the riparian reserve boundaries is possible for purposes of restoration. For example, if the previous harvest and planting conducted prior to implementation of riparian reserves has left the area in a high density young stand condition, then thinning might be used to accelerate development of large streamside trees. It should be noted that riparian buffers occur on the nonfederal lands in this landscape, but we have not mapped them in Fig. 4a, and their widths are smaller. In addition to riparian reserves, small areas of federal lands (<~40 ha) have been set aside to preserve unique botanical areas and old-growth forest-associated northern spotted owls (these small set-aside reserves are among the federal reserve areas in Fig. 4a).

A recent additional assessment of endemic salamanders associated with rocky soils in this landscape has resulted in another consideration for biodiversity conservation in the region. This region is part of a larger biogeographic zone considered to have the highest species richness of salamanders in the US Pacific Northwest (Bury and Bury 2005). The Siskiyou Mountains salamander (*Plethodon stormi*) is a rare species in the Applegate Watershed; its conservation has been of prime concern for federal land managers (Clayton et al. 2005). Habitat associations of this species have been studied (N. Suzuki, unpublished data, e.g., Welsh et al. 2007) and suitable habitat has been mapped along with all known sites (Fig. 4c). To advance combined timber and biodiversity concerns for the watershed, a subset of known sites has been proposed as "high priority" for species management, areas where conservation of the salamander would be of utmost importance. Selection of salamander conservation sites are intended to anchor the species within this portion of its range, to preserve the current species' distribution. It is assumed salamander occupancy is retained at some level outside these anchors for connectivity, which is supported by the prevalence of their rocky soil habitat throughout the area and knowledge that these animals can occur in suboptimal conditions.

A primary objective to the development of salamander conservation sites was to produce a pattern of well-distributed sites throughout the larger Applegate Watershed. To accomplish this, sites were evaluated within the context of an intermediate spatial scale, sixth field watersheds (i.e., regional sub-watershed hydrologic unit designations with catchment areas ranging from ~4,000 to 16,000 ha in western OR; Fig. 4d), chosen due to its existing use in forest management and aquatic resource planning in the area. Within sixth field watersheds, salamander sites were selected by a host of criteria. These included occurrence of sites within or adjacent to existing reserves (Fig. 4a), sites representing a range of fire risk areas and fire management areas (Fig. 4b), sites in or adjacent to areas thought to have optimum salamander habitat conditions (Fig. 4c), and locations both central and peripheral to the boundaries of the sixth field watersheds. Additional species were considered in this selection process, to include locations of other biota within priority salamander sites.

This Applegate Watershed design has several key elements described by Spies et al. (2006) for the dry provinces of the NWFP. The spatial extent of the large block of federal reserves at the southern boundary is reduced (it is less than the ~80% prescribed for the entire NWFP), yet may provide significant conservation benefits (Stritthold and DellaSala 2001). Fire risk has been modeled for the landscape (Fig. 4b) and can be used to integrate timber harvest priorities to reduce fuels, using approaches such shaded-fuel breaks and ladder fuel management. Furthermore, human communities at risk of fire are identified with the Wildland-Urban Interface, and are areas where higher priority fuels reduction treatments can be planned. Habitat islands for botanical, owl and salamander species have been identified throughout the area, and these may be managed. Specific management considerations include: retention of legacy forest attributes (e.g., large trees and dead wood); restoration (e.g., thinning of young stands) of late-successional and old-growth habitat conditions (i.e., for owl, salamander, and riparian areas); prescribed fire in some of the botanical zones; and retention of some canopy closure to maintain cool, moist surface microclimates and to avoid ground disturbing activities in priority salamander sites. Commodity production can be a priority in the intervening federal matrix, with commercial plantations having reduced biodiversity conservation objectives occurring on the nonfederal blocks (Fig. 4a). The outcome of these multiple priorities and concerns is a managed plantation landscape (nonfederal and federal managed lands) with small species-conservation areas anchoring habitat and species concerns throughout the landscape. While this design is conceptual, most elements are being implemented at this time. Monitoring and adaptive management is needed to advance the efficacy of such integrated matrix management to meet the diverse conservation and timber objectives of this ecosystem.

Regulatory approaches and incentives for biodiversity conservation

Habitat Conservation Plans for nonfederal lands under ESA

While a wide variety of approaches are currently available to potentially maintain or enhance biodiversity on nonfederal lands through innovative management practices (e.g., Carey and Curtis 1996; Franklin et al. 2002; Hartley 2002), landowners often are reluctant to invest for biodiversity conservation without clear economic benefit (Loehle et al. 2002). Consequently, federal and state regulations have been developed to provide minimum standards for conservation of biodiversity on forest plantations.

Under the ESA, the development of a Habitat Conservation Plan (HCP) provides regulatory incentives for nonfederal landowners to protect populations and habitats of threatened and endangered species. An important benefit of an HCP to nonfederal landowners is that they can obtain an incidental take permit for federally protected species in exchange for developing an HCP on their land (Noss et al. 1997). An incidental take permit allows a landowner to unintentionally harm individuals or modify habitats of endangered species while landowners continue forest management activities (Harding et al. 2001). Hence, this provision protects landowners from prosecution while they are attempting to balance management for species, habitats, and commodities on their lands.

Some landowners have successfully incorporated an HCP into their forest management practices at landscape scales (Loehle et al. 2002). However, without landowners' conscientious efforts to protect populations and habitats of endangered species, the HCP as a regulatory incentive presents several limitations. To enhance its effectiveness in endangered species conservation on nonfederal lands, several issues need to be remedied. First, nonfederal landowners are not required to address the recovery of endangered species in an HCP (Shilling 1997). HCPs are intended to maintain populations and habitats above the baseline conditions, which are often determined by the initial population and habitat conditions upon which the agreement was signed (Noss et al. 1997). As a result, an HCP does not particularly encourage landowners to improve habitat quality, increase populations, or to create new habitats for listed species on their land.

Second, the majority of HCPs lack monitoring programs to track population trends, and many existing monitoring programs are insufficient to evaluate the HCP's success (Kareiva et al. 1999). Although the development of an HCP is based upon the best available science, the process does not encourage landowners to incorporate adaptive management to address scientific uncertainties and modify management based on newly discovered information because a ''no surprise'' policy guarantees landowners that they would not be required to incur financial burden beyond the signed agreement (Wilhere 2002).

Third, concern is raised when an HCP is applied to the conservation of multiple species. Multi-species HCPs frequently fail to address adequate conservation measures for each species when they include species with no confirmed occurrence and distribution information in planning areas as well as little-known species (Rahn et al. 2006). Multi-species recovery plans appear to be less effective than single species recovery plans partly due to the lack of special attention to the ecological requirements of each species (Boersma et al. 2001; Taylor et al. 2005). Developing and implementing an HCP that covers a large number of species without clearly identifying species distributions, conditions of populations and habitats of species on a nonfederal property is essentially developing the plan in the absence of credible scientific information to help species recovery; such an approach might seriously jeopardize the persistence of endangered species (Noss et al. 1997; Kareiva et al. 1999; Reichhardt 1999; Rahn et al. 2006; Harding et al. 2001).

Programmatic HCP

Instead of each landowner developing an individual HCP, a state or local government can organize a group of stakeholders and develop a programmatic Habitat Conservation Plan to mitigate a group of similar management activities (e.g., forest management practices) proposed across a broad landscape as a whole (USDI and USDC 1996). Participants involved in the process are issued ''Certificates of Inclusion,'' which permit incidental take of species (USDI and USDC 1996).

For example, the Forest Practices Habitat Conservation Plan for the conservation of aquatic ecosystems by the State of WA was designed to cover five federally listed threatened or endangered salmonid species, 48 other fish and seven amphibian species across 3.7 million ha of nonfederal forestland in the state over the next 50 years (WSDNR 2005a). The plan was approved by the federal agencies in 2006. The foundation of the Forest Practices Habitat Conservation Plan is the state forest practices act and rules, updated to meet the recommendations of a multiple-stakeholder review, known as Forest and Fish report of 1999 (Call 2005). By meeting the requirements of the state forest practices act and rules, nonfederal landowners in WA are guaranteed an incidental take permit for species covered in the plan for 50 years, and can engage in forest management activities on their lands without any further legal restrictions under the ESA (WSDNR 2005a). An adaptive management process is used to determine whether changes or adjustments in forest practices rules and guidance are necessary to achieve program goals, performance target, or resource objectives (WSDNR 2005a). The Cooperative Monitoring Evaluation and Research committee was formed by resource and science experts who represent landowners, forest industry, environmental interests, state and federal agencies, and tribal governments (WSDNR 2005a). The committee develops and oversees research and monitoring programs as well as provides science-based technical advice during the adaptive management process. This framework for monitoring and adaptive management through stakeholder participation in the Forest Practices Habitat Conservation Plan could potentially be used as a model to remedy the general lack of coordinated monitoring and adaptive management in the HCP process.

Safe Harbor and Candidate Conservation Agreements under ESA

To supplement the passive nature of an HCP in protecting threatened and endangered species, the US Fish and Wildlife Service and the National Marine Fisheries Service developed two voluntary conservation programs that are intended to promote conservation of federally listed threatened and endangered species and rare species with a high likelihood of becoming listed under the ESA in the foreseeable future. Safe Harbor Agreements offer private landowners incentives to create or enhance potential future habitats (Safe Harbor) for species in exchange for no further future legal restrictions, allowing incidental take of endangered species and their habitat (US Fish and Wildlife Service 2001). Similarly, landowners can develop species conservation plans for rare species that are at risk of becoming listed under the ESA and make agreements, known as Candidate Conservation Agreements, with the US Fish and Wildlife Service or the National Marine Fisheries Service (US Fish and Wildlife Service 2002). In exchange, the landowners are assured of no further legal obligations in the event that the species become listed under the ESA; this process is called Candidate Conservation Agreements with Assurances. Once the term of Safe Harbor Agreements or Candidate Conservation Agreements ends, landowners are allowed to resume land use activities that may reduce the condition of habitat or populations as long as they maintain the baseline conditions agreed upon in the initial plan. Currently, only one nonfederal landowner of a plantation forest (144 acres = 58 ha) has entered a Safe Harbor Agreement with the US Fish and Wildlife Service in the Pacific Northwest Region (US Fish and Wildlife Service 2006). No nonfederal landowner of plantation forests has entered a Candidate Conservation Agreement in western OR and WA at this time. Consequently, the impacts of these programs are still uncertain relative to biodiversity conservation in the region's plantation forests.

State forest practices rules

Nonfederal landowners must comply with state forest practices rules in OR (ODF 2006) and WA (WFPB 2002). Hence, state forest practices rules have a great impact, in terms of area of nonfederal land, on how landowners manage their forests. To conserve biodiversity or ecological function of nonfederal forest lands, state forest practices rules in OR and WA primarily focus on retention of structural habitat elements, namely green trees, snags, and down wood, at stand scales and conservation of riparian and aquatic habitat for fish and other public resources. Improvement of structural habitat and protection of riparian areas are considered as key conservation strategies for many forest wildlife species in the region (e.g., Olson et al. 2001, 2007b). For example, in OR, nonfederal landowners are required to retain two snags or two green trees at least 30 feet (9.1 m) in height and 11 inches (28 cm) or greater in Diameter at Breast Height (DBH) and two pieces of live or dead down wood for each acre (0.4 ha) of land to enhance habitat complexity at stand scales (ORS527.676; ODF 2006). A similar structural habitat rule applies to landowners in WA (WAC222-30-020 [11]; WFPB 2002). The establishment of a Riparian Management Area, which includes retention of no-harvest riparian buffers and provides guidelines for forest management prescriptions in managed riparian buffers, is required in both states (OAR 629-635, ODF 2006; WAC 222-30-021, WFPB 2002; Olson et al. 2007b).

The riparian strategy in western WA requires landowners to establish a Riparian Management Zone comprised of: 1) a 50-foot (15 m) wide no-harvest buffer next to a stream; 2) a 10- to 100-foot (3 to 30 m) wide partial-harvest buffer next to the no-harvest buffer; and 3) a 22- to 67-foot (6.7 to 20 m) wide harvest buffer next to the partial-harvest buffer, on each side of a fish-bearing stream (WAC 222-30-021, WFPB 2002). Buffer widths for the partial-harvest buffer and harvest-buffer vary depending on stream size and forest site productivity. For non-fish-bearing streams, a no-harvest riparian buffer of either 50 feet (15 m) or 56 feet (17 m), depending on site sensitivities, is required only to protect selected portions (at least 50% of the stream length) of a stream on each side. Nonfederal riparian strategies in both OR and WA are less protective relative to similar strategies for federal lands, which include a no-harvest riparian buffer of at least 300 feet (91 m) on each side of fish-bearing streams and of at least 150 feet (46 m) on each side of permanently flowing non-fish-bearing streams (USDA and USDI 1994).

Small forest landowners in WA have an option to participate in a Forestry Riparian Easement Program (WAC 222-21, WFPB 2002). In this program, landowners receive a minimum of 50% of fair market stumpage value for those trees that are left unharvested, as mandated by the state rule to preserve riparian function, in exchange for leasing the volume of unharvested timber to the sate as a riparian buffer for 50 years (WSDNR 2005b). This program recognizes the contribution of small forest landowners to the conservation of riparian habitat.

As a state regulation addressing conservation of biological resources at a broad spatial scale, the WA forest practices rules require statewide analysis of watersheds, known as Washington Watershed Analysis, by dividing the state into watershed administrative units, each of which is approximately 4,047 to 20,234 ha in size to protect and restore public resources, including fish, water, and capital improvements of the state or its political subdivisions as well as cultural resources (WAC 222-22, WFPB 2002). Washington Watershed Analysis is a collaborative effort among resource scientists, landowners, agencies, tribes, the public, and other stakeholders. An interdisciplinary team of experts assesses resource conditions and identifies sensitive areas within each watershed. Forest management plans are developed for each watershed, and site-specific prescriptions are developed in cooperation with field managers, agency personnel, and landowners (WFPB 1997).

There also are likely unintended negative consequences of state forest practices rules to the conservation of biodiversity. For example, timber harvest prescriptions in riparian management zones in WA require retention of at least 20 trees/acre (~50 trees/ha) in timber-harvest buffer, preferably conifer with DBH > 12 inches (30 cm), and also encourage hardwood-to-conifer conversion through thinning from below in partial-harvest buffer (WAC 222-30-021, WFPB 2002). These prescriptions to maintain conifer dominance in riparian management zone are intended to maintain the recruitment of large conifer debris to enhance stream habitats (WSDNR 2005a). However, the operation to selectively remove hardwood would reduce habitat heterogeneity and possibly biodiversity because hardwood patches typically maintain a high species diversity of various groups of organisms (Harris 1984; Gomez and Anthony 1996, 1998; Neitlich and McCune 1997; Pabst and Spies 1998; Hagar 2007). Furthermore, under the reforestation stocking standard in OR (OAR 629-610-0020, ODF 2006), landowners are required to stock each forest stand with a fixed minimum number or basal area per acre of seedlings, saplings, or trees > 11 inches (28 cm) in DBH of acceptable species, well distributed throughout the stand. This minimum tree stocking requirement fundamentally limits landowners' options to enhance heterogeneity within stands during the reforestation process, and could potentially reduce spatial heterogeneity of forest stands across the landscape, although it was intended to promote a viable reforestation.

Forest certification

Guidelines in state forest practices rules are passive in nature because they are intended to lower impacts of management rather than to create particular types of structure to enhance habitat for species. For landowners who are willing to contribute more than the minimum requirement, options are available to pursue a wide variety of voluntary forest certification programs. Forest certification is a process in which forest management practices are evaluated by an independent certification organization based on a set of ecological, social, and economic standards (Society of American Foresters Study Group 1995). It provides a voluntary incentive for landowners who desire to be recognized for their management practices toward sustainable forestry. Although most forest certifications requires some conservation measures for biodiversity, their standards and guidelines are highly variable among certification organizations.

The American Tree Farm System is a voluntary certification program with the largest participation by small non-industrial landowners in the US. Its standards and guidelines encourage landowners to conserve biodiversity and maintain or enhance habitat for native fish, wildlife, and plant species (Standard 6, American Forest Foundation 2002). However, their performance measures and indicators for the Biodiversity Standard (Standard 6) do not provide a list of specific elements of habitat or biodiversity to be considered in the development of the landowner's conservation and management plan. The standards and guidelines provide great latitude for landowners to decide what constitutes biodiversity on their land. For example, landowners are required to manage forests to maintain or enhance habitat for fish, wildlife, and plant species that are ''desired by owner'' (Performance Measure 6.2 and Indicator 6.2.1), whereas opportunities to protect rare species and special habitat features are considered and addressed in the landowner's management plan only where such opportunities are practical (Indicator 6.1.1). The lack of strong language and specific goals and guidelines to conserve biodiversity leads to uncertainty that American Tree Farm System certification would provide landowners incentives to conserve biodiversity or manage habitat beyond what is already required by the State forestry practices rules. Because landowners can manage habitat for their own desired species, they may choose to manage for game species with some tangible recreational values (e.g., trout, deer, and elk), and may not encourage management and conservation of habitat for rare and endangered species, which often do not present tangible values. Other limitations with the American Tree Farm System may include lack of a requirement to monitor species. It also may be ineffective for implementing conservation practices at broad spatial scales because the certification is intended for small landowners.

An increasing number of industrial forests with large landholdings in OR and WA has been certified in recent years by the Sustainable Forestry Initiative (SFI), a third-party forest certification developed by the American Forest & Paper Association, an industry trade group based in the US (Fletcher et al. 2001). One of the SFI's land management objectives outlines the use of stand- and landscape-level measures to enhance wildlife habitat and to promote conservation of biodiversity, including forest flora and fauna, and aquatic fauna (Objective 4, SFI 2005). Under this biodiversity objective, landowners are required to facilitate programs to conserve biodiversity, including species, habitat, ecological or natural communities at both the stand and landscape level; protect threatened and endangered species; and locate and protect known sites of imperiled species and communities (Performance Measures 4.1-4.2, SFI 2005). However, SFI does not provide specific guidelines nor criteria for these conservation programs to be acceptable or successful. Furthermore, landowners are not required to address landscape-level conservation measures not only when credible scientific data are absent but also when landscape-level conservation measures are inconsistent with landowners management objectives and where practical (Objective 4, Indicator 5, SFI 2005). Consequently, SFI may potentially certify forests owned by landowners who have no intention of developing landscape-level conservation measures because of incompatibility of such measures with their land use or resource production objectives.

Under the SFI, the stand-level retention of habitat elements, such as snags, mast-producing trees, down wood, den trees, and nest trees, is based on regionally appropriate science (Objective 4, Indicator 4, SFI 2005). When compared with OR Forest Practices rules, requirements to satisfy the biodiversity objectives under SFI did not considerably exceed those already required under OR Forest Practices rules (Fletcher et al. 2001). Therefore, SFI certification may not have a significant positive impact on conservation of biodiversity in OR and WA beyond the impact from State Forest Practices Rules and HCPs under ESA.

Forest certification administered by the Forest Stewardship Council (FSC), a nonprofit-worldwide organization, is by far the most comprehensive certification program and provides more detailed criteria on conservation of biodiversity as well as other environmental and socio-economic concerns than SFI and OR practices rules (Fletcher et al. 2001). Under the FSC Principle 6, landowners are required to conserve biodiversity and its associated values, including water resources, soils, and unique and fragile ecosystems, and maintain the ecological functions and the integrity of the forest, such as structure and composition (Principle 6, FSC 2000). FSC differs from SFI in at least the following key points, outlined in their Pacific Coast regional forest stewardship standard (FSC 2005; SFI 2005).

First, the FSC Pacific Coast regional forest stewardship standard requires an environmental impact assessment at every relevant spatial scale from the stand or on-site facility where trees are harvested and processed to the entire landscape of the ownership. Biodiversity and ecosystem characteristics considered in the environmental impact assessment encompass structural, compositional, and functional elements. Landowners are asked to provide descriptions of ecological processes, common plant and animal species and their habitats, rare plant community types, rare species and their habitats, water resources, and soil resources (6.1.a, FSC 2005) and to compare a wide variety of measures of habitat complexity and spatial heterogeneity between current and historic variability of forest conditions, including composition and distribution of tree species, tree age-classes, structural habitat elements, habitat patches, forest seral stages, and other identifiable forest ecological types (AC6.1.3, FSC 2005).

Second, implementation of monitoring and adaptive management is considered as an integral element under the FSC certification. Monitoring of management activities and of environmental impacts is required by the FSC for large and/or intensively managed forests (Principle 8, FSC 2000, 2005); furthermore, an adaptive management process is used to revise management plans based on the monitoring results (Adaptive management, 8.4, FSC 2005). Elements of biodiversity to be periodically monitored included composition and observed changes in the flora and fauna (8.2, FSC 2005), specifically the changes in conditions of populations and habitats of threatened species relative to recovery goals, major habitat elements, and occurrence of rare species (8.2c, FSC 2005).

Third, FSC requires conservation of habitats for rare, threatened, or endangered species, and forests with rich biodiversity, such as old-growth forests in the US Pacific Northwest. This is implemented by designation of ''High Conservation Value Forests'' (Principle 9, FSC 2000, 2005), for which landowners are required to develop management plans to maintain or enhance ecological and biological values of High Conservation Value Forests. FSC does not require landowners to establish High Conservation Value Forests as no-harvest reserves and allows timber harvest in High Conservation Value forest harvest to the extent that such operation assure both quality and area of High Conservation Value forests for a long term (9.2, FSC 2005). To maintain and enhance areas and quality of these forests, annual monitoring is conducted to assess effectiveness of landowners' measures on conservation attributes (9.2, FSC 2005).

Fourth, under the principle of ''Plantation Forestry'' (Principle 10, FSC 2000, 2005), FSC addresses management standards at stand and landscape scales with specific guidelines to enhance spatial heterogeneity, stand complexity, and connectivity of forest habitats. At landscape scales, landowners are required to address the spatial arrangement among stands of different ages and rotation periods, wildlife corridors, and riparian zones that follow the pattern of forest stands found in the natural landscape characteristic of the region (10.2, FSC 2005). Also, they need to incorporate the spatial and functional relationship of their plantation to the surrounding area's natural forests, late-seral forests, and long-rotation forests into a management plan (10.1.b, FSC 2005). At forest stand scales, landowners are required to practice uneven-aged forest management using long rotation periods (>80 years) for a portion (30–50%) of their land to promote late-seral forest habitat; furthermore, they are required to enhance quality of early- and mid-seral wildlife habitat by maintaining structural and compositional diversity (10.5a, FSC 2005).

Among the state forest practices rules and three forest certification programs we assessed, FSC certification most comprehensively addressed fundamental elements of conservation of biodiversity and provided the most detailed criteria for each conservation element. A previous assessment similarly found that FSC addressed environmental and socio-economic issues better than SFI and OR forest practices rules (Fletcher et al. 2001). One of the strengths of FSC certification under the Pacific Coast regional forest stewardship standard is monitoring and adaptive management of large and/or intensively managed forests to advance biodiversity conservation. Furthermore, the FSC Pacific Coast regional forest stewardship standard provides clear guidelines that specifically address management for spatial heterogeneity and connectivity among forest stands across a landscape. However, it is still too early to tell whether these strict and ideal conservation standards of FSC will be successfully administered by landowners to yield significant positive contributions to the conservation of biodiversity. FSC's strict conservation standards may discourage landowners from choosing FSC certification.

For example, currently 13 nonfederal forests in 299,575 ha in OR and WA are certified by FSC (FSC 2006), whereas at least 16 forests with at least 3,254,997 ha are certified by the third-party certification option under SFI (SFI 2006).

Conclusions

Over the last couple of decades, a wide variety of innovative management approaches has been proposed to balance timber production and conservation of biodiversity in the US Pacific Northwest. Many of these approaches integrate ecological principles of natural disturbances into improvement of habitat heterogeneity at stand scales, landscape scales, or both. Among such approaches, reserve-matrix approach of the NWFP is the very first to comprehensively address forest management and conservation of biodiversity at multiple spatial scales.

A better integration of historical disturbance regimes into management of matrix lands appears to reduce the sharp contrasts in stand conditions between reserves and matrix by encouraging the development of larger late-successional patches and more variable tree canopy cover relative to conventional reserve-matrix approach; however, volume of timber production may be reduced. On the other hand, in disturbance-prone landscapes, options to actively manage entire landscapes may be necessary to reduce risk of catastrophic disturbance events, restore natural forest stand conditions, and protect and promote habitats of rare species, while producing timber throughout the process. On a landscape of predominantly young forest stands managed for timber production, one approach for conservation of biodiversity would be to use active management to provide full representation of forest stands in different structural stages of development across landscapes. Hence, the choice of landscape management approaches needs to consider the balance among conservation of biodiversity, restoration of ecosystems, timber production, and characteristics of disturbance in systems. At stand scales, landowners can promote heterogeneous habitat patterns and associated biota by planting tree seedlings at irregular spacing, thinning at various densities within or between stands, extending stand rotation age, artificially creating snags and down wood, and retaining structural legacies either through variable retention harvest in either aggregated or dispersed pattern.

There are no state laws or rules that explicitly require conservation of biodiversity on nonfederal lands in OR and WA. Current state forest rules may be lacking at least in the following points to ensure conservation of biodiversity on nonfederal lands. First, state forest practices rules generally do not address conservation or management of habitat at broad spatial scales (e.g., spatial configuration of various stand types), even though many industrial forests are large enough to consider landscape-level management guidelines. Second, some state forest practices rules, such as minimum tree stocking requirements (OAR 629-610-0020, ODF 2006) and riparian conservation strategy to selectively maintain conifer over hardwood trees (WAC 222-30-021, WFPB 2002), may have unintended negative consequences to reduce biodiversity by limiting landowners' options to enhance habitat heterogeneity. Some of the inherent limitations of the state forest practices rules may be the tendency to require landowners to meet a minimum standard for conservation and the inability to encourage landowners for continuous, incremental improvement of habitat for biodiversity on their land.

Under the ESA, the development of a HCP provides regulatory incentives for nonfederal landowners to protect populations and habitats of threatened and endangered species. To enhance effectiveness of an HCP as a recovery strategy of endangered species on nonfederal lands, the following measures could be considered: (1) develop economic incentives in addition to the incidental take permit for landowners who incrementally improve or enhance habitat and population conditions for endangered species, beyond baseline conditions; (2) establish quantitative baseline measurements on distribution, population level, and habitat conditions within the planning area for all species considered in the HCP, and develop quantitative goals for the recovery of population or habitat conditions for each species (Kareiva et al. 1999; Harding et al. 2001); (3) monitor population trends and habitat conditions, and integrate monitoring results into an adaptive management process (Wilhere 2002); (4) coordinate quantitative measurements of endangered species population trends and habitat conditions among landowners, agencies, and other stakeholders as part of monitoring and adaptive management processes (Noss et al. 1997; Kareiva et al. 1999); (5) develop multi-species HCPs based on a species-specific conservation strategy for each species and limit species covered under the HCP to only those with credible quantitative baseline information on species distribution, population levels, and habitat conditions within the planning area to help minimize adverse management impacts (Rahn et al. 2006).

Successful approvals of multi-species state-wide programmatic HCPs by federal agencies may encourage more states to develop similar programmatic multi-species HCPs and potentially replace individual HCPs. The potential future trend toward multi-species programmatic HCPs would shift the nature of the HCP process from a fine-filter conservation approach for a specific species at a specific site to a coarse-filter conservation approach for many species over a broad landscape. Multi-species programmatic HCPs at a state level may be able to remedy some of the limitations of individual HCPs, such as the lack of coordinated monitoring and adaptive management processes. On the other hand, there also are concerns over the implication of multiple-species state-wide programmatic HCPs for the conservation of biodiversity. First, effectiveness of adaptive management to adjust to a higher conservation standard would be limited if minimum or some low standards were used as an initial mitigation measure for potentially disturbing management activities (e.g., no-harvest riparian buffers cannot be widened overnight once a riparian area is harvested following the current forest practices rules). Second, previous multi-species conservation approaches have been showed to be ineffective (Boersma et al. 2001; Taylor et al. 2005). Third, negative consequences of management activities could potentially spread over the entire state. Fourth, nonfederal landowners are allowed to continue management activities without making any change for the duration of the HCP under the no surprise policy of the ESA. Hence, failure of multi-species programmatic HCPs at a state level could have significant negative consequences on conservation of biodiversity across the landscape over a long period of time.

Among the state forest practices rules and three forest certification programs we reviewed, FSC certification most comprehensively addressed fundamental elements of conservation of biodiversity and provided the most detailed criteria for each conservation element. However, positive contributions of FSC certification to conservation of biodiversity may be limited to relatively small land areas because of its tough conservation requirements for landowners' management activities. Consequently, current conservation standards on nonfederal lands would largely remain lower than those on federal lands. It is likely that any removal of conservation measures on federal lands due to a policy change would not be compensated by the current level of conservation efforts on nonfederal lands. Furthermore, future changes in strategies for biodiversity conservation on federal lands may occur without coordination with nonfederal lands because of the differences in regulatory and voluntary incentives between ownerships.

Acknowledgments We thank D. Clayton, R. Nauman, H. Welsh, E. Reilly, S. Morey, B. Devlin, and L. Ollivier for conservation planning in the Applegate Watershed., and E. Reilly for help assembling GIS layers that were used in Fig. 4. Kathryn Ronnenberg assisted with graphics and editing. Funding and support was provided by the Aquatic and Land Interactions Program of the US Forest Service, Pacific Northwest Research Station and Oregon State University.

References

American Forest Foundation (2002) Standards of sustainability for forest certification: including performance measures and field indicators (2004–2008 AFF standard). Washington, DC, USA, available at http://www.treefarmsystem.org. Cited 25 October 2006

Barbour RJ, Haynes RW, Martin JR, Lee DC, White R, Bormann BT (2006) Context for the Northwest Forest Plan. In Haynes RW, Bormann BT Lee DC, Martin JR (Tech Eds) Northwest Forest Plan: the first 10 years (1994–2003): synthesis of monitoring and research results. Gen Tech Rep PNW-GTR-651, US Department of Agiculture, Forest Service, Pacific Northwest Research Station, pp 11–22

Boersma PD, Kareiva P, Fagan WF, Clark JA, Hoekstra JM (2001) How good are endangered species recovery plans? BioScience 51:643–649

Bordelon MA, McAllister DC, Holloway R (2000) Sustainable forestry Oregon style. J Forestry 98(1):26–34

Bormann BT, Lee DC, Kiester AR, Spies TA et al (2006) Synthesis: interpreting the Northwest Forest Plan as more than the sum of its parts. In: Haynes RW, Bormann BT, Lee DC, Martin JR (Tech Eds) Northwest Forest Plan: the first 10 years (1994–2003): synthesis of monitoring and research results. Gen Tech Rep PNW-GTR-651, US Department of Agriculture, Forest Service, Pacific Northwest Research Station, pp 23–48

Bury RB, Bury GW (2005) Biogeographic patterns. In: Jones LLC, Leonard WP, Olson DH (eds) Amphibians of the Pacific Northwest. Seattle Audubon Society, Seattle, WA, pp 14–16

Call R (2005) Collaborative processes: lessons learned from the Timber Fish Wildlife, Chelan, and Forest and Fish Agreements. Thesis, University of Washington

Carey AB, Curtis RO (1996) Conservation of biodiversity: a useful paradigm for forest ecosystem management. Wildl Soc Bull 24:610–620

Cissel JH, Swanson FJ, Grant GE, Olson DH, Gregory SV, Garman SL, Ashkenas LR, Hunter MG, Kertis JA, Mayo JH, McSwain MD, Swetland SG, Swindle KA, Wallin DO (1998) A landscape plan based on historic fire regimes for a managed forest ecosystem: the Augusta Creek study. Gen Tech Rep PNW-GTR-422. US Department of Agriculture, Forest Service, Pacific Northwest Research Station, Portland, OR, p 82

Cissel JH, Swanson FJ, Weisberg PJ (1999) Landscape management using historical fire regimes: Blue River, Oregon. Ecol Appl 9:1217–1231

Cissel JH, Anderson P, Olson DH, Puettmann K, Berryman S, Chan S, Thompson C (2006) BLM density management and riparian buffer study: establishment report and study plan. US Department of Interior, US Geological Survey, Scientific Investigations Report 2006–5087, p 144

Clayton D, Olson DH, Nauman RS (2005) Conservation assessment for the Siskiyou Mountains salamander (*Plethodon stormi*), Version 1.3. USDA Forest Service, Region 6, and USDI Bureau of Land Management, Oregon, Interagecy Special-Status and Sensitive Species Program. Portland, OR. Available at: http://www.or.blm.gov/ISSSP/Conservation_Planning-and-Tools.htm, accessed 27 October 2006

Curtis RO (1997) The role of extended rotations. In: Kohm KA, Franklin JF (eds) Creating a forestry for the 21st century. Island Press, Washington, DC, pp 165–170

Curtis RO, DeBell DS, Harrington CA, Lavender DP, St Clair JB, Tappeiner JC, Walstad JD (1998) Silviculture for multiple objectives in the Douglas-fir region. Gen Tech Rep PNW-GTR-435. US Department of Agriculture, Forest Service, Pacific Northwest Research Station, Portland, OR, p 123

DellaSala DA, Reid SB, Frest TJ, Strittholt JR, Olson DM (1999) A global perspective on the biodiversity of the Klamath-Siskiyou ecoregion. Nat Areas J 19:300–319

Duncan N, Ed. (1999) Management recommendations for survey and manage mollusks, Version 2.0. US Department of Agriculture, Forest Service, and US Department of Interior, Bureau of Land Management, Portland, OR. Available at: http://www.blm.gov/or/plans/surveyandmanage/MR/TM23Species/TerrestrialMollusk.pdf. Cited 30 October 2006

Epstein RA (1997) Babbitt v Sweet Home Chapters of Oregon: the law and economics of habitat preservation. Supreme Court Econ Rev 5:1–57

Fletcher R, Adams P, Radosevich S (2001) Comparison of two forest certification systems and Oregon legal requirements. College of Forestry, Oregon State University, Corvallis, OR, Available at: http://www.oregonsolutions.net/documents/forestry/certification_comp.pdf. Cited 20 October 2006

Forest Stewardship Council (FSC) (2000) Principles and criteria for forest stewardship, Forest Stewardship Council. Washington, DC, Available at http://www.fscus.org/images/documents/FSC_Principles_Criteria.pdf. Cited 15 October 2006

Forest Stewardship Council (FSC) (2005) Revised final Pacific coast (USA) regional forest stewardship standard version 9.0, Available at http://www.fscus.org/images/documents/2006_standards/pcwg_9.0_NTC.pdf. Cited 12 October 2006

Forest Stewardship Council (FSC) (2006) Certificates in US/ certified forests. Forest Stewardship Council. Available via DIALOG. http://www.fscus.org/certified_companies/index.php?num=*&type=forests. Cited 18 October 2006

Franklin JF, Berg DR, Thornburgh DA, Tappeiner JC (1997) Alternative silvicultural approaches to timber harvesting: variable retention harvest systems. In: Kohm KA, Franklin JF (eds) Creating a forestry for the 21st century. Island Press, Washington, DC, pp 111–139

Franklin JF, Spies TA, Pelt RV, Carey AB, Thornburgh DA, Berg DR, Lindenmayer DB, Harmon ME, Keeton WS, Shaw DC, Bible K, Chen J (2002) Disturbance and structural development of natural forest ecosystems with silvicultural implications, using Douglas-fir forests as an example. Forest Ecol Manage 155:399–423

Gomez DM, Anthony RG (1996) Amphibian and reptile abundance in riparian and upslope areas of five forest types in western Oregon. Northwest Sci 70:109–119

Gomez DM, Anthony RG (1998) Small mammal abundance in riparian and upland areas of five seral stages in western Oregon. Northwest Sci 72:293–302

Hagar JC (2007) Wildlife species associated with non-coniferous vegetation in Pacific Northwest conifer forests: A review Forest Ecol Manage (in press). doi: 10.1016/j.foreco.2007.03.054

Hansen AJ, Garman SL, Weigand JF, Urban DL, McComb WC, Raphael MG (1995) Alternative silvicultural regimes in the Pacific Northwest: simulations of ecological and economic effects. Ecol Appl 5:535–554

Harding EK, Crone EE, Elderd BD, Hoekstra JM, McKerrow AJ, Perrine JD, Regetz J, Rissler LJ, Stanley AG, Walters EL, NCEAS Habitat Conservation Plan working group (2001) The scientific foundations of Habitat Conservation Plans: a quantitative assessment. Conserv Biol 15:488–500

Harris LD (1984) The fragmented forest: island biogeography theory and the preservation of biotic diversity. University of Chicago Press, Chicago

Hartley MJ (2002) Rationale and methods for conserving biodiversity in plantation forests. Forest Ecol Manage 155:81–95

Hayes JP, Chan SS, Emmingham WH, Tappeiner JC, Kellogg LD, Bailey JD (1997) Wildlife response to thinning young forests in the Pacific Northwest. J Forestry 95:28–33

Hunter ML Jr (1993) Natural fire regimes as spatial models for managing boreal forests. Biol Conserv 65:115–120

Kareiva P, Andelman S, Doak D, Elderd B, Groom M, Hoekstra J, Hood L, Frances J, Lamoreux J, LeBuhn G, McCulloch C, Regetz J, Savage L, Ruckelshaus M, Skelly D, Wilbur H, Zamudio K, NCEAS Habitat Conservation Plan Working Group (1999) Using science in habitat conservation plans, National Center for Ecological Analysis and Synthesis, University of California, Santa Barbara, CA and American Institute of Biological Science, Washington, DC

Loehle C, MacCracken JG, Runde D, Hicks L (2002) Forest management at landscape scales; solving the problems. J Forestry 100(6):25–33

McComb WC, Spies TA, Emmingham WH (1993) Douglas-fir forests: managing for timber and mature-forest habitat. J Forestry 91:31–42

Molina N, Hussey T, Mulder B (2006a) Key management implications of the Northwest Forest Plan. In: Haynes RW, Bormann BT, Lee DC, Martin JR (Tech eds) Northwest Forest Plan: the first 10 years (1994–2003): synthesis of monitoring and research results. Gen Tech Rep PNW-GTR-651, US Department of Agiculture, Forest Service, Pacific Northwest Research Station, pp 243–277

Molina R, Marcot BG, Lesher R (2006b) Protecting rare, old-growth, forest-associated species under the survey and manage program guidelines of the northwest forest plan. Conserv Biol 20:306–318

Neitlich PN, McCune B (1997) Hotspots of epiphytic lichen diversity in two young managed forests. Conserv Biol 11:172–182

Noss RF, O'Connell MA, Murphy DD (1997) The Science of conservation planning: habitat conservation under the Endangered Species Act. Island Press, Washington, DC

Olson DH, Rugger C (2007) Preliminary study of the effects of headwater riparian reserves with upslope thinning on stream habitats and amphibians in western Oregon. For Sci 53:331–342

Olson DH, VanNorman KJ, Huff RD (2007a) The utility of strategic surveys for rare and little known species under the US federal Northwest Forest Plan. Gen Tech Rep PNW-GTR-708, US Department of Agiculture, Forest Service, Pacific Northwest Research Station, Portland

Olson DH, Hagar JC, Carey AB, Cissel JH, Swanson FJ (2001) Wildlife communities in westside and high montane forest. In: Johnson DH, O'Neil TA (eds) Wildlife-habitat relationships in Oregon and Washington. Oregon State University Press, Corvallis, OR, pp 187–212

Olson DH, Anderson PD, Frissell CA, Welsh HH Jr, Bradford DF (2007b) Biodiversity management approaches for stream-riparian areas: perspectives for Pacific Northwest headwater forests, microclimates, and amphibians. Forest Ecol Manage (in press). doi: 10.1016/j.foreco.2007.03.053

Oregon Department of Forestry (ODF) (2001) Northwest Oregon State Forest Management Plan: final draft. Oregon Department of Forestry, Salem, Oregon, USA, Available at: http://www.oregon.gov/ODF/STATE_FORESTS/nwfmp.shtml. Cited 25 October 2006

Oregon Department of Forestry (ODF) (2006) Oregon Department of Forestry Forest Practice Administrative Rules and Forest Practices Act: chapter 629 forest practice administration. Oregon Department of Forestry, Salem, Oregon, USA, Available at: http://egov.oregon.gov/ODF/PRIVATE_FORESTS/docs/fp/FPArulebk.pdf. Cited 25 October 2006

Palik B, Mitchell RJ, Pecot S, Battaglia M, Pu Mou (2003) Spatial distribution of overstory retention influences resources and growth of longleaf pine seedlings. Ecol Appl 13:674–686
Pabst RJ, Spies TA (1998) Distribution of herbs and shrubs in relation to landform and canopy cover in riparian forests of coastal Oregon. Can J Bot 76:298–315
Public Land Foundation (2005) Forest plan revisions and future for O&C forests. Available via DIALOG http://www.publicland.org/forestPlansRevised.htm. Cited 30 October 2006
Rahn ME, Doremus H, Diffendorfer J (2006) Species coverage in multispecies habitat conservation plans: where's the science? Bioscience 56(7):613–619
Reeves GH, Williams JE, Burnett K, Gallo K (2006) The aquatic conservation strategy of the Northwest Forest Plan. Conserv Biol 20:319–329
Reichhardt T (1999) 'Inadequate science' in US habitat plans. Nature 397:287
Rundio DE, Olson DH (2007) Influence of headwater site conditions and riparian buffers on terrestrial salamander response to forest thinning. For Sci 53:320–330
Rykken JJ, Moldenke AR, Olson DH (2007) Headwater riparian forest-floor invertebrate communities associated with alternative forest management practices. Ecol Appl 17:1168–1183
Shilling F (1997) Do habitat conservation plans protect endangered species? Science 276(5319):1662–1663
Society of American Foresters Study Group (1995) Forest certification. J Forestry 93(4):6–10
Spies TA, Turner MG (1999) Dynamic forest mosaics. In: Hunter JR (eds) Maintaining biodiversity in forest ecosystems. Cambridge University Press, Cambridge, pp 95–160
Spies TA, Hemstrom MA, Youngblood A, Hummel S (2006) Conserving old-growth forest diversity in disturbance-prone landscapes. Conserv Biol 20:351–362
Strittholt JR, DellaSala DA (2001) Importance of roadless areas in biodiversity conservation in forested ecosystems: Case study of the Klamath-Siskiyou Ecoregion of the United States. Conserv Biol 15:1742–1754
Strittholt JR, DellaSala DA, Jiang H (2006) Status of mature and old-growth forests in the Pacific Northwest. Conserv Biol 20:363–374
Sustainable Forestry Initiative (SFI) (2005) Sustainable Forestry Initiative 2005–2009 Standard, Washington, DC, USA, Available at http://www.aboutsfb.org/generalPDFs/SFBStandard2005–2009.pdf. Cited 12 October 2006
Sustainable Forestry Initiative (SFI) (2006) Sustainable Forestry Initiative participants that have completed 3rd party certification, Sustainable Forestry Initiative. Available via DIALOG http://www.sfiprogram.org/generalPDFs/SFI_Certification_List_Website.pdf. Cited 18 October 2006
Taylor MFJ, Suckling KF, Rachlinski JJ (2005) The effectiveness of the Endangered Species Act: a quantitative analysis. Bioscience 55(4):360–367
Thomas JW, Raphael MG, Anthony RG et al (1993) Viability assessments and management considerations for species associated with late-successional and old-growth forest of the Pacific Northwest. The Report of the Scientific Analysis Team, US Department of Agriculture, p 530
Thomas JW, Franklin JF, Gordon J, Johnson KN (2006) The Northwest Forest Plan: origins, components, implementation experience, and suggestions for change. Conserv Biol 20:277–287
US Department of the Interior (USDI), Fish and Wildlife Service and US Department of Commerce (USDC), National Oceanic and Atmospheric Administration/National Marine Fisheries Service (1996) Habitat Conservation Planning and Incidental Take Permit Processing Handbook. Available via DIALOG. http://www.fws.gov/endangered/hcp/hcpbook.html. Cited 23 March 2007
US Department of Agriculture (USDA) and US Department of Interior (USDI) (1993) Forest ecosystem management assessment team (FEMAT): an ecological, economic, and social assessment. Portland, OR
US Department of Agriculture (USDA) and US Department of Interior (USDI) (1994) Record of decision on management of habitat for late-successional and old-growth forest related species within the range of the northern spotted owl [Northwest Forest Plan]. Portland, OR
US Fish and Wildlife Service (2001) Safe Harbor Agreements for private property owners: questions and answers. Endangered Species Program, US Fish and Wildlife Service, Arlington, VA, USA, Available at: http://www.fws.gov/northflorida/Documents/FWS-safeharborqa.pdf. Cited 23 October 2006
US Fish and Wildlife Service (2002) Candidate Conservation Agreements with Assurances for non–federal property owners. Endangered Species Program, US Fish and Wildlife Service, Arlington, VA, USA. Available at: http://www.fws.gov/endangered/listing/cca.pdf. Cited 8 October 2006
US Fish and Wildlife Service (2006) US Fish and Wildlife Service Conservation Plans and Agreements Database. Available via DIALOG. http://ecos.fws.gov/conserv_plans/servlet/gov.doi.hcp.servlets.PlanSelect. Cited 25 October 2006

Washington Forest Practices Board (WFPB) (1997) Standard methodology for conducting watershed analysis manual, version 4.0, Washington Forest Practices Board, Olympia, WA, USA. Available at: http://www.dnr.wa.gov/forestpractices/watershedanalysis/manual. Cited 20 October 2006

Washington Forest Practices Board (WFPB) (2002) Forest Practices Rules, Title 222 WAC, Washington Forest Practices Board, Olympia, WA, USA. Available at: http://www.dnr.wa.gov/forestpractices/rules. Cited 15 October 2002

Washington State Department of Natural Resources (WSDNR) (2005a) Final Forest Practices Habitat Conservation Plan. Available via DIALOG. http://www.dnr.wa.gov/htdocs/agency/federalassurances/final_fphcp/index.html. Cited 17 March 2007

Washington State Department of Natural Resources (WSDNR) (2005b) Forestry Riparian Easement Program: conserving Washington's forested stream sides. Small Forest Landowner Office, Washington State Department of Natural Resources, Olympia, WA, USA, Available at http://www.dnr.wa.gov/sflo/frep/frepfaqs.pdf. Cited 15 October 2006

Welsh HH Jr, Stauffer H, Clayton DR, Ollivier LM (2007) Strategies for modeling habitat relationships of uncommon species: an example using the Siskiyou Mountains salamander (*Plethodon stormi*). Northwest Sci 81:15–36

Wessell SJ (2005) Biodiversity in managed forests of western Oregon: species assemblages in leave islands, thinned, and unthinned forests, Thesis, Oregon State University

Wilhere GF (2002) Adaptive management in habitat conservation plans. Conserv Biol 16(1):20–29

Impact of four silvicultural systems on birds in the Belgian Ardenne: implications for biodiversity in plantation forests

Gaëtan du Bus de Warnaffe · Marc Deconchat

Originally published in the journal Biodiversity and Conservation, Volume 17, No 5, 1041–1055.
DOI: 10.1007/s10531-008-9364-x

Abstract Uneven-aged management of conifer plantations is proposed as a way to increase the value of these forests for the conservation of bird diversity. To test this assumption, we compared the impact of four common silvicultural systems on bird communities, defined by cutblock size (large in even-aged silvicultural systems/smaller in uneven-aged silvicultural systems) and tree species composition (spruce/beech) in the Belgian Ardenne where beech forests have been replaced by spruce plantations. The abundances of bird species were surveyed in young, medium-aged and mature stands in 3–5 forests per silvicultural system (66 plots in all). The effect of silvicultural systems on bird species richness, abundance and composition were analysed both at the plot and at the silvicultural system levels. In plots of a given age, beech stands were richer in species. The composition of bird species at the plot level was explained by stand age and tree composition, but weakly so by stand evenness. For the silvicultural systems, bird species richness was significantly higher in even-aged and in beech forests, and bird species composition depended on the silvicultural system. This study emphasises the importance of maintaining native beech stands for birds and suggests that uneven-aged management of conifer plantations does not provide a valuable improvement of bird diversity comparatively with even-aged systems.

Keywords Silvicultural system · Biodiversity · Bird communities · Silvicultural cycle · Coniferous plantation

G. du Bus de Warnaffe · M. Deconchat (✉)
UMR1201 Dynamiques forestières dans l'espace rural, INRA, Chemin de Borde Rouge, BP 52627, 31326 Castanet-tolosan, France
e-mail: marc.deconchat@toulouse.inra.fr

G. du Bus de Warnaffe
Unité des Eaux et Forêts, Université catholique de Louvain, Place Croix du Sud 2 bte 9, 1348 Louvain-la-Neuve, Belgium

E.G. Brockerhoff et al. (eds.), *Plantation Forests and Biodiversity: Oxymoron or Opportunity?*
DOI: 10.1007/978-90-481-2807-5_6

Introduction

The replacement of native broadleaf stands by uniform conifer plantations is a matter of concern for biodiversity conservation (Lack 1933, 1939; Ledant et al. 1983; Laiolo et al. 2004) and this question needs detailed analysis. Bird species composition is affected by tree species composition (e.g. Moss 1978; Müller 1987; Bersier and Meyer 1994, 1995; Hansen 1995) with few species associated with conifers while some are more associated with broadleaf species. Bird species composition is also influenced by vertical and horizontal vegetation structure that is determined by tree growth in the stand (Wigley and Roberts 1997; Lertzmann and Fall 1998) and the silviculture (Bellamy et al. 1996; Jokimäki and Huhta 1996; Drapeau et al. 2000). The size of the disturbance created by harvesting operations (cutbock size) defines different silvicultural systems and is known to influence biodiversity (Attiwill 1994; Chesson and Pantastico-Caldas 1994; Schnitzer and Carson 2001). In most of the cases, planted conifers are managed with large cutblocks (>2 ha) that are considered as unfavorable for bird diversity conservation (Ledant et al. 1983).

To improve the value of planted conifer forests for bird diversity, alternative silvicultural systems based on varying the areas where mature trees are harvested have been proposed (Kerr 1999). To test this idea, the differences in bird diversity between cutblock sizes in planted conifer forests have to be compared to similar differences in the original broadleaf forests. The Belgian Ardenne has the particularity of containing within a restricted region, four main silvicultural systems, including conifer plantations and broadleaf forest, and both forests managed by small and large cutblocks. In the forest manager's terminology, the large cutblock sizes are typical of the "even-aged" silvicultural system, while smaller cutblock sizes are typical of the "uneven-aged" silvicultural system used in this part of Europe (Kerr 1999).

Silvicultural systems have to be characterized by considering the whole silvicultural cycle. Moreover, as biodiversity can be influenced considerably by stand age, the effect of silvicultural systems can only be understood by considering the whole cycle (du Bus de Warnaffe 2002). Yet the age of the stand should be seen as a stage rather than an absolute age, since the effect of the absolute age on birds depends on the composition of the stand. Three stages can be identified in managed forests: a short one just after logging when low vegetation is dominant, a medium-aged stage when trees grow rapidly and induce a closed canopy, a long mature stage when trees have commercial dimensions and induce a high canopy with an overstorey (du Bus de Warnaffe and Lebrun 2004). An over-mature stage with collapsing and senescent trees can be identified in forests where harvesting does not occur (Fuller and Moreton 1987). Different silvicultural systems can be compared for each stage, or by gathering the stages over space, using a space-for-time substitution. Two spatial levels must therefore be considered: the plot, which only considers one stage, and a larger spatial and temporal scale integrating the complete silvicultural cycle of a silvicultural system (Huston 1999).

The hypothesis tested in this paper is that uneven-aged conifer planted forests have a higher value for bird conservation than even-aged conifer planted forests. This difference was tested with a sampling design including several stages of forest development and was compared with the same design in natural beech forests. These comparisons help to identify the impacts on biodiversity of the silvicultural systems applied to a large part of the forests in Europe and may provide guidance to mitigate their consequences on biodiversity conservation.

Materials and methods

Study region

The study was conducted in the Belgian Ardenne, between Namur and Luxembourg (Fig. 1). The historical land-use types in this region are pastures and broadleaf woodlands, which now account for 20 and 40% of the region (Paquet et al. 2006). They have been partly transformed into commercial conifer plantations (30% of the area) over the last 150 years (Devillez and Delhaise 1991). The elevation of our study plots ranged from 320 to 560 m, mean annual rainfall from 1,050 to 1,200 mm yr^{-1} and mean annual temperatures from 7.3 to 7.8°C (Weissen et al. 1994). All study plots comprised plantations established on *Luzulo-Fagetum* or *Luzulo-Quercetum* vegetation types, according to Noirfalise (1984) and Rameau et al. (2000) phytosociological systems, on flat or very gently sloping ground with acid and moderately dry soils (*Dystric cambisol*) (FAO 1990). The main tree species are native, mostly Norway spruce (*Picea abies* (L.) Karst), beech (*Fagus sylvatica* L.) and oaks (*Quercus petraea* (Mattme.) Liebl. and *Quercus robur* L.), with few introduced species, mostly Douglas fir (*Pseudotsuga menziesii* (Mirb.) Franco). Rotation length is typically 60–80 years for spruce, which is usually planted, and 120–150 years for beech, which is usually natural. Logging is done by clearcut on cutblocks with sizes ranging from 0.1 ha to more than 2 ha.

In even-aged systems, all the tree of a stand (>1 ha) are of the same age at a given time. In this system, logging is applied on large areas (cutblocks) by clearcutting. Even-aged systems result from planted forests for conifer tree and for beech tree from naturally regenerated forests managed to produce timber wood. In uneven-aged systems, the trees of different ages are mixed on smaller areas (<0.5 ha), logging is done by cutting mature trees on small cutblocks, as younger trees remain we do not be considered it as a clearcut. Uneven-aged conifer silvicultural system has developed from even-aged planted forests where small logging areas have been used rather than the large typical clearcuts. Forest managers consider it as a way to improve the sustainability of planted conifer forests.

Sampling design

The study compared four important silvicultural systems in the Belgian Ardenne:

(1) Even-aged conifer (EC): planted forests with greater than 80% cover of Norway spruce logged by clearcut on large cutblocks (>2 ha);

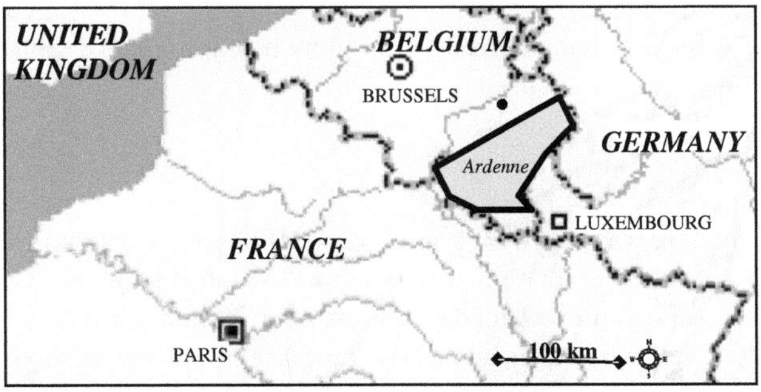

Fig. 1 Study area: ecological limits of the Belgian Ardenne (gray area with solid lines)

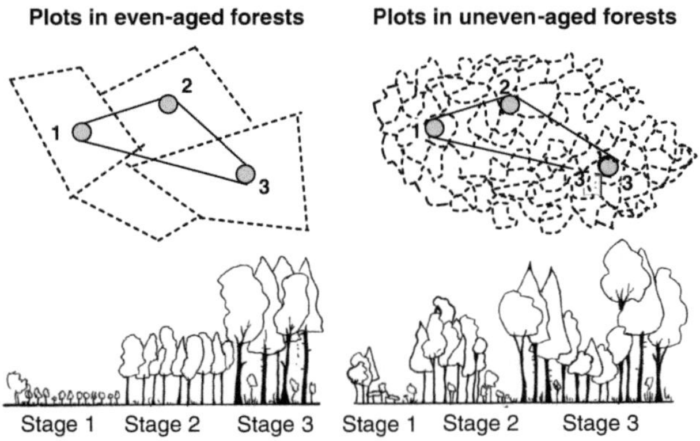

Fig. 2 Scheme of the location of the plots in even-aged and uneven-aged forest areas, in aerial and field views. Broken lines define stands of different ages and/or tree composition, circles represent bird counting zones in plots (25 m circle) and solid line convex hulls define the silvicultural system (about 15 ha)

(2) Even-aged Beech (EB): naturally regenerated forests with greater than 80% cover of Beech logged by clearcut on large cutblocks (>2 ha);
(3) Uneven-aged conifer (UC): planted forests with greater than 80% cover of Norway spruce logged on small cutblocks (<0.5 ha) producing a mix of trees of different ages;
(4) Uneven-aged Beech (UB): naturally regenerated forests with greater than 80% of Beech logged by small cutblocks (<0.5 ha).

We selected three to six forests per silvicultural system, these forests comprised at least 15 ha corresponding to the silvicultural system as defined above and managed for at least two rotations with the same system (du Bus de Warnaffe and Dufrêne 2004). The size of the cutblocks and the composition of each forest were determined by GIS analysis of 1/10,000 aerial photographs, and checked on site.

Plots were selected in three non-overlapping stages covering the silvicultural cycles of each silvicultural system (Fuller and Moreton 1987; Hansen 1995; Lertzmann and Fall 1998): regeneration stage (stage 1: trees 3–10 years old), medium-aged stage (stage 2: 20–40 years old for conifer, 30–60 years old for beech), and mature stage (stage 3: 50–80 years old for conifer, 80–140 years old for beech). The stage in uneven-aged systems was defined according to the time since the last logging, it is similar to the age of the oldest trees at a given time. The plots were separated by at least 200 m. A set of three plots belonging to these three stages in the same forest and the same silvicultural system defined a silvicultual cycle since it included the tree stages (Fig. 2).

The sampling was thus characterized by 54 plots belonging to 18 silvicultural systems (Tables 1, 2).

Bird data

The bird survey method was based on point counts (Bibby et al. 1985; Frochot and Roché 1990; Petty and Avery 1990) within a maximum 25 m fixed radius visually estimated. Singing birds were surveyed by trained observers over 20 min periods in each plot, twice during the breeding season (April and early June 2000), to record both sedentary and migrant species, and to reduce the bias associated with differences in detectability. The data were collected in the first 4 h after dawn, avoiding rainy and windy days. According to the

Table 1 Number of plots

Silvicultural system	EC	EB	UC	UB	Total
Number of plots					
In stage 1	4	3	6	5	18
In stage 2	4	3	6	5	18
In stage 3	4	3	6	5	18
Total number of plots	12	9	18	15	54

Silvicultural systems are defined by cutblock size and tree species composition of the forest: EC, even-aged conifer; EB, even-aged beech; UC, uneven-aged conifer; UB, uneven-aged beech. Each silvicultural system contains three stages. See text and Fig. 2 for details

Table 2 Major characteristics of plots in each class (see Table 1 for codes)

Silvicultural system	Stage	Tree species	Altitude (m)	Mean dbh (cm)	Basal area (m^2/ha)	Cutblock size (ha)
EC	1	PA	380–520	2–7	1–7	4–12
	2	PA, PM	320–490	20–27	34–41	–
	3	PA, PM	320–520	43–50	47–53	–
EB	1	FS	410–540	2–6	1–5	3–6
	2	FS	380–540	28–44	21–31	–
	3	FS	380–460	43–59	22–28	–
UC	1	PA, PM	420–580	2–8	18–27	0.02–0.45
	2	PA, PM	420–580	22–36	32–40	–
	3	PA, PM	420–580	41–52	31–42	–
UB	1	FS, QP, QR	410–500	1–8	11–20	0.03–0.25
	2	FS	350–500	23–33	17–28	–
	3	FS, QP, QR	350–500	34–52	21–29	–

All plots were situated on flat or very slightly sloping ground, on acid brown and moderately dry soils. Dbh (diameter at breast height) and basal area were measured on 0.20 ha. Cutblock size was measured for stage 1. Tree species: PA, *Picea abies*; PM, *Pseudotsuga menziesii*; FS, *Fagus sylvatica*; QP, *Quercus petraea*; QR, *Quercus robur*

territorial behavior of most of the bird species in spring, the fixed radius of the plots and the experience of the surveyors, we assumed to have comparable lists, but not necessarily exhaustive, of the bird species living in the plots (Buckland et al. 2001; Kery and Schmid 2004). All recorded species were used for the analyses, except over-flying birds, such as raptors and corvids, which were discarded. The abundance was estimated as two individuals (a pair) for each bird heard singing and one individual for each bird that was only seen or heard calling (not singing). The highest abundance recorded on the two dates was used as abundance index (Frochot and Roché 1990). For silvicultural systems, abundance of each species was the sum of the abundance index in the three plots (stages 1, 2 and 3).

Data analysis

ANOVA was used to test for differences in species richness and abundance between the datasets defined by the silvicultural systems and the stages: three-way ANOVA for the plot analysis (cutblock size, tree species composition, growth stage) and two-way ANOVA for the analysis of silvicultural systems (cutblock size, tree species composition) (Sokal and Rohlf 2000). Interactions between factors were included in the model and post-hoc tests, after Bonferonni correction, were used to identify significant differences between the means.

We used a linear ordination [Correspondence Analysis (Hill 1974)] to reduce the bird community data (in presence–absence) to a smaller set of dimensions, allowing us "to describe the strongest patterns in species composition" (McCune and Grace 2002). The result is an ordination of the species and the samples along axes computed as the solution of linear equations linking (1) the species space, where each sample is a coordinate, and (2) the sample space, where each species is a coordinate. Correspondence analysis can be interpreted as a summary of the departure of the observed contingency table (species by sample) from a null hypothesis of independence between species and samples, estimated by the χ^2 distance (Couteron et al. 2003). Several orthogonal axes can be computed and can be interpreted as follow: the closer the species on the axes, the more similar their distributions in samples; the closer the samples, the more similar their bird species composition. The higher values along the axes indicate samples or species with composition or distribution, respectively, more different from the mean composition or distribution of the whole sample (Balent and Courtiade 1992). Samples and species ordinations can be displayed on the same plan: the closer a species and a sample, the higher the probability, estimated from χ^2 distance, to have this species observed in this sample (Pelissier et al. 2003). This ordination of the samples, based on the covariations and associations among the species, was constrained by the silvicultural system classes to measure their influence on the bird species communities. This so called "between-group analysis" can be seen as a discriminant analysis adapted to species survey data and is a special case of the Canonical Correspondence Analysis (CCA) with only one explaining qualitative factor. It allows us to test the influence of qualitative variables on the structure of a species community (McCune and Grace 2002). The results are displayed as factorial plans where the separation of the sample classes is maximized according to their species composition (Thioulouse et al. 1997). A permutation test measured the departure of the observed structure from a random distribution of the species and gave a significance level of the difference between groups. All calculations were performed with R software (R development core team 2006) and with ade4 package (Chessel et al. 2004).

Results

A total of 44 species were found but 10 were recorded only once. The most abundant species were Chaffinch (*Fringilla coelebs*), Robin (*Erithacus rubecula*), Wren (*Troglodytes troglodytes*), and Wood pigeon (*Columba palumbus*) (Table 3).

Species richness and abundance

Plot analysis (each stage)

We found 3–20 species per plot. Tree species composition was the only factor with a significant effect on bird species richness ($F = 10.8353$; df = 1; $P = 0.0020$). The mean species richness in beech plots (14.12 ± 4.38; $n = 24$) was higher than in conifer plots (10.50 ± 3.59; $n = 30$). The variability of the species richness was higher in beech plots than in conifer plots (Fig. 3). In beech plots, the mean species richness of small cutblocks size (uneven-aged system) was not significantly different from the richness in the larger cutblocks, but in both cases, intermediate stages 2 had a lower species richness that masked the higher differences observed with stages 1 and 3 (Fig. 3). When considering only these two stages, uneven-aged plots had higher bird species richness than in even-aged plots, the few cases with extremely low values may explain why these differences were not significant.

Table 3 List of bird species observed in Belgian Ardenne

Code	Scientific name	Beech even-aged	Beech uneven-aged	Beech	Conifer even-aged	Conifer uneven-aged	Conifer	Total
ATRI	*Anthus trivialis*	3	1	4	2	0	2	6
ACAU	*Carduelis cannabina*	0	0	0	0	1	1	1
CCAR	*Carduelis carduelis*	0	0	0	1	1	2	2
CSPI	*Carduelis spinus*	0	0	0	0	0	0	0
CBRA	*Certhia brachydactyla*	1	4	5	0	1	1	6
CFAM	*Certhia familiaris*	0	2	2	0	4	4	6
CCOC	*Coccothraustes coccothraustes*	1	4	5	2	0	2	7
CPAL	*Columba palumbus*	3	4	7	3	4	7	14
CCAN	*Cuculus canorus*	2	1	3	4	0	4	7
DMAJ	*Dendrocopos major*	3	5	8	0	4	4	12
DMED	*Dendrocopos medius*	0	3	3	0	0	0	3
DMIN	*Dendrocopos minor*	1	0	1	0	0	0	1
DMAR	*Dryocopus martius*	3	3	6	0	1	1	7
ECIT	*Emberiza citrinella*	1	0	1	1	0	1	2
ERUB	*Erithacus rubecula*	3	5	8	4	6	10	18
FCOE	*Fringilla coelebs*	3	5	8	4	6	10	18
GGLA	*Garrulus glandarius*	2	5	7	2	5	7	14
LCUR	*Loxia curvirostra*	1	0	1	1	1	2	3
NCAR	*Nucifraga caryocatactes*	0	0	0	1	0	1	1
PATE	*Parus ater*	3	2	5	4	5	9	14
PCAE	*Parus caeruleus*	1	5	6	0	2	2	8
PCRI	*Parus cristatus*	0	2	2	3	2	5	7
PMAJ	*Parus major*	3	5	8	3	1	4	12
PMON	*Parus montanus*	0	3	3	0	0	0	3
PPAL	*Parus palustris*	3	4	7	1	3	4	11
PCOL	*Phylloscopus collybita*	3	3	6	3	4	7	13
PSIB	*Phylloscopus sibilatrix*	3	4	7	0	2	2	9
PTRO	*Phylloscopus trochilus*	3	1	4	4	1	5	9
PCAN	*Picus canus*	1	0	1	0	0	0	1
PMOD	*Prunella modularis*	3	1	4	4	4	8	12
PPYR	*Pyrrhula pyrrhula*	1	1	2	0	1	1	3
RIGN	*Regulus ignicapillus*	0	0	0	3	3	6	6
RREG	*Regulus regulus*	1	3	4	4	6	10	14
STOR	*Saxicola torquata*	1	0	1	0	0	0	1
SEUR	*Sitta europaea*	3	5	8	0	2	2	10
SVUL	*Sturnus vulgaris*	1	0	1	0	0	0	1
SATE	*Sylvia atricapilla*	3	4	7	4	6	10	17
SBOR	*Sylvia borin*	1	0	1	1	1	2	3
SCOM	*Sylvia communis*	0	0	0	1	0	1	1
TTRO	*Troglodytes troglodytes*	3	4	7	4	6	10	17
TMER	*Turdus merula*	3	5	8	4	6	10	18
TPHI	*Turdus philomelos*	3	5	8	4	5	9	17
TPIL	*Turdus pilaris*	0	1	1	1	0	1	2
TVIS	*Turdus viscivorus*	3	4	7	4	4	8	15

Note: Scientific names of the following species have changed in 2007: *Parus ater* is now *Periparus ater*, *Parus caeruleus* is *Cyanistes caeruleus*, *Parus cristatus* is *Lophophanes cristatus*, *Parus montanus* is *Poecile montana*, *Parus palustris* is *Poecile palustris*, *Regulus ignicapillus* is *Regulus ignicapilla* and *Saxicola torquata* is *Saxicola rubicola*

Total is the total number of forests (at silvicultural system level) where a given species was observed; Beech and Conifer are, respectively, the number of beech or conifer forests where a given species was observed (beech + conifer = Total); the same for even-aged and uneven-aged columns, splited according to tree species composition of the forests

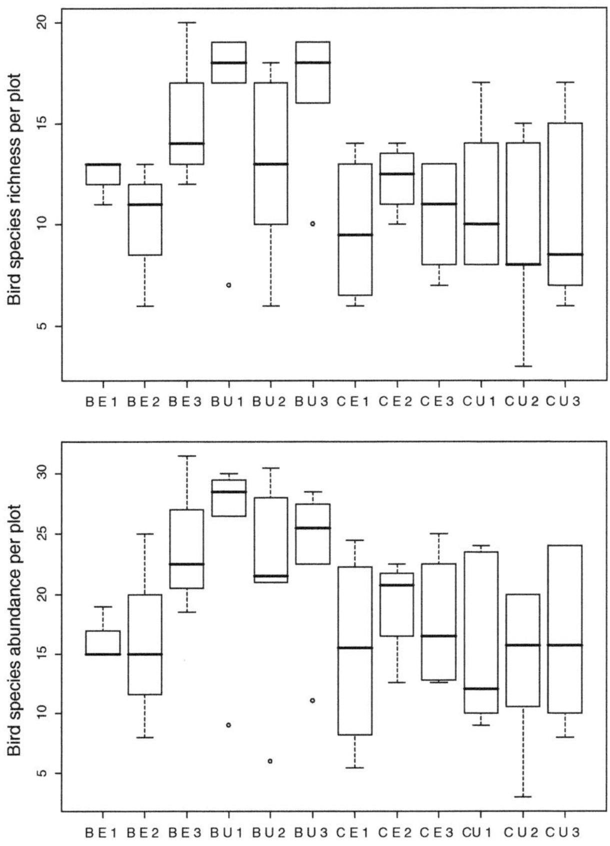

Fig. 3 Box plots of the bird species richness (top) and abundance (bottom) in sample plots according to tree species composition (B: beech or C: conifer), cutblock size (E: even-aged or U: uneven-aged) and stages (1, 2 or 3)

In conifer plots, no particular differences were identified between species richness according to cutblock size and stage.

Total bird abundance was highly correlated to species richness ($r^2 = 0.86$, $P < 0.001$). As for bird species richness, tree species composition was the only factor explaining a significant part of bird abundance variability ($F = 7.4354$; df = 1; $P = 0.0092$); the highest mean abundance was in beech plots (21.46 ± 7.75; $n = 24$), the lowest in conifer plots (16.03 ± 6.31; $n = 30$) (Fig. 3). In conifer plots, no clear pattern was observed for abundance, nor for species richness. In beech plots, on the other hand, the pattern was different. In even-aged plots, mature stands had clearly a higher abundance (but few samples with extremely low values) than younger stages. Conversely, the highest value was for first stage in uneven-aged plots, but with lower differences with the other stages comparatively with even-aged plots.

Silvicultural system analysis (stages 1 + 2 + 3 pooled together)

We found 12–27 species in the silvicultural systems. Cutblock size ($F = 6.8983$; df = 1; $P = 0.0176$) and tree composition ($F = 4.7767$; df = 1; $P = 0.0431$) significantly explained bird species richness variability, with the highest bird species richness in even-aged beech forests (24.67 ± 2.08; $n = 3$) and the lowest in uneven-aged conifer forests (16.33 ± 4.88; $n = 6$) (Fig. 4). The difference of bird species richness between beech and conifer was

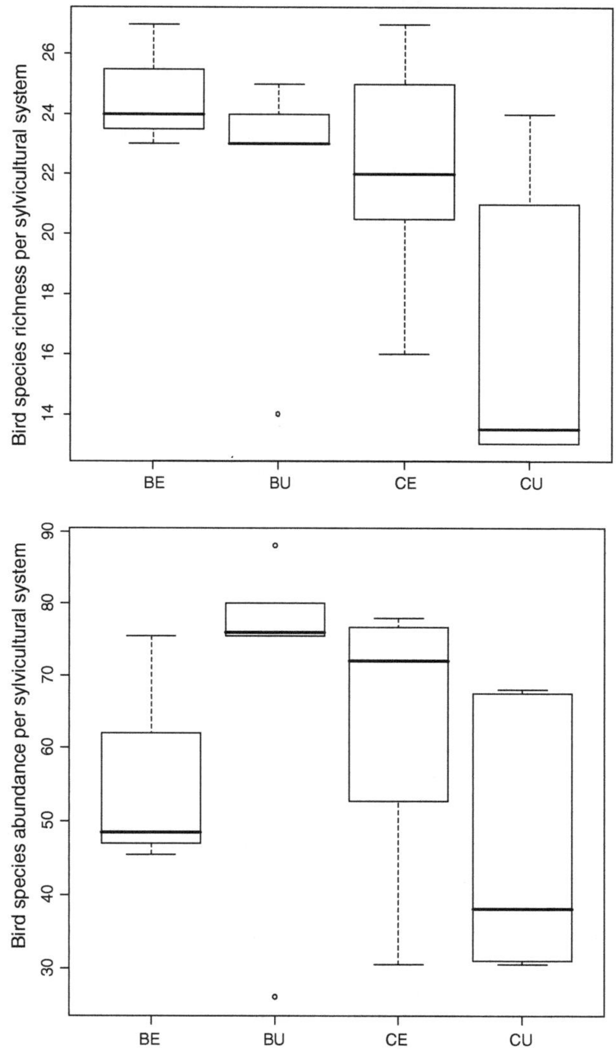

Fig. 4 Box plots of the bird species richness (top) and abundance (bottom) in silvicultural systems according to tree species composition (B: beech or C: conifer) and cutblock size (E: even-aged or U: uneven-aged)

higher in uneven-aged forests than in even-aged ones. No factor explained a significant part of bird species abundance, however, it can be noticed that in beech forests, the abundance was higher in uneven-aged forests than in even-aged, while the differences were less visible in conifer forests (Fig. 4).

Species composition

Plot analysis

The first stages of the even-aged silvicultural systems were significantly ($P < 0.001$) separated from the others groups along the first axis of the between group analysis (Fig. 5). The beech plots and the conifer plots of the older stages were separated along the second axis,

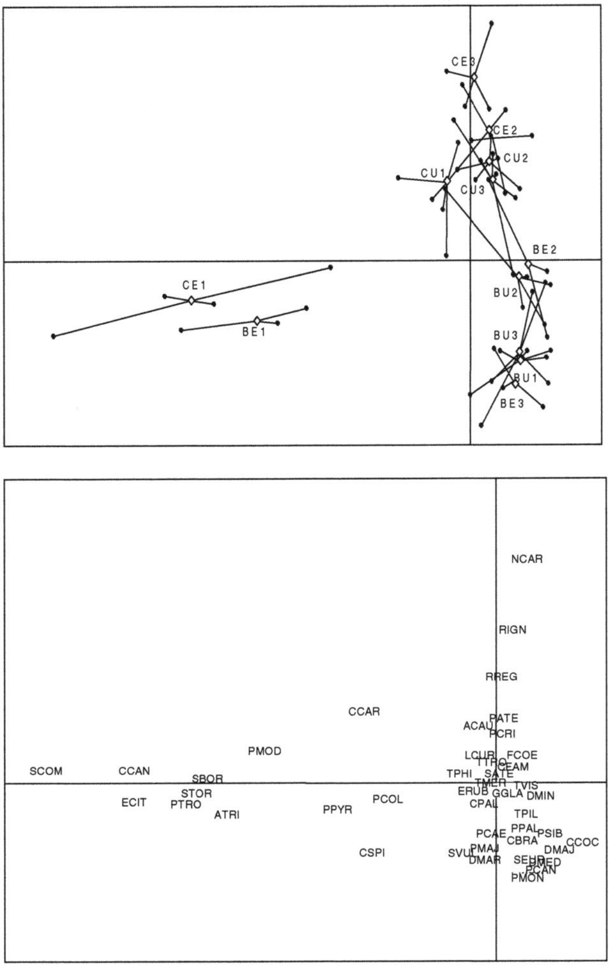

Fig. 5 Scatterplot of the between group analysis of the bird community data at the sample plot level. Top plot: sample plots (black dots) are linked to the mean position (diamond) of their silvicultural system and stage identified by the following code: EC = Even-aged Conifer; EB = Even-aged Beech; UC = Uneven-aged Conifer; UB = Uneven-aged Beech, the final number indicating the stage (see text and Fig. 2). Bottom plot: Ordination of the bird species on the same axes. The code of the species is based on the first genus letter and the three letters of the scientific name (Table 3). Their positions have been slightly modified for a better readability

but the even-aged and uneven-aged plots in these groups were not separated. *Sylvia communis*, *Emberiza citrinella* and *Carduelis cannabina* were positively associated with even-aged first stage plots, while *Regulus* species and *Nucifraga caryocatactes* were positively associated with coniferous plots. No particular species seemed to be associated with the beech plots, since most of the species close to the position of these plots were close to the origin of the factorial plan and thus were common in most of the samples (Fig. 5).

Silvicultural systems analysis

Conifer plots were separated ($P < 0.001$) from the other ones on the first axis of the between group analysis (Fig. 6). Even-aged beech forests were separated from the other

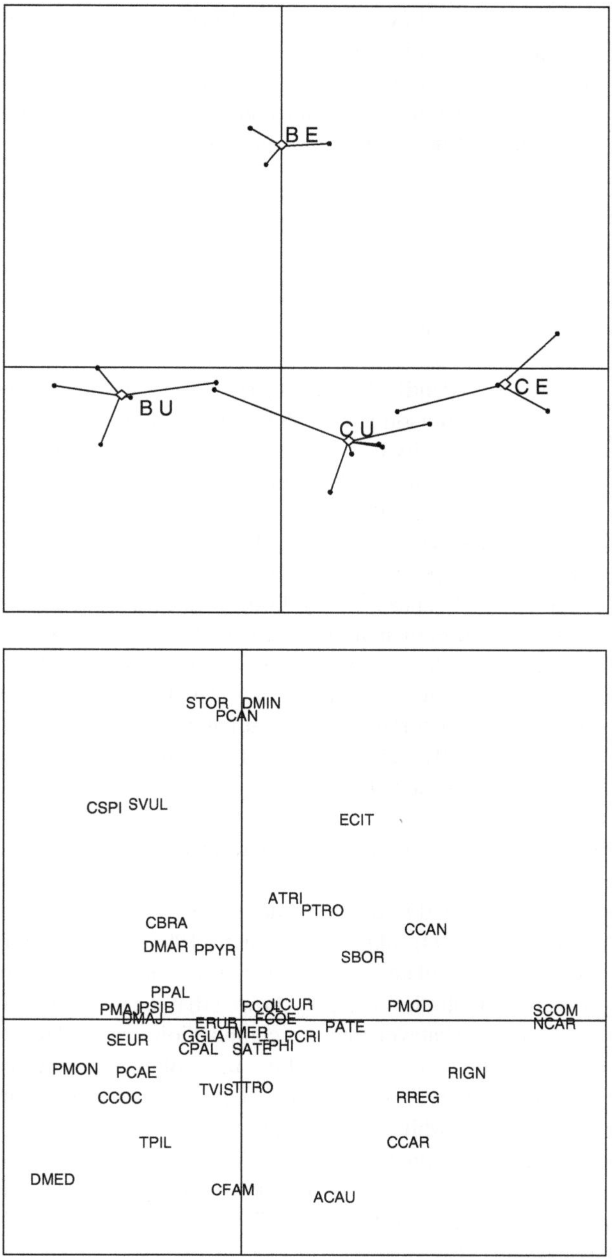

Fig. 6 Scatterplot of the between group analysis of the bird community data at the silvicultural system level. Top plot: forest (black dots) are linked to the mean position (diamond) of their silvicultural system identified by the following code: EC = Even-aged Conifer; EB = Even-aged Beech; UC = Uneven-aged Conifer; UB = Uneven-aged Beech (see text and Fig. 2). Bottom plot: Ordination of the bird species along the same axes. The code of the species is based on the first genus letter and the three letters of the scientific name (Table 3). Their positions have been slightly modified for a better readability

groups on the second axis, with a lower variability of their composition (distribution along axes). Cutblock size showed a significant effect ($P < 0.001$) in beech forests. *Regulus* species, *Nucifraga caryocactactes*, *Sylvia communis* and *Carduelis carduelis* were associated with the coniferous forests; *Dendrocopos minor*, *Saxicola torquata* and *Picus canus* were associated with the even-aged beech, *Dendrocopos medius* was associated with uneven-aged beech forests.

Discussion

Stand composition: conifer vs. beech

Although some studies have identified little impact of tree species composition on bird communities (Müller 1987; Patterson et al. 1995; Donald et al. 1998), most authors have found, as we have, a greater diversity in broadleaf forests compared with coniferous forests of similar stages, at plot level (Moss 1978; James and Wamer 1982; Bibby et al. 1985; Lebreton et al. 1987; Lebreton and Choisy 1991; Baguette et al. 1994; Solonen 1996; Gjerde and Saetersdal 1997) as well at larger levels including the whole silvicultural cycle (Jokimäki and Huhta 1996; Drapeau et al. 2000). Conifer forests seem to attract only few bird species, as suggested by Drapeau et al. (2000). In the Ardenne region, the total bird species richness in mature conifer plantation is estimated to be 43 species with a mean richness per plot of 13 species, while the estimations for beech forests are 44 species for the total richness and 16 species for the mean richness per plot (Paquet et al. 2006). However, historical factors may play an important role in that pattern because Norway spruce stands have been planted for about only 150 years in Belgium and thus bird communities may have not adapted yet to this new habitat.

Stand structure: even-aged vs. uneven-aged silvicultural systems

Bird community was more related to the dominant tree species (beech vs. conifer) than to cutblock size (Baguette et al. 1994; Jokimäki and Huhta 1996; Kirk and Hobson 2001). The only differences identified were related to the first stages, especially with bird composition: the between group analysis at plot level clearly showed that the bird composition of the first stage of even-aged systems, whatever the tree composition, was different from the other stages (Fig. 5). The species more associated to first stages of even-aged forests were mainly species known to be able to live in open habitats, having adapted to the practice of large clear-cuts (Paquet et al. 2006), while smaller logged areas in uneven-aged forests seem to have fewer associated species. Paquet et al. (2006) demonstrated that the species associated to the open areas in forest are not intermediate between the typical bird communities from agricultural habitat and forests, but were "specific" and contributed to 38.6% to the conservation value in large open areas.

These results do not confirm the main hypothesis of the paper, that uneven-aged management of planted conifer forests improves their bird conservation value. However, this conclusion should be moderated by the spatial dimension of even-aged and uneven-aged silvicultural systems, which has not been taken into account in this study (Picket et al. 1989; Kotliar and Wiens 1990). Even-aged system produces a coarser grain spatial pattern of heterogeneity than uneven-aged systems, with larger patches of even-aged trees. This induces edge-effects that may also have an influence on bird species distribution (Deconchat and Balent 2001).

Stand age: bird diversity according to silvicultural stages

Bird species composition in even-aged stage 1, i.e. large cutblocks, had a sharp contrast with stage 2 and 3. It was characterized by species known to be associated with open habitat conditions (Haila et al. 1980; Fuller and Moreton 1987; Baguette et al. 1994; Jacob 1996). Though Bibby et al. (1985) suggested that very few species require large clear-cuts, a number of species preferred even-aged stage 1 as it has been already noticed by Paquet et al. (2006). We were surprised not to obtain a higher species richness in stage 1 than in stages 2 and 3, as did a number of authors (Müller 1987; Bersier and Meyer 1994, Patterson et al. 1995; Jokimäki and Huhta 1996; Fuller and Green 1998). Indeed, species richness can greatly vary in young stands in plantations (Frochot 1971; Bibby et al. 1985), as well as in natural forest, even under strong disturbances such as coppice clear-cuts (Deconchat and Balent 2001).

Some authors have identified differences in bird communities between medium-aged and mature stands (Fuller and Moreton 1987; Lebreton and Pont 1987) with some bird species associated with old and senescent trees. We did not identify such a pattern, probably because of the intensive silviculture practiced in the Ardenne, based on short rotations, high densities and systematic removal of diseased and dead trees. At stage 3 in our study area, trees were not very large (Table 2), and hollow or dead trees were rare, which makes a difference with the same stage observed in less intensive contexts.

Conclusion

The results confirm the strong impact of tree species composition on bird species richness, abundance and composition. In the Belgian Ardenne, the massive introduction of spruce plantations has allowed some new species to breed (e.g. *Nucifraga caryocatactes*) but their bird diversity is clearly of lower conservation value than in the beech forests they have supplanted (Ledant et al. 1983). The conversion of even-aged conifer plantations to uneven-aged management (Schütz 2001), which has been proposed as a way to improve biodiversity in conifer forests, does not seem to improve their ability to shelter richer or more diverse bird community than in even-aged plantations. Even-aged management in beech forest was suspected to have negative impacts on biodiversity (Paquet et al. 2006). This opinion is not supported by the results of our study. We observed that the first stage after clear-felling on large zones seemed to offer temporary habitats for species also inhabiting fallow areas and extensive meadows (Delvaux 1998; Paquet et al. 2006), and species richness and composition do not differ much in simple (even-aged) and more complex (uneven-aged) canopies of the same age (stage 2 or 3).

Acknowledgements We are most grateful to J.P. Jacob and his team for the bird sampling, to Ph. Lebreton for methodological and scientific advice, to the Ministry of the Walloon Region for their financial support for the project and finally, to all the field engineers and technicians (DNF) who allowed us to carry out this study. H. Thomas and J. Willm helped to improve the manuscript. We thank the reviewers of a first version of this paper.

References

Attiwill MP (1994) The disturbance of forest ecosystems: the ecological basis for conservative management. For Ecol Manage 63:247–300

Baguette M, Deceuninck B, Muller Y (1994) Effects of spruce afforestation on bird community dynamics in a native broadleaved forest area. Acta Oecol Oecol Gen 15(3):275–288

Balent G, Courtiade C (1992) Modelling bird communities/landscape patterns relationships in a rural area of South-Western France. Landsc Ecol 6(3):195–211

Bellamy PE, Shelley A, Hinsley A, Newton I (1996) Factors influencing bird species numbers in small woods in south-east England. J Appl Ecol 33:249–262

Bersier LF, Meyer DR (1994) Bird assemblages in mosaic forests—the relative importance of vegetation structure and floristic composition along the successional gradient. Acta Oecol Oecol Gen 15:561–576

Bersier LF, Meyer DR (1995) Relationships between bird assemblages, vegetation structure and floristic composition of mosaic patches in riparian forests. Rev Ecol 50:15–32

Bibby CJ, Phillips B, Sedon JE (1985) Birds of restocked conifer plantations in Wales. J Appl Ecol 22:619–633

Buckland ST, Anderson DR, Burnham KL, Laake JL, Borchers DL, Thomas L (2001) Introduction to distance sampling. Oxford University Press, Oxford, UK, 432 pp

Chessel D, Dufour A-B, Thioulouse J (2004) The ade4 package-I-one-table methods. R News 4:5–10

Chesson P, Pantastico-Caldas M (1994) The forest architecture hypothesis for diversity maintenance. Trends Ecol Evol 9(3):79–80

Couteron P, Pelissier R, Mapaga D, Molino J-F, Teillier L (2003) Drawing ecological insights from a management-oriented forest inventory in French Guiana. For Ecol Manage 172(1):89–108

Deconchat M, Balent G (2001) Vegetation and bird community dynamics in fragmented coppice forests. Forestry 74(2):105–118

Delvaux A (1998) Espèces sensibles cherchent mise à blanc d'accueil. Forêt Wallonne 34:11–17

Devillez F, Delhaise C (1991) Histoire de la forêt wallonne. Forêt Wallonne 13:2–12

Donald PF, Fuller RJ, Evans AD, Gough SJ (1998) Effects of forest management and grazing on breeding bird communities in plantations of broadleaved and coniferous trees in western England. Biol Conserv 85:183–197

Drapeau P, Leduc A, Giroux JF, Savard JPL, Bergeron Y, Vickery WL (2000) Landscape-scale disturbance and changes in bird communities of boreal mixed-wood forests. Ecol Monogr 70(3):423–444

du Bus de Warnaffe G (2002) Impact des systèmes sylvicoles sur la biodiversité : une approche comparative en Ardenne – Réaction de la flore vasculaire, des coléoptères carabidés et de l'avifaune chanteuse à la structure de l'habitat forestier, à plusieurs échelles spatiales. Doctoral thesis, University of Louvain, Belgium, 132 pp

du Bus de Warnaffe G, Dufrêne M (2004) To what extent can management variables explain species assemblages? A study with carabid beetles in forests. Ecography 27(6):701–714

du Bus de Warnaffe G, Lebrun P (2004) Effects of forest management on carabid beetles in Southern Belgium: implications for biodiversity conservation. Biol Cons 118(2):219–234

FAO (1990) FAO-Unesco soil map of the world. Revisited legend. Soils bulletin 60, FAO, Rome, p 119

Frochot B (1971) Ecologie des oiseaux forestiers de Bourgogne et du Jura. Doctoral thesis, University of Dijon, France - CNRS Document A.0.5264

Frochot B, Roché J (1990) Suivi de populations d'oiseaux nicheurs par la méthode des indices ponctuels d'abondance IPA. Alauda 58:29–35

Fuller RJ, Green GH (1998) Effects of woodland structure on breeding bird populations in stands of coppiced lime *Tilia cordata*) in western England over a 10-year period. Forestry 71:199–217

Fuller RJ, Moreton BD (1987) Breeding bird populations of kentish sweet chestnut (*Castanea sativa*) coppice in relation to age and structure of the coppice. J Appl Ecol 24:13–27

Gjerde I, Saetersdal M (1997) Effects on avian diversity of introducing spruce plantations in the native pine forests of western Norway. Biol Conserv 79:241–250

Haila Y, Järvinen O, Väisänen RA (1980) Effects of changing forest structure on long-term trends in bird populations in SW Finland. Ornis Scand 11:12–22

Hansen AJ (1995) Bird habitat relationships in natural and managed forests in the West Cascades of Oregon. Ecol Appl 5(3):555–569

Hill MO (1974) Correspondence analysis: a neglected multivariate method. Appl Stat 23:340–354

Huston MA (1999) Local processes and regional patterns: appropriate scales for understanding variation in the diversity of plants and animals. Oikos 86:393–401

Jacob JP (1996) Avifaune nicheuse de clairières en forêt de Soignes. Aves 33:221–228

James FC, Wamer NO (1982) Relationships between temperate forest bird communities and vegetation structure. Ecology 63:159–171

Jokimäki J, Huhta E (1996) Effects of landscape matrix and habitat structure on a bird community in northern Finland: a multi-scale approach. Ornis Fenn 73:97–113

Kerr G (1999) The use of silvicultural systems to enhance the biological diversity of plantation forests in Britain. Forestry 72:191–205

Kery M, Schmid H (2004) Monitoring programs need to take into account imperfect species detectability. Basic Appl Ecol 5(1):65–73

Kirk D, Hobson KA (2001) Bird-habitat relationships in jackpine boreal forests. Biol Conserv 147:217–143
Kotliar NB, Wiens JA (1990) Multiple scales of patchiness and patch structure a hierarchical framework for the study of heterogeneity. Oikos 59(2):253–260
Lack D (1933) Habitat selection in birds with special reference to the effects of afforestation on the Breckland avifauna. J Anim Ecol 2:239–262
Lack D (1939) Further changes in the Breckland avifauna caused by afforestation. J Anim Ecol 8:277–285
Laiolo P, Caprio E, Rolando A (2004) Can forest management have season-dependent effects on bird diversity? Biodivers Conserv 13(10):1925–1941
Lebreton P, Choisy JP (1991) Avifaune et altérations forestières III: Incidences des aménagements forestières – substitutions Quercus/Pinus en milieu subméditerranéen (with english summary). Ecologie 22:213–220
Lebreton P, Pont B (1987) Avifaune et altérations forestières I: l'avifaune des boisements résineux du Haut-Beaujolais, considérations générales (with english summary). Acta Oecol Oecol Gen 8:227–235
Lebreton P, Broyer J, Pont B (1987) Avifaune et altérations forestières II: L'avifaune des boisements résineux du Haut-Baujolais; relations structurales végétation-avifaune (with english summary). Rev Ecol 4:71–81
Ledant J-P, Jacob J-P, Devillers P (1983). Protégeons nos oiseaux. Animaux menacés en Wallonie. Région Wallonne, Jambes, Duculot, Gembloux, 325 pp
Lertzman K, Fall J (1998) From forest stand to landscape: spatial scale and the roles of disturbances. In: Peterson L, Parker VT (eds) Ecological scale, theory and applications. Complexity in ecological systems. Columbia University Press, New York, pp 339–367
McCune B, Grace JB (2002) Analysis of ecological communities. Gleden beach, Oregon, MjM Software Design, 300 pp
Moss B (1978) Diversity of woodland song-bird populations. J Anim Ecol 47:521–527
Müller Y (1987) L'avifaune nicheuse des deux successions écologiques du pin sylvestre et du hêtre dans les Vosges du Nord (with english summary). Acta Ecol Oecol Gen 8:185–189
Noirfalise A (1984) Les stations forestières de Belgique. Presses agronomiques de Gembloux, Belgium, p 235
Paquet JY, Vandevyvre X, Delahaye L, Rondeux J (2006) Bird assemblages in a mixed woodland-farmland landscape: the conservation value of silviculture-dependant open areas in plantation forest. For Ecol Manage 227(1–2):59–70
Patterson IJ, Ollason JG, Doyle P (1995) Bird populations in uplands spruce plantations in northern Britain. For Ecol Manage 79:107–131
Pelissier R, Couteron P, Dray S, Sabatier D (2003) Consistency between ordination techniques and diversity measurements: two strategies for species occurrence data. Ecology 84(1):242–251
Petty SJ, Avery MI (1990) Bird census methods and techniques. In: Forest bird communities. A review of the ecology and management of forest bird communities in relation to silviculture practices in the British uplands. Forestry Commission, Edinburgh, pp 12–16
Picket STA, Kolasa J, Armesto JJ, Collins SL (1989) The ecological concept of disturbance and its expression at various hierarchical levels. Oikos 54:129–136
R Development Core Team (2006) R: a language and environment for statistical computing. R Foundation for Statistical Computing, Vienna, Austria. URL: http://www.R-project.org. ISBN 3-900051-07-0
Rameau JC, Gauberville C, Drapier N (2000) Gestion forestière et biodiversité. Identification et gestion intégrée des habitats et espèces d'intérêt communautaire, partie Wallonie et Grand-Duché de Luxembourg. ENGREF, ONF & IDF, Paris
Schnitzer AS, Carson WP (2001) Treefall gaps and the maintenance of species diversity in a tropical forest. Ecology 82(4):913–919
Schütz JP (2001) Opportunities and strategies of transforming regular forests to irregular forests. For Ecol Manage 151:87–94
Sokal RR, Rohlf FJ (2000) Biometry, the principles and practice of statistics in biological research, 3rd edn. WH Freeman & co, New York, pp 887
Solonen T (1996) Patterns and variations in the structure of forest bird communities in southern Finland. Ornis Fenn 73:12–26
Thioulouse J, Chessel D, Dolédec S, Olivier JM (1997) ADE-4 Multivariate analysis and graphic display software. Stat Comp 7:75–83
Weissen F, Bronchart L, Piret A (1994) Fichier écologique des essences et Guide de boisement des stations forestières de Wallonie. Groupe inter-universitaire « définition de l'aptitude des stations ». FSAGx-UCL-ULB-Ulg, Namur, Belgium
Wigley TB, Roberts TH (1997) Landscape-level effects of forest management on faunal diversity in bottomland hardwoods. For Ecol Manage 90:141–154

The early effects of afforestation on biodiversity of grasslands in Ireland

Erika Buscardo · George F. Smith · Daniel L. Kelly · Helena Freitas · Susan Iremonger · Fraser J. G. Mitchell · Saoirse O'Donoghue · Anne-Marie McKee

Originally published in the journal Biodiversity and Conservation, Volume 17, No 5, 1057–1072.
DOI: 10.1007/s10531-007-9275-2

Abstract The target rate of afforestation in Ireland over the next 30 years is 20,000 ha per year, which would result in an increase of the forest cover from the current 10% to 17%. In order to promote sustainable forest management practices, it is essential to know the composition and conservation value of habitats where afforestation is planned and the effects of subsequent planting upon biodiversity. The objectives of this study were to investigate changes in vegetation composition and diversity of grasslands 5 years after afforestation with Sitka spruce (*Picea sitchensis*) and determine the primary ecological and management factors responsible for these changes. Species cover, environmental and management data were collected from 16 afforested and unplanted improved and wet grassland site pairs in Ireland. Our results indicate that 5 years after tree planting, there were significant changes in richness, composition, and abundance of species. Competitive and vigorous grasses were more abundant in planted than in unplanted sites, as were generalist species found in both open and wooded habitats, while small-stature shade-sensitive species were less abundant. Vascular plant species richness and Shannon's diversity index were higher in unplanted wet grassland, than in the planted sites. Bryophyte species richness was higher in planted improved grassland than in unplanted sites. The differences were primarily the result of the exclusion of grazing, ground preparation, changes in nutrient management and drainage for afforestation. Drainage ditches provided a temporary habitat for less competitive species, but the overall effect of drainage was to reduce the diversity of species dependent on wet conditions. Variance partitioning showed differences in the relative influences of environmental and management variables on biodiversity in the two habitats, probably due to the greater pre-afforestation grazing pressure and fertilisation levels in improved grasslands. The differences in biodiversity

---

E. Buscardo · G. F. Smith · D. L. Kelly · S. Iremonger · F. J. G. Mitchell · S. O'Donoghue · A.-M. McKee
BIOFOREST Project, Department of Botany, School of Natural Science, Trinity College Dublin, Dublin 2, Ireland

E. Buscardo (✉) · H. Freitas
Centre for Functional Ecology, Department of Botany, University of Coimbra, 3001-456 Coimbra, Portugal
e-mail: erikatea@ci.uc.pt

E.G. Brockerhoff et al. (eds.), *Plantation Forests and Biodiversity: Oxymoron or Opportunity?*
DOI: 10.1007/978-90-481-2807-5_7

between planted and unplanted grasslands indicate that afforestation represents a threat to semi-natural habitats where distinctive and highly localised plant communities could potentially occur.

Keywords Afforestation · Biodiversity · Grassland · Ordination · Plantation · Variation partitioning

Abbreviations

CCA	Canonical correspondence analysis
Dbh	Diameter at breast height (1.3 m)
TVE	Total variation explained
IG	Improved grassland
WG	Wet grassland
TI	Total inertia

Introduction

In most European countries, large areas of land are being converted from agriculture to forestry (Elemans 2004; Wulf 2004). In Ireland, afforestation of agricultural land is progressing at one of the highest rates in Europe (MCPFE 2003). The government objective is to increase the country's forest cover from its current level of approximately 10% to 17% by 2030 to create a forest industry of sufficient scale to meet economic and social targets (Department of Agriculture Food and Forestry 1996; Forest Service 2000). This proposed large-scale and rapid expansion of forestry will alter the landscape radically in many parts of the country, with significant implications for biodiversity.

The net effects of afforestation will depend on the biodiversity of the new forest and on that of the habitat it replaces. Afforestation of improved grasslands may be of net benefit to biodiversity, particularly in intensively managed landscapes. However, it is likely that afforestation is mostly being carried out on the less agriculturally productive land (Heritage Council 1999). The most ecologically interesting grasslands are semi-natural communities that are long established and usually species-rich (Byrne et al. 1997). Recent survey work in eastern Ireland shows that semi-natural grasslands exist as isolated fragments within an intensively farmed landscape (Byrne et al. 1997). Most semi-natural grasslands lack formal nature conservation designations that preclude afforestation; therefore, sites of high biodiversity value could potentially be planted. Changes in grassland management, including afforestation, may induce corresponding changes in the structure of the vegetation or plant species composition (Alard et al. 1994). The conservation of semi-natural grasslands will depend therefore on the awareness at the planning stage of the effects the proposed forest will have on them and the surrounding area.

Most research on the effects of afforestation on understorey plant communities focuses on the mature stages of plantation development and emphasises the role of forest structure, particularly canopy cover (Hill and Jones 1978; Hill 1979; Sykes et al. 1989; Wallace et al. 1992; Wallace and Good 1995; Fahy and Gormally 1998; Ferris et al. 2000; French 2005). In contrast, early changes in composition and structure of vegetation following afforestation have received comparatively little attention. The vegetation of planted areas is influenced by the cessation of agricultural practices, such as grazing, burning, liming and fertiliser application, and by forest establishment operations, notably land drainage, ploughing or other ground preparation, and tree planting (Wallace et al. 1992). For example, cessation of grazing often results in the loss of several species from plant communities in the face of competition from a restricted range of aggressive grasses such as *Molinia caerulea* and *Festuca rubra* (Welch and Rawes 1964; Ball 1974). Other vegetation changes may be due to some of the changes in soil properties reported from pastures converted to plantations (Parfitt et al. 1997; Alfredsson et al. 1998; Jobbàgy and Jackson 2003, 2004; Farley and Kelly 2004). After plantation, grassland communities are progressively replaced by species characteristic of shady environments, notably bryophytes, and species of disturbed soils, especially rushes and grasses (Hill and Jones 1978; Hill 1979; Wallace et al. 1992; Ferris et al. 2000; French 2005). Afforestation of upland vegetation replaces specialised associations which are present at both extremes of the fertility gradient with less distinctive communities in which generalist species preponderate (Wallace et al. 1992), and it is likely that afforestation has a similar impact on lowland grasslands. However, the extent to which changes in forest structure or changes in land management at the early stages of afforestation are responsible remains unclear. Better understanding of the relative importance of different factors influencing understorey plant diversity will inform decisions on whether to plant particular habitats or sites. It will also improve biodiversity management in existing forests and inform efforts to restore open habitats in former plantation forests.

Given recent increases in the rate of afforestation, the paucity of previous studies and the potential for significant biodiversity loss, there is a need for research to determine the effects of land use change from grassland to coniferous plantation forest. The objective of this study was to investigate the effects of afforestation on the composition and structure of species assemblages and the relationships of the changes observed with ecological and management factors. The patterns of variation in biodiversity and soil properties were assessed in 16 recently afforested sites and 16 non-afforested equivalent habitats in two vegetation types, improved grasslands and unimproved wet grasslands.

Materials and methods

Study sites

Sixteen pairs of sites were selected in two habitat types that are commonly used for afforestation: eight improved agricultural grasslands and eight unimproved wet grasslands (Fig. 1). The first is characterised by intense management, subjected to fertilisation (ground rock phosphate, artificial N and N-P-K fertiliser) and heavily grazed or used for silage making, while the second is used for generally less intensive livestock grazing and is subjected to a lower fertilisation rate than improved grasslands. The site locations were selected to encompass the geographic and environmental range occupied by these grassland types in Ireland. The grasslands were defined according to a broad classification scheme developed by the Heritage Council (Fossitt 2000). Improved grasslands were typically dominated by *Lolium perenne* and *Trifolium repens*. Wet grasslands included oligotrophic sites characterised by *Molinia caerulea, Juncus acutiflorus* and *Agrostis canina* and more base-rich sites characterised by such species as *Carex hirta, Iris pseudacorus* and *Filipendula ulmaria*. Each pair of sites consisted of one pre-afforestation site and one post-afforestation site, which were geographically (usually adjacent) and environmentally matched. The planted sites were comprised of 5 year old forest of Sitka spruce (*Picea sitchensis*), the most commonly planted species in Ireland (Forest Service

Fig. 1 Map of Ireland showing the geographical location of the 16 site pairs. ○ indicates improved grasslands, ▲ indicates unimproved wet grasslands

2004), often mixed with a smaller amount of Japanese larch (*Larix kaempferi*). The improved grasslands were mostly on brown earth soils, with two site pairs on brown podzolics. Wet grasslands were located on gley soils with mull humus.

Data sampling procedure

Fieldwork was conducted during the summers of 2002 and 2004. In each site three plots (10 m × 10 m) were surveyed. All plots were orientated with the corners at the four cardinal points of the compass. Within each 10 m × 10 m plot, two smaller subplots measuring 2 m× 2 m were placed at opposite corners perpendicular to the slope. In afforested sites, plots were placed so that the 4 m^2 plots did not include any drainage ditches. Within each 100 m^2 plot the presence of all vascular plant species was recorded as well as the presence of bryophytes forming patches more than 50 cm^2. Species that dominated the vegetation were noted, but no other abundance data were collected. The average height and percent cover of vegetation in each of several layers/growth-form categories were recorded: saplings, shrubs (as defined by growth form), field layer (herbaceous vegetation, including ferns) and ground layer (mosses and liverworts). Also recorded were the percent cover of bare soil or rock, percent cover of litter and percent cover of standing water. For every 4 m^2 subplot a list of vascular plant and bryophyte species was made and their percentage cover estimated to the nearest 5%. Below 5% two additional cover-abundance categories were distinguished: (a) 3% and (b) 0.5%. For all bryophytes only species forming patches more than 5 cm^2 were recorded. For each 2 m × 2 m subplot, we recorded height and percentage cover of each of the four vegetation layers listed above and height and diameter at breast height (dbh—1.3 m) of each sapling occupying 5% or more of the plot (dbh was not recorded for trees < 2 m tall).

The nomenclature of vascular plants, mosses and liverworts follows Stace (1997), Smith (2004) and Paton (1999) respectively.

For each 10 m × 10 m plot the following environmental and management data were recorded: latitude, longitude, slope, aspect (degrees), elevation, soil drainage (3-point scale: poorly-, moderately- or well-drained), grazing intensity (4-point scale: from 0, no sign of mammal grazing to 3, heavy grazing), recreational use (3-point scale: 0, no sign of human impact; 1, sign of low-impact human presence; 2, high impact human recreational use), presence or absence of drainage ditches and other aspects of site management. Aspect was transformed to a linear scale such that southwest aspect was assigned a value of 1, northeast aspect was assigned a value of 0 and northwest and southeast aspects had a value of 0.5:

$$\text{If } 0 \leq x \leq 225 \rightarrow x' = \frac{|x - 45|}{180}; \text{ if } 226 \leq x \leq 360 \rightarrow x' = \frac{405 - x}{180}$$

where x = aspect and x' = transformed aspect.

Soil samples were collected from each 10 m× 10 m plot to a depth of 10 cm. Each sample was comprised of nine evenly spaced subsamples. Soil pH was determined on field-moist samples using a glass electrode on a 2:1 distilled water:soil suspension. Soil samples were air-dried and analysed for: total N by Indophenol Blue method (Berthelot 1859) with colorimetric detection at 630 nm, total orthophosphate P by the Molybdenum Blue method (Murphy and Riley 1962) with colorimetric detection at 880 nm, and total Ca, Mg and K by Atomic Absorption Spectroscopy using a Varian SpectrAA 400. Organic matter content was determined by loss on ignition. Soil element percentages were converted to mg/L after calculation of bulk density following the method of Jeffrey (1970). Adjacent to one of the 100 m^2 plots in each site, a 30 cm deep soil pit was dug, the soil profile was sketched and the soil type determined according to the Irish classification (Gardiner and Radford 1980).

Data analyses

The floristic data were analysed using Canonical Correspondence Analysis (CCA) (ter Braak 1986) on 4 m^2 subplots (82 and 126 species for improved grasslands and unimproved wet grasslands, respectively) employing mean abundance data (obtained by averaging abundances of the two 4 m^2 subplots located in each 100 m^2 plot). Separate CCAs were conducted for the two grassland types. All the planted tree species were eliminated from the vegetation records prior to analysis. Multivariate analyses were performed using CANOCO version 4.5 (ter Braak and Smilauer 2002). Forward selection was used to select significant explanatory variables. Only those significant at the $p < 0.05$ level were included. Significant variables were divided into two groups: environmental variables and management variables. To quantify the contribution of each class of variables in explaining species composition, variation partitioning was performed using partial CCA (Borcard et al. 1992; Økland and Eilertsen 1994). Partial ordination is an important tool for studying residual variation after 'factoring out' one or more variables (Legendre and Legendre 1998; ter Braak and Smilauer 2002). This method not only allows statistical testing, but also provides a quantification of the variation explained by the different sets of variables (Økland and Eilertsen 1994). Three CCAs (the first constrained by all significant environmental and management variables, the second constrained by environmental variables only and the third constrained by management variables only) and two partial CCAs (the first constrained by environmental variables with management variables as covariables and the second constrained by management variables with environmental variables as covariables) were performed on each habitat data set (4 m^2 subplots). Significances of the CCAs were tested using a Monte Carlo permutation test (with 499 permutations for both explanatory variables and final results) against the null hypothesis of no difference between planted and unplanted sites. Default options were employed, including the reduced model in permutation tests.

The following diversity indices were calculated for each 4 m^2 subplot: vascular plant species richness, bryophyte species richness, Shannon diversity index (H') (Pielou 1975) and the evenness index (J'), an index which describes the dominance structure of species (Smith and Wilson 1996). The mean of the two subplots was the experimental unit. Species richness was also calculated for each 100 m^2 plot. Species frequency in unplanted and planted sites (using presence/absence in 100 m^2 plots) was compared using the McNemar test (Sokal and Rohlf 1995) followed by the Hochberg procedure for controlling experimentwise error rates for multiple independent tests (Hochberg 1988). To test the differences in species richness, Shannon diversity index and evenness between unplanted and planted sites, paired *t*-tests were used. Soil data in planted and unplanted sites were compared using paired *t*-tests and non-parametric Wilcoxon's paired signed-ranks tests. Statistical testing was carried out using SPSS version 12.0 (SPSS 2003).

Results

Species frequency

The major aspect of the change in species frequency after afforestation of improved grassland was the increase in competitive grasses such as *Arrhenatherum elatius* and *Festuca rubra* and the drastic decrease of species of neutral pasture, including small-stature grassland herbs such as *Bellis perenne*, *Cerastium fontanum*, *Ranunculus repens*, *Taraxacum officinale* agg., *Trifolium repens* and grasses such as *Lolium perenne* (Table 1). *Rubus fruticosus* was completely absent from the unplanted grasslands (but abundant in the surrounding hedges) and occurred in 13 of the planted plots.

Plantations on wet grassland had higher frequencies of species of wet, disturbed conditions, such as *Carex viridula* and *Riccardia chamedryfolia* than unplanted wet grasslands (Table 2). Tall, highly competitive species like *Deschampsia cespitosa* and *Rubus fruticosus* agg. were more frequent in planted than unplanted sites, while low-growing herbs like *Ranunculus flammula*, *R. repens* and *Trifolium repens* were more frequent in unplanted plots.

Table 1 Significant changes in species frequency (%) between unplanted and afforested site pairs in improved grassland 10 m × 10 m plots

Species	UP	P	p-value
Arrhenatherum elatius	0	41.67	***
Festuca rubra	33.33	83.33	*
Potentilla anglica	0	33.33	*
Rubus fruticosus	0	54.17	*
Bellis perennis	58.33	12.5	*
Cardamine spp.	50	0	**
Cerastium fontanum	79.17	16.67	***
Lolium perenne	100	20.83	***
Ranunculus acris	41.67	0	*
Ranunculus repens	91.67	33.33	**
Rumex obtusifolius	50	8.33	*
Taraxacum officinale agg.	87.5	37.5	*
Trifolium repens	83.33	16.67	***

$N = 24$, *UP* unplanted, *P* planted

p-values are the results of the McNemar test (*$p < 0.05$, **$p < 0.01$, ***$p < 0.001$), adjusted to control for the experimentwise error rate

Table 2 Significant changes in species frequency (%) between unplanted and afforested site pairs in unimproved wet grassland 10 m × 10 m plots

Species	UP	P	p-value
Carex viridula	8.33	54.17	*
Deschampsia cespitosa	16.67	70.83	**
Kindbergia praelonga	8.33	50	*
Riccardia chamedryfolia	0	29.17	*
Rubus fruticosus	4.17	58.33	**
Ranunculus flammula	41.67	12.5	*
Ranunculus repens	91.66	54.17	*
Senecio jacobaea	33.33	0	*
Trifolium repens	66.67	29.17	*

$N = 24$, *UP* unplanted, *P* planted

P-values are the results of the McNemar test (*$p < 0.05$, **$p < 0.01$), adjusted to control for the experimentwise error rate

In both the habitat groups there were no significant differences in mean vascular plant species richness in 100 m^2 plots between planted and unplanted sites according to paired t-tests. Vascular plant species richness was significantly higher in unplanted 4 m^2 subplots of unimproved wet grassland (Table 3). Bryophyte species richness in 100 m^2 plots was significantly higher in planted improved grassland than in unplanted sites. Evenness was somewhat higher in improved grassland planted subplots ($p = 0.07$), and Shannon's diversity index was significantly lower in wet grassland planted plots.

No species in the Irish Red Data Book for vascular plants (Curtis and McGough 1988) or bryophytes listed as rare by Holyoak (2003) were found in the habitat or plot surveys.

Table 3 Mean values ± SEM of vascular plant species richness (VSR), bryophyte species richness (BSR), Shannon's diversity index (H′) and evenness index (J′) in 100 m^2 plots and 4 m^2 subplots in unplanted and planted sites in improved (IG) and wet grasslands (WG)

	Improved grasslands			Unimproved wet grasslands		
	UP	P	*p*-value	UP	P	*p*-value
100 m^2						
VSR	17.1 ± 2.0	14.5 ± 1.8	0.298	28.9 ± 1.7	25.8 ± 2.3	0.320
BSR	**0.2 ± 0.1**	**0.96 ± 0.2**	**0.020**	2.4 ± 0.6	4.1 ± 0.8	0.167
4 m^2						
VSR	10.1 ± 1.4	8.0 ± 0.9	0.261	**17.7 ± 1.5**	**10.9 ± 0.9**	**0.006**
BSR	0.5 ± 0.2	0.4 ± 0.2	0.336	2.4 ± 0.6	1.6 ± 0.4	0.485
H′	1.45 ± 0.17	1.44 ± 0.14	0.962	**2.37 ± 0.08**	**2.02 ± 0.12**	**0.023**
J′	0.57 ± 0.42	0.67 ± 0.3	0.070	0.74 ± 0.01	0.73 ± 0.24	0.415

UP unplanted, *P* planted

$N = 8$ site pairs in each group. *p*-values indicate the results of paired *t*-tests between site pairs, significant differences among planted and unplanted sites are indicated in bold type

However, some unplanted wet grassland communities were of high conservation value. Four sites contained communities referable to the EU Habitats Directive Annex I habitat 'Molinia meadows on calcareous, peaty or clayey-silt-laden soils (6410)' (European Commission 1999).

Ordination and variance partitioning

Twelve explanatory variables in the improved grassland subplots and eleven in the unimproved wet grassland subplots significantly explained variation in species abundances (Table 4). The most important variables accounting for variation in improved grasslands were sapling height, elevation, shrub cover and drainage ditches. The largest fractions of variation explained by single variables in unimproved wet grasslands were observed for northing, shrub cover, ground layer cover, elevation, easting and sapling height.

The differences in improved grassland plant community composition between planted and unplanted sites were driven by a management gradient along the first axis (Fig. 2). Axis 1 explained 13% of the total variability in the species data. It was positively correlated with sapling height, field layer height and litter cover, and negatively correlated with grazing. Axis 2 explained 7.6% of the variability and was most closely associated with elevation. There was a contrast between species typical of grassland characterised by intensive management (e.g. *Lolium perenne*, *Taraxacum officinale* agg., *Bellis perenne*, *Trifolium* spp., *Cynosurus cristatus* and *Cerastium fontanum*) on the left hand side of the ordination diagram in correspondence with the unplanted plots, and competitive grasses (e.g. *Agrostis stolonifera*, *Arrhenatherum elatius*, *Deschampsia cespitosa* and *Elytrigia repens*) and species that characterise abandoned pastures and field margins (e.g. *Rubus fruticosus* agg., *Urtica dioica*) on the right hand side of the ordination diagram, in association with the planted plots. Bryophyte species such as *Fissidens bryoides*, *Thuidium tamariscinum* and *Lophocolea bidentata* were also found in planted sites, related to management variables and to variables indirectly related to afforestation and grazing cessation (i.e. field height and shrub cover).

Table 4 Variation explained by significant environmental and management variables in CCAs of mean species abundances in 4 m^2 subplots in improved and unimproved wet grasslands

Variable	Set	Fraction of TI	
		IG	WG
Ground layer cover	E	0.030	0.048
Field layer height	E	0.026	–
Shrub height	E	–	0.034
Shrub cover	E	0.047	0.059
Litter cover	E	0.030	–
Easting	E	0.021	0.044
Northing	E	–	0.063
Slope	E	0.036	–
Aspect	E	–	0.027
Elevation	E	0.062	0.055
K	E	–	0.023
Mg	E	0.034	–
Sapling height	M	0.114	0.042
Sapling cover	M	0.030	–
Grazing	M	0.030	0.027
Drainage ditches	M	0.039	0.029

TI total inertia, *IG* improved grassland, *WG* unimproved wet grassland, *M* management variable, *E* environmental variable

Significant variables were included by forward selection ($P \leq 0.05$)

Planted and unplanted wet grassland plots from the same site pairs were separated along a management gradient, but the separation between them was not as defined as in improved grassland. One extreme of the management gradient included the more heavily grazed unplanted plots, whereas afforested plots containing drainage ditches and taller saplings were located towards the other extreme (Fig. 3). Axes 1 and 2 explained the same amount of variation in the species data equal to approximately 8%. Axis 1 was positively correlated with northing and negatively correlated with drainage ditches while there were not variables which were specifically correlated with axis 2. Species like *Agrostis stolonifera* and *Festuca rubra* were positively related to the plantation management variables. Biogeographical and soil variables, such as northing and K concentration, were relatively more important in structuring plant communities in wet grasslands than in improved grasslands.

Variation partitioning between the environmental and management data sets is shown in Table 5. Environmental variables counted for 75% of the total variation explained in unimproved wet grassland subplots and approximately 55% in improved grassland subplots. The contribution of management factors was similar between the two data sets and equal to approximately 25% of the TVE. In unimproved wet grassland plots, the small amount of variation accounted by both of the two sets of explanatory variables together indicated that the interaction between management and environmental variables was weak. In contrast, shared variation between the two sets of explanatory variables was considerable in improved grassland subplots and accounted for 20% of total variation explained.

Soils

Total nitrogen and calcium concentrations in improved grasslands were significantly higher in unplanted than in planted sites (Table 6). Soil pH was lower in the planted sites. There

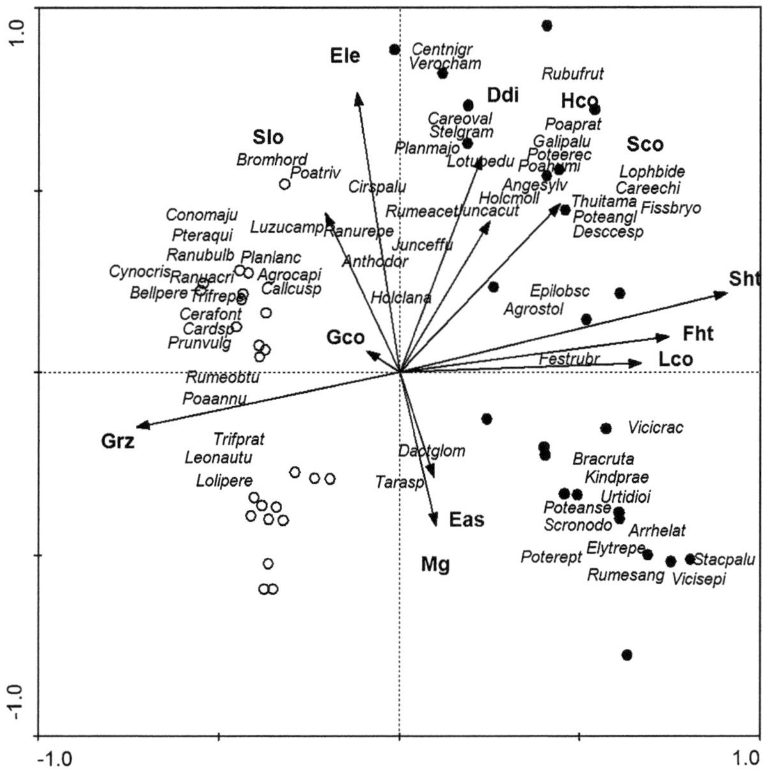

Fig. 2 Samples-species-explanatory variables triplot from CCA (axes 1 and 2) performed on improved grassland 4 m^2 subplots: species fit range is between 3 and 100% (i.e. 82 species were reduced to 64). ○ = unplanted plots; • = planted plots; Gco = ground layer cover; Fht = field layer height; Hco = shrub cover; Lco = litter cover; Eas = easting; Slo = slope; Ele = elevation; Mg = magnesium; Sht = sapling height; Sco = sapling cover; Grz = grazing; Ddi = drainage ditches; agrocapi = *Agrostis capillaries*; agrostol = *Agrostis stolonifera*; angesylv = *Angelica sylvestris*; anthodor = *Anthoxanthum odoratum*; arrhelat = *Arrhenatherum elatius*; bellpere = *Bellis perennis*; bracruta = *Brachythecium rutabulum*; bromhord = *Bromus hordeaceus*; callcusp = *Calliergonella cuspidate*; cardsp = *Cardamine* sp.; careechi = *Carex echinata*; careoval = *Carex ovalis*; centnigr = *Centaurea nigra*; cerafont = *Cerastium fontanum*; cirspalu = *Cirsium palustre*; conomaju = *Conopodium majus*; cynocris = *Cynosurus cristatus*; dactglom = *Dactylis glomerata*; desccesp = *Deschampsia cespitosa*; elytrepe = *Elytrigia repens*; epilobsc = *Epilobium obscurum*; kindprae = *Kindbergia praelonga*; festrubr = *Festuca rubra*; fissbryo = *Fissidens bryoides*; galipalu = *Galium palustre*; holclana = *Holcus lanatus*; holcmoll = *Holcus mollis*; juncacut = *Juncus acutiflorus*; junceffu = *Juncus effusus*; leonautu = *Leontodon autumnalis*; lolipere = *Lolium perenne*; lophbide = *Lophocolea bidentata*; lotupedu = *Lotus pedunculatus*; luzucamp = *Luzula campestris*; planlanc = *Plantago lanceolata*; planmajo = *Plantago major*; poaannu = *Poa annua*; poahumi = *Poa humilis*; poaprat = *Poa pratensis*; poatriv = *Poa trivialis*; poteangl = *Potentilla anglica*; poteanse = *Potentilla anserine*; poteerec = *Potentilla erecta*; poterept = *Potentilla reptans*; prunvulg = *Prunella vulgaris*; pteraqui = *Pteridium aquilinum*; ranuacri = *Ranunculus acris*; ranubulb = *Ranunculus bulbosus*; ranurepe = *Ranunculus repens*; rubufrut = *Rubus fruticosus* agg.; rumeacet = *Rumex acetosa*; rumeobtu = *Rumex obtusifolius*; rumesang = *Rumex sanguineus*; scronodo = *Scrophularia nodosa*; stacpalu = *Stachys palustris*; stelgram = *Stellaria graminea*; tarasp = *Taraxacum* sp.; thuitama = *Thuidium tamariscinum*; trifprat = *Trifolium pratense*; trifrepe = *Trifolium repens*; urtidioi = *Urtica dioica*; verocham = *Veronica chamaedrys*; vicicrac = *Vicia cracca*; vicisepi = *Vicia sepium*

were also significant differences in organic carbon and litter cover percentage between planted and unplanted improved grasslands. No significant differences were found in soil properties between planted and unplanted wet grasslands, although the results indicate

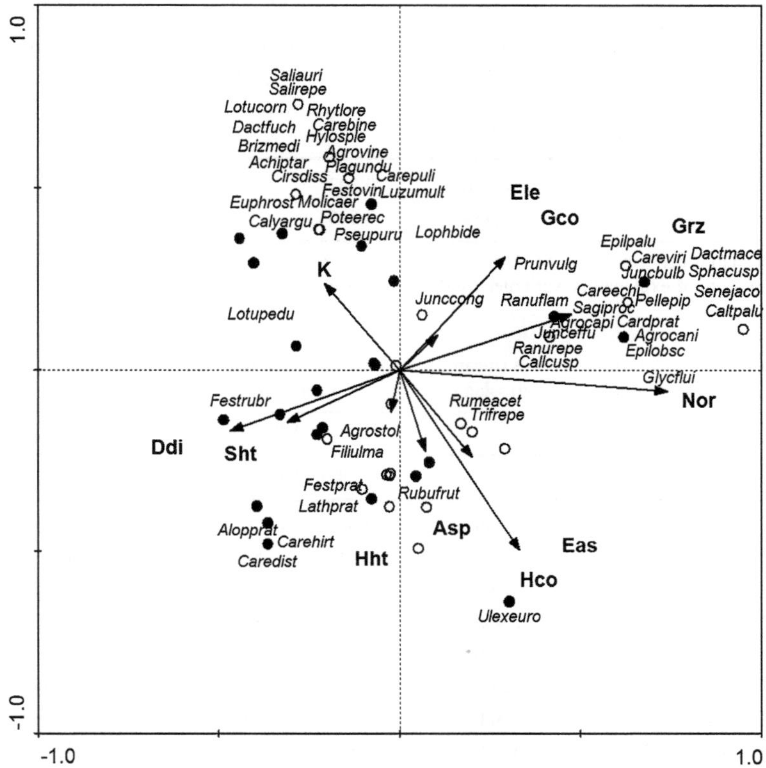

Fig. 3 Samples-species-explanatory variables triplot from CCA (axes 1 and 2) performed on unimproved wet grassland 4 m^2 subplots: species fit range is between 7 and 100% (i.e. 126 species were reduced to 55). ○ = unplanted plots; • = planted plots; Gco = ground layer cover; Hht = shrub height; Hco = shrub cover; Eas = easting; Nor = northing; Asp = aspect; Ele = elevation; K = potassium; Sht = sapling height; Grz = grazing; Ddi = drainage ditches; achiptar = *Achillea ptarmica*; agrocaca = *Agrostis canina* subsp. *canina*; agrocapi = *Agrostis capillaris*; agrostol = *Agrostis stolonifera*; agrovine = *Agrostis vinealis*; alopprat = *Alopecurus pratensis*; brizmedi = *Briza media*; callcusp = *Calliergonella cuspidata*; caltpalu = *Caltha palustris*; calyargu = *Calypogeia arguta*; cardprat = *Cardamine pratensis*; carebine = *Carex binervis*; caredist = *Carex disticha*; careechi = *Carex echinata*; carehirt = *Carex hirta*; carepuli = *Carex pulicaris*; careviri = *Carex viridula*; cirsdiss = *Cirsium dissectum*; dactfuch = *Dactylorhiza fuchsii*; dactmace = *Dactylorhiza maculata* supsp. *ericetorum*; epilobsc = *Epilobium obscurum*; epilpalu = *Epilobium palustre*; euphrost = *Euphrasia rostkoviana*; festovin = *Festuca ovina*; festprat = *Festuca pratensis*; festrubr = *Festuca rubra*; filiulma = *Filipendula ulmaria*; glycflui = *Glyceria fluitans*; hylosple = *Hylocomium splendens*; juncbulb = *Juncus bulbosus*; junccong = *Juncus conglomeratus*; junceffu = *Juncus effusus*; lathprat = *Lathyrus pratensis*; lophbide = *Lophocolea bidentata*; lotucorn = *Lotus corniculatus*; lotupedu = *Lotus pedunculatus*; luzumult = *Luzula multiflora*; molicaer = *Molinia caerulea*; pellepip = *Pellia epiphylla*; plagundu = *Plagiomnium undulatum*; poteerec = *Potentilla erecta*; prunvulg = *Prunella vulgaris*; pseupuru = *Pseudoscleropodium purum*; ranuflam = *Ranunculus flammula*; ranurepe = *Ranunculus repens*; rhytlore = *Rhytidiadelphus loreus*; rubufrut = *Rubus fruticosus* agg.; rumeacet = *Rumex acetosa*; sagiproc = *Sagina procumbens*; saliauri = *Salix aurita*; salirepe = *Salix repens*; senejaco = *Senecio jacobaea*; sphacusp = *Sphagnum cuspidatum*; trifrepe = *Trifolium repens*; ulexeuro = *Ulex europaeus*

possible trends in calcium, nitrogen, organic carbon and litter cover percentage (Table 6). The difference in soil pH between unplanted and planted wet grasslands was almost significant ($p = 0.06$).

Table 5 Variation partitioning of species data in 4 m^2 plots in improved grasslands (IG) and wet grasslands (WG) between environmental (E) and management (M) explanatory variables using partial CCA

Data set				Fraction of TI			
	n_E	n_M	TVE	X	E	M	S
IG	8	4	0.498	0.502	0.273	0.125	0.100
WG	8	3	0.454	0.546	0.342	0.102	0.010

S variation shared between E and M, *X* unexplained variation, n_E, n_M number of significant variables from the *E* and *M* sets of explanatory variables used in each ordination, *TI* total inertia

Table 6 Comparison of soil properties (mean ± SEM) in unplanted (UP) and planted sites (*P*) of improved grassland (IG) and unimproved wet grassland (WG) ($n = 8$ for each grassland type)

	IG			WG		
	UP	*P*	*p*-value	UP	*P*	*p*-value
N (mg/L)[a]	**3,844 ± 154**	**3,235 ± 164**	**0.008**	4,470 ± 239	4,229 ± 133	0.188
P (mg/L)[a]	734 ± 67	673 ± 43	0.394	552 ± 38	546 ± 27	0.896
K (mg/L)[a]	6,709 ± 848	7,494 ± 795	0.120	4,833 ± 529	5,240 ± 506	0.348
Ca (mg/L)[b]	**8,063 ± 3641**	**2,935 ± 854**	**0.036**	2,119 ± 377	1,582 ± 250	0.090
Mg (mg/L)[a]	2,458 ± 360	2,260 ± 329	0.276	1,675 ± 129	1,751 ± 161	0.484
pH[a]	**6.59 ± 0.30**	**6.04 ± 0.29**	**0.044**	5.76 ± 0.09	5.50 ± 0.14	0.060
Organic C (mg/L)[b]	**8,728 ± 1001**	**6,360 ± 557**	**0.030**	11,184 ± 1,001	10,724 ± 1120	0.483
Litter cover (%)[b]	**2.79 ± 0.70**	**14.29 ± 2.45**	**0.012**	6.67 ± 1.52	8.88 ± 1.08	0.164

Significant differences among planted and unplanted sites are indicated in bold type

Statistical tests: [a]*T*-test and the [b]non-parametric Wilcoxon's paired signed-ranks test

Discussion

Changes in vegetation following afforestation result from a series of factors, some of which are imposed quickly whilst others develop gradually; these may act simultaneously or sequentially (Sykes et al. 1989). The substantial differences in species richness, relative abundance and species composition observed between planted and unplanted sites were primarily due to removal of grazing, changes in nutrient management and drainage for afforestation. Wallace and Good (1995) studied the effects of afforestation on upland plant communities in England and found that in the early stages (<6 years), stands tend to have poorly developed woodland flora; although cover may be high, the species present on mineral soils tend to be those characteristic of unmanaged sites. Sykes et al. (1989) reported the effects of reduced grazing and ground preparation following afforestation of upland grasslands. They found that grasses and rushes increase most in frequency, while species of wet habitats and small stature species become less frequent. These changes are similar to what we have found in lowland situations in this study.

In afforested improved grasslands, vigorous and competitive grasses commonly found in both wooded and unwooded habitats were significantly more frequent and less competitive ruderal species, which generally have open habitat affinities, were significantly less frequent than in unplanted sites. Analogous floristic changes were recorded in grasslands after the exclusion of grazing in other studies (Welch and Rawes 1964; Ball 1974; Stránská 2004). The first response to release from heavy grazing is a change in the structure of the community to one of greater height and more abundant tussocks; this is accompanied by a change in the competitive balance between species, with the gradual loss of some species, increase in others and invasion of new species (Ball 1974). The higher bryophyte species richness found in planted improved grasslands suggests they have colonised the new habitat under the shade of the taller ungrazed grasses. Some colonising species may be present in the soil bank or may invade from neighbouring vegetation (Stránská 2004). Surrounding habitats, such as hedgerows and other ecological corridors like rivers and road banks, represent a potential resource of species which may be able to colonise the newly afforested areas. In this study, a decrease in grazing pressure allowed highly competitive species, such as *Rubus fruticosus* and *Arrhenatherum elatius*, to colonise the grassland from neighbouring hedgerows. In many cases, this may lead to a species-poor community mostly dominated by vigorous grasses. However, improved grasslands are themselves species-poor habitats, and therefore afforestation of these habitats will have little short-term impact on biodiversity.

In unimproved wet grasslands, changes were similar but less marked if compared with improved grasslands, because of generally lower pre-afforestation grazing pressures and fertilisation. Drainage contributed to the decrease in species preferring wet conditions and also to the increases in competitive species. Drainage ditches were responsible for the discrepancies between plot scales in species richness differences between planted and unplanted wet grasslands. They were included in 100 m^2 plots, but not 4 m^2 plots, and provided new, reduced-competition habitats for bryophytes and pioneer vascular plant species such as *Riccardia chamaedryfolia* and *Carex viridula*. This accounts for the lack of a difference in species richness at the 100 m^2 scale while species richness at the 4 m^2 was lower in planted sites due to greater competitive pressures.

Variation between the floristic composition of unplanted and planted grassland is related to both natural variability and forestry induced changes, including changes in drainage, ground preparation and shading by trees (Wallace and Good 1995). In this study, the relative importance of explanatory variables differed between improved grassland and unimproved wet grassland plots. Environmental variables explained a higher fraction of the variation in unimproved wet grasslands (75.3%) than in improved grasslands (54.8%). This difference is best explained by the higher variation in environment among unimproved wet grassland sites. The low percentage of shared variation (2.2%) between environmental and management variables in unimproved wet grasslands showed that the two sets of explanatory variables were weakly correlated. On the other hand, in improved grasslands environmental variables accounted for 25% of the variation, and the two sets of variables shared 20% of the total variance explained, showing that the effects of afforestation and the consequent cessation of grazing on vegetation communities were apparently greater in improved grasslands than in wet grasslands. This finding is reflected by differences in soil properties which were found only for improved grasslands, where plantations had lower total nitrogen, calcium and organic carbon and higher soil acidity. Although it is possible that some of these differences in soil chemistry between paired sites were present prior to afforestation, studies examining the shift of native grassland to forest reveal various degrees of surface soil acidification after conifer establishment (Page 1968; Hornung 1985; Reich et al. 2005).

The decline in pH may be associated with the loss of exchangeable cations, particularly Ca, organic acid inputs and increased soil respiration (Page 1968; Hornung 1985; Jobbágy and Jackson 2003), and accumulation of cations in tree biomass (Hornung 1985; Parfitt et al. 1997). In the present study, where the trees were still small, it is unlikely that the acidification was due to tree nutrient uptake. The pH decrease may therefore be linked to lower soil Ca after afforestation and/or to organic acid inputs which enter the soil predominantly from the surface, causing maximum acidification in the topsoil. As litter percentage cover increased significantly under *Picea sitchensis* differences in pH levels between unplanted and planted sites in this study could also have been caused by differences in quantity and quality of litter, litter decomposability and microbial activity.

Lower nitrogen in afforested sites could be explained by several factors including the accumulation of nitrogen in tree biomass which is expected to be greatest within the first 15 years of plantation (Gholz et al. 1985), denitrification (loss of total N) and leaching of N due to enhanced mineralisation/nitrification (Parfitt et al. 1997). In this study, most improved grassland sites had previously been fertilised annually with synthetic N-P-K fertilisers and/or animal manure. Fertiliser applications ceased after afforestation, leading to a temporal gradient of decreasing fertility which may represent a further explanation for the observed difference in total nitrogen.

Concentrations of soil organic C in topsoil were significantly lower under Sitka spruce compared with unplanted improved grassland. The causes of changes in soil organic carbon when land use is changed may be ascribed to soil disturbance through ploughing or ripping, clearing of the original vegetation and control of competing vegetation in the initial years (Turner and Lambert 2000). Other important factors affecting soil carbon are previous land uses, climate, soil texture, site management, type of forest established (Polglase et al. 2000), and low production of detrital materials (e.g. litterfall, root turnover) in the early stages of stand development (Parfitt et al. 1997).

The absence of significant differences between unplanted and planted sites in unimproved wet grasslands might be due to the greater organic matter content and consequent buffering capacity of the wet soils, to the clay texture, to water logging conditions which could have delayed the effects of conversion between grasslands and plantation, and to a lower fertilisation rate of pre-afforested sites. In addition, variability in soil chemistry among wet grassland site pairs probably obscures differences between planted and unplanted sites.

Conclusions

The initial effects of afforestation appeared to be largely the results of three factors: exclusion of grazing, forestry drainage and changes in nutrient management. The impacts of afforestation on vegetation and soil properties of the studied habitats were greater in improved grasslands than in wet grasslands. Species sensitive to shading declined under competitive pressure from vigorous grasses and other aggressive colonist species. Drainage ditches provided a temporary habitat for new pioneer and less competitive species, but the overall effect of drainage contributed to a reduction in species richness and diversity. Planted improved grasslands had lower total nitrogen, calcium, organic carbon and pH than unplanted grasslands; these differences may be the result of ground preparation for afforestation and the effects of conifer litter, but it is impossible to rule out pre-existing differences. Any changes may have been lessened in wet grassland by several factors, e.g. soil texture and poor drainage. Since afforestation of these sites occurred only 5 years ago—a short period compared with the age of a mature forest—more changes may be expected to occur in the next few years. The tree saplings were still small, and the planted habitats were still grasslands rather than woodlands. When the trees become more mature and form a closed canopy, the understorey vegetation will be almost completely eliminated if it is a densely shading species such as Sitka spruce (Hill 1979; French 2005). Because of its short- and long-term effects, we conclude that afforestation will have a detrimental effect on semi-natural habitats. Such habitats, such as semi-natural wet grasslands, should not be afforested, unless similar habitats are abundant in the landscape. On the other hand, the effect of afforestation on improved and semi-improved grasslands will be neutral or positive, particularly in landscapes that contain little semi-natural woodland habitat.

Acknowledgements This research is a product of the BIOFOREST project, which is funded by the Council for Forest Research and Development (COFORD) and the Environmental Protection Agency (EPA) under the Irish National Development Plan 2000–2006. The authors acknowledge Dr Laura French, Ms Linda Coote, Ms Aoife Delaney and Ms Jacqueline Bolli for assistance in the field and with database management, and two anonymous referees for valuable comments on this paper. Thanks are also due to the Portuguese Foundation for Science and Technology (FCT) for grant aiding E.B.'s work (SFRH/BM/12697/2003).

References

Alard D, Bance JF, Frileux PN (1994) Grassland vegetation as an indicator of the main agro-ecological factors in a rural landscape: consequences for biodiversity and wildlife conservation in central Normandy (France). J Environ Manage 42:91–109

Alfredsson H, Condron LM, Clarholm M, Davis MR (1998) Changes in soil acidity and organic matter following the establishment of conifers on former grassland in New Zealand. For Ecol Manage 112:245–252

Ball ME (1974) Floristic changes on grasslands and heaths on the isle of Rhum after a reduction or exclusion of grazing. J Environ Manage 2:299–318

Berthelot MPE (1859) Violet d'aniline. Repert Chim Appl 1:284

Borcard D, Legendre P, Drapéau P (1992) Partialling out the spatial component of ecological variation. Ecology 73:1045–1055

Byrne C, Kelly DL, Jeffrey DW (1997) An experimental study of floristic composition in semi-natural grassland in eastern Ireland. In: Proceedings of the international occasional symposium of the European Grassland Federation, Warszawa, Łomża, Poland, 19–23 May 1997

Curtis TGF, McGough HN (1988) The Irish red data book. 1. Vascular plants. Stationery Office, Dublin

Department of Agriculture Food, Forestry (1996) Growing for the future: a strategic plan for the development of the forestry sector in Ireland. Stationery Office, Dublin

Elemans M (2004) Light, nutrients and the growth of the herbaceous forest species. Acta Oecol 26:197–202

European Commission (1999) Interpretation Manual of European Union Habitats. EUR 15/2. DG Environment

Fahy O, Gormally M (1998) A comparison of plant and carabid beetle communities in an Irish oak woodland with a nearby conifer plantation and clearfelled site. For Ecol Manage 110:263–273

Farley KA, Kelly EF (2004) Effects of afforestation of a páramo grassland on soil nutrient status. For Ecol Manage 195:281–290

Ferris R, Peace AJ, Humphrey JW, Broome AC (2000) Relationships between vegetation, site type and stand structure in coniferous plantations in Britain. For Ecol Manage 136:35–51

Forest Service (2000) Forest biodiversity guidelines. Department of the Marine and Natural Resources, Dublin

Forest Service (2004) Forestry statistics. Department of Agriculture and Food, Dublin. http://www.agriculture.gov.ie/forestry/files/standard.xls

Fossitt JA (2000) A guide to habitats in Ireland. The Heritage Council, Dublin

French L (2005) Ground flora communities in Ireland's plantation forests: their diversity, structure and composition. PhD thesis, University of Dublin, Trinity College, Dublin

Gardiner MJ, Radford T (1980) Soil associations of Ireland and their land use potential. An Foras Taluntais, Dublin

Gholz HL, Fisher RF, Pritchett WL (1985) Nutrient dynamics in slash pine plantation ecosystems. Ecology 66:647–659

Heritage Council (1999) Forestry and the National Heritage. Policy paper. Kilkenny, Ireland

Hill MO, Jones EW (1978) Vegetation changes resulting from afforestation of rough grazing in Caeo forest, south Wales. J Ecol 66:433–456

Hill MO (1979) The development of a flora in even-aged plantations. In: Ford ED, Malcolm DC, Atterson J (eds) The ecology of even-aged forest plantations. Institute of Terrestrial Ecology, Cambridge, pp 175–192

Hochberg Y (1988) A sharper Bonferroni procedure for multiple tests of significance. Biometrika 75:800–802

Holyoak D (2003) The distribution of bryophytes in Ireland. Broadleaf Books, Dinas Powys

Hornung M (1985) Acidification of soils by trees and forests. Soil Use Manage 1:24–28

Jeffrey DW (1970) A note on the use of ignition loss as a means for the approximate estimation of soil bulk density. J Ecol 58:297–299

Jobbágy EG, Jackson RB (2003) Pattern and mechanisms of soil acidification in the conversion of grasslands to forests. Biogeochemistry 64:205–229

Jobbágy EG, Jackson RB (2004) The uplift of soil nutrients by plants: biogeochemical consequences across scales. Ecology 85:2380–2389

Legendre P, Legendre L (1998) Numerical ecology. Elsevier Science BV, Amsterdam

MCPFE (2003) State of Europe's forests 2003. MCPFE Liason Unit, Vienna

Murphy J, Riley JP (1962) A modified single solution method for the determination of phosphate in natural waters. Anal Chim Acta 27:31–36

Økland RH, Eilertsen O (1994) Canonical correspondence analysis with variation partitioning: some comments and an application. J Veg Sci 5:117–126

Page G (1968) Some effects of conifer crops on soil properties. Commonw For Rev 47:52–62

Parfitt RL, Percival HJ, Dahlgren RA, Hill LF (1997) Soil and solution chemistry under pasture and radiata pine in New Zealand. Plant Soil 191:279–290

Paton JA (1999) The liverwort flora of the British Isles. Harley Books, Colchester

Pielou EC (1975) Ecological diversity. Wiley, New York

Polglase PJ, Paul KI, Khanna PK, Nyakuengama JG, O'Connell AM, Grove TS, Battaglia M (2000) Change in soil carbon following afforestation or reforestation. National carbon accounting system. Technical Report No. 20. Australian Greenhouse Office, Canberra

Reich PB, Oleksyn J, Modrzynski J, Mrozinski P, Hobbie SE, Eissenstat DM, Chorover J, Chadwick OA, Hale CM, Tjoelker MG (2005) Linking litter calcium, earthworms and soil properties: a common garden test with 14 tree species. Ecol Lett 8:811–818

Smith AJE (2004) The moss flora of Britain and Ireland. Cambridge University Press, Cambridge

Smith B, Wilson JB (1996) A consumer's guide to evenness indices. Oikos 76:70–82

Sokal RR, Rohlf FJ (1995) Biometry, 3rd edn. WH Freeman and Co, New York

SPSS (2003) SPSS 12.0 for Windows. SPSS, Chicago

Stace C (1997) New flora of the British Isles. Cambridge University Press, Cambridge

Stránská M (2004) Successional dynamics of *Cynosurus* pasture after abandonment in Podkrkonoší. Plant Soil Env 50:364–370

Sykes JM, Lowe VPW, Briggs DR (1989) Some effects of afforestation on the flora and fauna of an upland sheepwalk during 12 years after planting. J Appl Ecol 26:299–320

ter Braak CJF, Smilauer P (2002) CANOCO reference manual and CanoDraw for Windows user's guide: software for Canonical Community Ordination (version 4.5). Microcomputer Power, Ithaca

ter Braak CJF (1986) Canonical correspondence analysis: a new eigenvector technique for multivariate direct gradient analysis. Ecology 67:1167–1179

Turner J, Lambert M (2000) Change in organic carbon in forest plantation soils in eastern Australia. For Ecol Manage 133:231–247

Wallace HL, Good JEG (1995) Effects of afforestation on upland plant communities and implications for vegetation management. For Ecol Manage 79:29–46

Wallace HL, Good JEG, Williams TG (1992) The effects of afforestation on upland plant communities: an application of the British National Vegetation Classification. J Appl Ecol 29:180–194

Welch D, Rawes M (1964) The early effects of excluding sheep from high-level grasslands in the north Pennines. J Appl Ecol 1:281–300

Wulf M (2004) Plant species richness of afforestations with different former use and habitat continuity. For Ecol Manage 195:191–204

Multi-scale habitat selection and foraging ecology of the eurasian hoopoe (*Upupa epops*) in pine plantations

Luc Barbaro · Laurent Couzi · Vincent Bretagnolle · Julien Nezan · Fabrice Vetillard

Originally published in the journal Biodiversity and Conservation, Volume 17, No 5, 1073–1087.
DOI: 10.1007/s10531-007-9241-z

Abstract Bird conservation can be challenging in landscapes with high habitat turnover such as planted forests, especially for species that require large home ranges and juxtaposition of different habitats to complete their life cycle. The eurasian hoopoe (*Upupa epops*) has declined severely in western Europe but is still abundant in south-western France. We studied habitat selection of hoopoes in pine plantation forests using a multi-scale survey, including point-counts at the landscape level and radio-tracking at the home-range scale. We quantified habitat use by systematically observing bird behaviour and characterized foraging sites according to micro-habitat variables and abundance of the main prey in the study area, the pine processionary moth (*Thaumetopoea pityocampa*). At the landscape scale, hoopoes selected habitat mosaics of high diversity, including deciduous woods and hedgerows as main nesting sites. At the home-range scale, hoopoes showed strong selection for short grassland vegetation along sand tracks as main foraging habitats. Vegetation was significantly shorter and sparser at foraging sites than random, and foraging intensity appeared to be significantly correlated with moth winter nest abundance. Hoopoe nesting success decreased during the three study years in line with processionary moth abundance. Thus, we suggest that hoopoes need complementation between foraging and breeding habitats to establish successfully in pine plantations. Hoopoe conservation requires the maintenance of adjacent breeding (deciduous woods) and foraging habitats (short swards adjacent to plantation edges), and consequently depends on the maintenance of habitat diversity at the landscape scale.

L. Barbaro (✉) · J. Nezan · F. Vetillard
UMR1202, Biodiversité, Gènes et Communautés, INRA, Cestas, 33612, France
e-mail: luc@pierroton.inra.fr

L. Couzi
Centre de Recherches sur la Biologie des Populations d'Oiseaux, Muséum National d'Histoire Naturelle, 55 rue Buffon, Paris 75005, France

V. Bretagnolle
Centre d'Etudes Biologiques de Chizé, CNRS, Beauvoir sur Niort 79360, France

E.G. Brockerhoff et al. (eds.), *Plantation Forests and Biodiversity: Oxymoron or Opportunity?*
DOI: 10.1007/978-90-481-2807-5_8

Keywords Bird conservation · Foraging · Habitat complementation · Home range · Mosaic landscapes · Pine plantations · Radio-tracking · *Thaumetopoea pityocampa* · *Upupa epops*

Introduction

The relevance of heterogeneous habitat mosaics for bird conservation has been recently highlighted with the emergence of the continuum model as a new paradigm in landscape ecology (Manning et al. 2004). This model considers landscape as a mosaic of habitats of different qualities instead of using a binary classification of habitat and non-habitat, as in the classical island biogeography theory (Kupfer et al. 2006). It also predicts that the effects of surrounding matrix habitats may be more important than processes occurring within habitat patches, as demonstrated for birds in different biogeographic areas (Wiens 1995; Wolff et al. 2002; Wethered and Lawes 2003; Tubelis et al. 2004). Edges between matrix and breeding habitats can have positive effects on bird populations because of differences in resource availability and microclimate at edges, and when food is taken outside the breeding habitat (McCollin 1998). In the latter case, species need the complementation of non-substitutable resources in the landscape mosaic to complete their life cycle (Dunning et al. 1992; Brotons et al. 2004; Ouin et al. 2004). In western Europe, habitat complementation at the landscape scale is probably essential for the conservation of several bird species that have been declining at least in part of their European range through the past decades (Burfield and van Bommel 2004): turtle dove (*Streptopelia turtur*, Browne and Aebischer 2003), wryneck (*Jynx torquilla*, Freitag 2004), woodlark (*Lullula arborea*, Bowden 1990), red-backed shrike (*Lanius collurio*, Virkkala et al. 2004) or linnet (*Carduelis cannabina*, Eybert et al. 1995).

Habitat complementation has important implications for bird conservation in heterogeneous landscape mosaics with high turnover in space and time, such as plantation forests (Barbaro et al. 2005; Paquet et al. 2006). Some species of particular conservation concern need the juxtaposition of breeding and foraging resources found in semi-natural habitat patches that may no longer be available in landscapes composed entirely of commercial plantations. The identification of key foraging habitats, especially when distinct from the main breeding habitat, consequently arises as a major issue in bird conservation management. For example, the presence of adjacent semi-natural grasslands is beneficial to farmland birds in mosaic forest-agricultural landscapes, both in northern and southern Europe (Preiss et al. 1997; Pons et al. 2003; Virkkala et al. 2004). Managed grasslands are suitable foraging habitats for open habitat specialists, but also for species such as the eurasian hoopoe (*Upupa epops epops*) nesting in wooded habitats and foraging on grassland seeds or invertebrates.

Hoopoes preferably inhabit farmlands with trees or walls where they nest in hollows, and open habitats with short sward structures where they forage on large ground-living insects (Kristin 2001). They also occur in cleared and thinned forests (Camprodon and Brotons 2006), and their bimodal distribution in bird-habitat ordination models suggest that they use multiple habitats (Preiss et al. 1997). The hoopoe is classified as declining in western Europe and France (Burfield and van Bommel 2004; Julliard and Jiguet 2005). Food quality and accessibility as well as the availability of suitable nesting cavities are major limiting factors (Martin-Vivaldi et al. 1999; Arlettaz et al. 2000). In western Europe, hoopoes occur in farmlands where they feed mainly on molecrickets (*Gryllotalpa gryllotalpa*) and Lepidoptera larvae (Fournier and Arlettaz 2001). They also inhabit pine plantations, where they specialize in pupae of the pine processionary moth (*Thaumetopoea pityocampa*), which is a serious forest pest (Battisti et al. 2000; Kristin 2001). In south-western France, hoopoes breed in oak forest fragments embedded within a landscape matrix of maritime pine (*Pinus pinaster*) plantation forests (Barbaro et al. 2007).

Here, we examine habitat selection by hoopoes at the landscape-scale (i.e., distribution of breeding pairs), at the home-range scale (i.e., habitat use of individual birds), and at the micro-habitat scale (i.e., selection of foraging sites). Specifically, we ask if (i) landscape mosaics occupied by hoopoes show significant differences in habitat composition compared to unoccupied ones; (ii) hoopoe behaviour is different according to habitat within home range; (iii) hoopoe select foraging sites with particular micro-habitat attributes; and (iv) hoopoe foraging intensity is positively related to pine processionary moth abundance.

Methods

Study area

The study took place in the Landes de Gascogne forest, south-western France, a region covering c.10,000 km² dominated by intensively managed maritime pine plantation forests. Climate, soil composition and current sylvicultural practices are described in Maizeret (2005). The distribution of breeding hoopoes was sampled at two nested scales within the study area. At the landscape-scale, the study site spans c.10000 ha (44°40′N to 44°44′N, 0°57′W to 0°46′W) and is composed of small (<5 ha) and isolated patches of oak (mainly *Quercus robur*) woodlands embedded in a matrix of pine plantations of different ages. At the home-range scale, we selected a part of the study site covering 180 ha, including pine stands, clearcuts and oak woodland patches bordered by large maize fields. Grasslands (with *Molinia caerulea, Pseudarrhenatherum longifolium, Agrostis curtisii* and *Ulex minor*) and heathlands (with *Pteridium aquilinum, Ulex europaeus, Erica cinerea, E. scoparia* and *Calluna vulgaris*) occur in recent clearcuts, firebreaks and sand track edges of the study area.

Bird surveys

The distribution of breeding hoopoes was surveyed at the landscape-scale in 2002–2003. Two observers performed 286 point-counts with unlimited distance using a sampling survey stratified by the main habitat types (see below). Points were established at least 400 m apart to avoid double counting (Sutherland et al. 2004). We conducted two 20-min visits before and after mid-May, within 5 h after sunrise and excluding rainy days.

Habitat use of individual birds in the 180 ha-area was investigated using the territory mapping method. Territory mapping is considered to be the standard method for birds showing territorial behaviour and not ranging widely (Bibby et al. 2000), such as the hoopoe (Kristin 2001). Between 2004 and 2006, hoopoes were monitored twice a week in the morning (9:00–11:00) or late afternoon (16:00–18:00) from mid-April to mid-July. We drove slowly along the dense network of tracks to survey the whole area, including the interior of pine stands typically distant of 50–100 m from the nearest track. Birds located inside the stands could be sighted from the tracks because stand understorey is regularly cut for management access. All stands were checked carefully with binoculars before driving along the tracks to locate hoopoes before they were disturbed and to avoid a bias on the detection probability of birds between habitats. Driving was used because hoopoes were more closely approached by car, and because territory mapping involves locating individual birds as precisely and rapidly as possible to avoid double counting the same birds that may have moved from their initial locations (Bibby et al. 2000). The location and behaviour of every recorded bird were mapped on a Geographic Information System (ArcView, ESRI, Redlands, CA, USA), except birds seen in flight and those showing any change in behaviour because of the observer's presence. The coupling of territory mapping with colour-marking and radio-tagging (see below) allowed us to attribute a large majority of sightings to known individuals, as well as to distinguish adults from fledglings and additional non-breeding individuals (Bibby et al. 2000).

Nesting success

In the 180 ha-area, we established 13 specific nestboxes in 2002 to monitor breeding parameters. Nestboxes were located in deciduous tree patches to mimic natural conditions, and because male hoopoes were expected to aggregate in such habitats to sing and visit cavities (Martin-Vivaldi et al. 2002). At the beginning of the study, a breeding population of hoopoes was already established in the area. They nested only in large hollows in deciduous trees (mostly oaks) because cavities are lacking in pine plantations. This population bred continuously in natural cavities during the study with 3–4 pairs from 2004 to 2006. The provision of nestboxes provided nesting opportunities for additional pairs, which fluctuated from 4 pairs in 2004 and 3 pairs in 2005 to 6 pairs in 2006. Consequently, the total density of breeding pairs varied from 1 pair/20–45 ha during the study. We monitored breeding parameters (laying date, clutch size, brood size at hatching and number of fledged youngs) for 4 pairs in 2004, 4 pairs in 2005 (including 2 second clutches) and 9 pairs in 2006 (including 3 second clutches). Nestboxes were checked at the critical periods of egg laying, hatching and fledging, and nesting success was calculated only for the 14 successful clutches, by dividing the number of fledglings by clutch size (Martin-Vivaldi et al. 1999).

Ringing and radio-tracking

From 2004 to 2006, we caught 61 fledglings (c.20 days old) and 20 adults, and ringed them with a metal ring and Darvic plastic colour rings using combinations that allowed visual re-identification of individuals. In addition, 15 birds were radio-tagged using tail-mounted 1.3 g-tags (Pip-tags, Biotrack, UK), i.e., <2% of body mass, with a life of c.6 weeks and a range of c.1–2 km. Tags were glued on to the central tail feathers, but 8 out of the 15 hoopoes removed the tags by pulling out the rectrice within 24 h following the capture. The remaining 7 birds kept their tags between 2 and 46 days, and 6 were followed long enough to obtain more than 10 direct relocations (Table 1). Radio-tagged birds were relocated every day by approximate triangulation based on signal strength, until the bird was sighted and its precise location mapped (Browne and Aebischer 2003). We used interval sampling at more than 30-min intervals between two consecutive fixes to achieve independence of locations and to avoid bias by relocating birds disturbed by the observer (Sierro et al. 2001). Direct fixes were completed with additional, associated fixes when the bird identity could be confirmed visually (either by colour rings or by direct sighting of birds flying from foraging sites to the nest). This allowed an improved sample size for birds that had lost their transmitters early, and to estimate home ranges for two additional birds that were not tagged in 2005 (Table 1).

Table 1 Radio-tracking parameters and home-range sizes for 17 hoopoes

Year	Sex	Age	Nestbox	Sighting period	Tracking days	Direct fixes[a]	Associated fixes[a]	Total fixes	Home range (ha)[b]	
									MCP	KER
2004	Fem	–	11	April 29–June 7	21	44	6	50	9.77	6.52
2004	Male	2nd year	5	April 9–June 2	2	2	28	30	7.52	7.79
2004	Male	2nd year	7	May 17–27	10	12	5	17	14.03	26.46
2004	Fem	–	7	April 29–May 27	10	15	4	19	17.37	24.27
2004	Male	2nd year	11	May 17–June 4	18	19	10	29	7.41	7.91
2005	Male	2nd year	–	June 2–3	1	3	1	4	–	–
2005	Male	2nd year	14	April 11–May 13	–	0	37	37	11.89	15.29
2005	Male	2nd year	7	April 12–July 20	–	0	40	40	12.29	16.90
2006	Male	2nd year	1	April 26–May 24	6	21	11	32	15.57	23.93
2006	Male	1st year	1	May 11–12	1	1	0	1	–	–
2006	Male	2nd year	14	April 28–July 20	1	1	39	40	16.63	22.47
2006	Male	2nd year	10	April 27–July 20	1	1	30	31	8.82	9.71
2006	Male	2nd year	11	April 27–June 14	1	1	27	28	9.46	10.89
2006	Fem	2nd year	11	April 26–May 31	12[c]	30	7	37	30.76	21.98
2006	Fem	2nd year	14	April 26–June 13	1	6	11	17	7.97	10.43
2006	Male	2nd year	9	April 21–June 8	1	3	13	16	9.25	9.88
2006	Male	2nd year	–	April 18–June 2	1	6	10	16	12.99	20.99

[a] Direct fixes were obtained by relocations of radio-tagged individuals and associated fixes by re-sightings of known individuals

[b] Estimates of home-range sizes calculated by minimum convex polygons (MCP) and fixed kernel density functions (KER)

[c] Transmitter failed after 12 days but was still on the bird when re-captured 46 days after

Habitat use and foraging ecology

Habitat maps were digitized on GIS from colour aerial orthophotographs at the scale 1:25,000. We used the following 7 habitat types, with field calibration: mature pine plantation (tree height >7 m), young pine plantation (tree height <7 m), deciduous woodland and hedgerow, shrubland and heathland, semi-natural grassland (including herbaceous firebreaks), hay meadow and crop (maize field). We calculated the percentage cover of each habitat and a set of landscape metrics within 400 m-radius buffers of 50.3 ha around point-counts using Fragstats software (McGarigal et al. 2002). Previous studies showed that the most significant landscape metrics related to bird distribution were mean patch size (in ha), edge density (total length of all edges between all habitat patches, in m ha^{-1}) and the Shannon index of habitat diversity (Barbaro et al. 2005). In the 180 ha-area, we measured micro-habitat variables in 1-m^2 quadrats located at 40 foraging sites and 40 random sites in May–June 2006. Foraging sites were located by direct observations of foraging hoopoes and the quadrats were centred on the empty cocoons left by the birds when extracting processionary moth pupae from the ground (Battisti et al. 2000). Control plots were established randomly within the same area using GIS tools to create random points. In both plots, we recorded the distance to the nearest occupied cavity, vegetation height, and percentage cover of the main plant species, bare ground, woody debris, litter, bryophyte, grass and shrub layers (Bowden 1990; Sutherland et al. 2004).

Prey availability

Previous observations of foraging hoopoes in the study area indicated that birds feed mostly on two prey species, which are typically extracted from the first cm of the ground (Kristin 2001): pupae of the pine processionary moth and adult field crickets (*Gryllus campestris*). As moth pupae seemed to be quantitatively the most important prey, we monitored moth populations by counting winter nests in tree crowns (Hodar and Zamora 2004; Battisti et al. 2005). Pine processionary moth larvae live gregariously and build a winter silk nest in the tree crown periphery. Density of winter nests is known to be maximal at pine stand edges because of female moth preference for trees standing out against clear sky (Démolin 1969). The short swards along stand edges allow the caterpillars to burrow themselves into the upper 5 cm of the soil for pupating, where they are exposed to hoopoe predation (Battisti et al. 2000). We assumed a significant relationship between moth nest density and below-ground pupae abundance per edge. To estimate moth abundance within the 180 ha-area, one observer counted all winter nests in the first two tree rows of pine stand edges in early spring of each study year. We did not sample the interior of pine stands because most feeding hoopoes used the herbaceous fringe between a track and a plantation, and food availability should only be measured in habitats where birds can actually forage (Wolda 1990).

Statistical analyses

For binary point-count data, we compared mean landscape attributes for occupied and unoccupied mosaics using two-sample *t*-tests, and Mann–Whitney *U*-tests when data had non-normal distributions. We used binomial Generalized Linear Models (GLM) with logit link to relate hoopoe occurrence and landscape variables. Stepwise backward model selection was performed with Akaike's Information Criterion using the 'stepAIC' procedure in R package (R Development Core Team 2006). For radio-tracking data, home-range sizes were calculated for the 15 individuals with more than 16 relocations (direct plus associated fixes). Among these birds, 8 had more than 30 relocations and thus allowed a reliable estimate of 70–80% of their maximum home-range area (Sutherland et al. 2004). We used two methods to estimate home-range sizes: minimum convex polygon (MCP) and fixed kernel density function with 95% of the fixes. For kernel functions, we used least-squares cross-validation for calculation of the smoothing parameter *H* (Worton 1989). For habitat use, we compared the proportion of relocations in each habitat within individual home ranges (used habitats) to habitat availability at two levels: within the 180 ha-area and within individual home ranges (Aebischer et al. 1993). For each individual, we calculated a forage ratio, using the B_{i1} index of Manly et al. (1972), by dividing the proportion of bird records in a given habitat by the proportion of habitat available, then by dividing the forage ratio for each habitat by the sum of forage ratios for all habitats (Sutherland et al. 2004). We tested for differences in mean forage ratios between habitats using the Kruskal–Wallis *H*-test. We then used compositional analysis to compared habitat use within home range to habitat availability within total study area, and within individual home ranges (Aebischer et al. 1993). Compositional analysis was performed using the R package 'adehabitat' with missing values replaced by 0.01%, and randomisation tests (1000 permutations) to assess for the significance of habitat selection (Calenge 2006). We compared the proportion of birds having a particular behaviour in each habitat to the proportion of each habitat available by means of χ^2 tests and non-parametric Kendall correlation coefficients (Robinet et al. 2003). Micro-habitat variables were log-transformed when necessary to improve normality and data recorded at foraging sites were compared with those measured at the nearest control plot using a paired *t*-test. We used one-way ANOVA to test for a year effect on prey abundance (number of moth nests per stand edge) and nesting success, and linear regression to relate the log-transformed number of foraging hoopoes to prey abundance. To test for between-year variation in the regression slopes between foraging intensity and prey abundance, we performed an ANCOVA with the year as factor, the number of foraging hoopoes as response variable and prey abundance as covariate.

Results

Habitat selection at the landscape scale

Hoopoes were recorded in half of the 286 point-counts. There was a significant effect of habitat type on hoopoe mean abundance, with significantly higher abundance in deciduous woodlands compared to all other habitats (ANOVA, F-ratio = 5.662, $P < 0.0001$). Landscape mosaics occupied by breeding pairs had significantly higher habitat diversity and smaller mean patch size than unoccupied mosaics, thus hoopoes tended to select the most heterogeneous parts of the landscape (Table 2). They avoided areas that contained a high proportion of mature pine plantations and favoured areas with more deciduous woods, hedgerows and meadows (Table 2). In addition, the stepwise GLM selection using AIC retained four variables: deciduous woodland, grassland and meadow covers and Shannon index of habitat diversity, but only the latter showed a significant effect on hoopoe occurrence (coefficient = 1.09 ± SE 0.34, z-value = 3.21, $P < 0.001$).

Habitat selection at the home-range scale

Home-range sizes were estimated for 15 hoopoes (Table 1 and Fig. 1), of which 13 were radio-tagged birds. They measured on average 12.78 ha (SD ±5.96; range 7.41–30.76 ha) when calculated with the MCP method, and 15.69 ha (±7.07; range 6.52–26.46 ha) when calculated with the kernel method, a difference which was not significant (Mann–Whitney test, U = 86.0, P = 0.27). Similarly, for the 6 individuals with more than 10 direct relocations,

Table 2 Mean ± SD values of landscape attributes measured within 50 ha-areas around point-counts for mosaics occupied or unoccupied by hoopoes (*t*-tests for landscape structure and Mann–Whitney *U*-tests for habitat cover

Landscape attributes	Occupied	Unoccupied	*t*- and *U*-tests	*P*
Landscape structure				
Edge density (m/ha)	229.73 ± 66.26	216.36 ± 71.87	−1.636	ns
Mean patch size (ha)	2.16 ± 1.23	2.61 ± 1.57	2.676	*
Shannon index	1.71 ± 0.37	1.54 ± 0.51	−3.355	**
Habitat cover (%)				
Mature pine	37.81 ± 20.66	45.93 ± 24.69	12152.5	**
Young pine	17.99 ± 17.06	17.11 ± 18.43	9674.5	ns
Deciduous wood	9.47 ± 12.11	8.05 ± 12.72	8681.5	*
Hedgerow	0.32 ± 0.74	0.14 ± 0.43	8469.5	***
Shrubland	15.96 ± 13.06	13.51 ± 13.82	8856.5	ns
Grassland	4.73 ± 5.62	6.24 ± 8.37	11023.0	ns
Meadow	2.97 ± 8.13	0.91 ± 3.12	8821.0	**
Crop	2.29 ± 8.20	2.00 ± 8.29	9396.5	ns

d.f. = 284, ***$P < 0.001$, **$P < 0.01$, *$P < 0.05$, ns = not significant

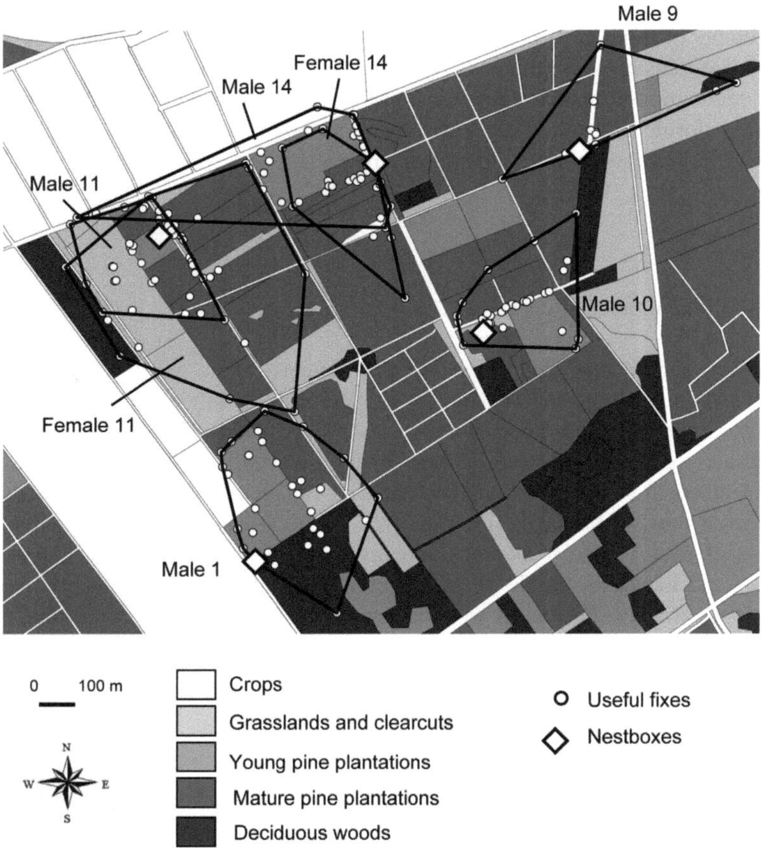

Fig. 1 Home ranges of 7 radio-tagged hoopoes in 2006 (white dots indicate fixes used to calculate home-range sizes by means of minimum convex polygons)

home-range sizes estimated with total fixes and with direct fixes only were not statistically different ($U = 15.5$, $P = 0.69$), and the estimates strictly identical for 4 birds. Home ranges overlapped largely within breeding pairs but generally not between pairs, with few exceptions (Fig. 1). Within home ranges, forage ratios differed significantly among habitats (Kruskal–Wallis test, $H = 49.0$, d.f. = 5, $P < 0.0001$). Mean forage ratios were higher for sand tracks and deciduous woodlands and hedgerows than for the other habitats, demonstrating positive selection of these two habitats as compared to their availability (Fig. 2). Compositional analysis showed that habitat selection for the 13 tagged hoopoes differed significantly from random at both levels of habitat availability within total study area ($\Lambda = 0.026$, $P < 0.001$) and individual home ranges ($\Lambda = 0.025$, $P < 0.001$). At the home range level, the ranking matrix of preferred used habitats gave the following order (>>> indicating significant differences): Sand tracks > Deciduous woods and hedgerows >>> Mature pines > Grasslands >>> Crops > Young pines.

Habitat use according to behaviour

A total number of 711 hoopoe observations were made between 2004 and 2006 ($n = 225$ in 2004, $n = 142$ in 2005 and $n = 344$ in 2006). Half of the birds were recorded in sand tracks

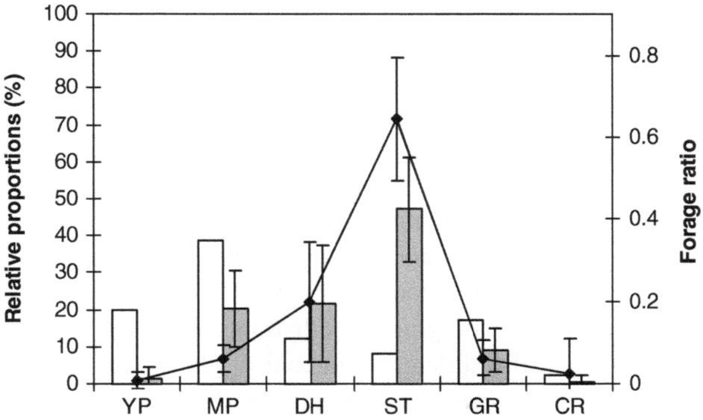

Fig. 2 Relative proportions (%) of habitat types available within the study area (white bars), mean (±SD) proportion of hoopoe relocations per habitat type within home ranges (grey bars) and mean (±SD) forage ratio per habitat type (black line). YP = young pine, MP = mature pine, DH = deciduous woods and hedgerows, ST = sand tracks, GR = grasslands, CR = crops

and their herbaceous edges (51%), 17% in mature pine plantations, 14% in oak woods, 9% in hedgerows and 8% in grasslands. Habitat use differed between years ($\chi^2 = 30.22$, d.f. = 8, $P < 0.0001$), with grasslands being less used in 2004 than in 2005–2006 and sand tracks more used in 2004–2005 than in 2006. The most common behaviour noted was roosting, either in a tree or on the ground (47% of sightings), then foraging (34%), singing (11%) and feeding chicks (8%). Hoopoe behaviour varied significantly among habitats ($\chi^2 = 455.43$, d.f. = 12, $P < 0.0001$) and among years ($\chi^2 = 60.09$, d.f. = 6, $P < 0.0001$), with more foraging birds in 2005–2006 than in 2004. Singing hoopoes were recorded in all wooded habitats, including pine plantations (Fig. 3a), while foraging birds were mainly recorded from sand tracks and secondly from grasslands (Fig. 3b). The proportion of birds having a particular behaviour in each habitat was compared to habitat availability in the 180 ha-area. We find significant habitat selection for all behaviour categories, according to the non-significant Kendall correlation coefficients ($\tau = -0.359$ for breeding behaviour, $\tau = 0.105$ for singing, $\tau = 0.200$ for foraging and $\tau = -0.200$ for roosting, all $P > 0.05$).

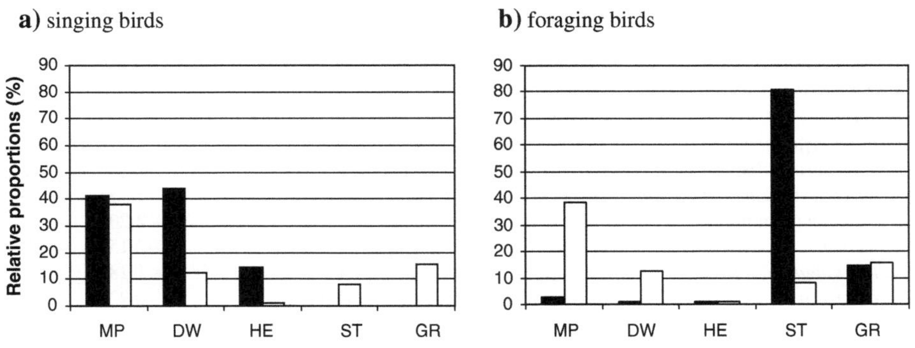

Fig. 3 Relative proportions (%) of hoopoe behaviour category per habitat type (black bars), compared to habitat availability (white bars) in the 180 ha study area: (**a**) singing birds and (**b**) foraging birds (data are pooled over the three study years). MP = mature pine, DW = deciduous woods, HE = hedgerows, ST = sand tracks, GR = grasslands

Small-scale selection of foraging habitats

The mean distance between foraging sites and nests was 271 ± 143 m (range 8–600 m), which was not significantly different from the distance measured between the randomly located plots and the nearest nest (Table 3). Foraging sites were all located in sand track edges adjacent to pine plantations, except one located in a mature pine plantation and four located in grasslands far from plantation edges. Vegetation was significantly shorter (7.4 cm ± 7.1) in foraging sites compared to control plots (23.8 cm ± 15.7). Bare ground (25.7% ± 26.7 versus 12.1% ± 29.3) and bryophytes (17.1% ± 18.2 versus 5.3% ± 13.2) had significantly higher cover in foraging sites than in control plots (Table 3). Vegetation composition at foraging sites also differed from control plots, with cover of bracken *Pteridium aquilinum*, gorse *Ulex europaeus* and deciduous shrubs being significantly higher in control plots, and cover of short annual graminoids, dicots and dwarf gorse *Ulex minor* being higher in foraging sites.

Prey abundance, foraging and nesting success

Moth abundance non-significantly decreased during the study (ANOVA, $F = 1.58$, d.f. = 2, $P = 0.21$, $n = 90$), from an average of 62.3 (SD ±38.2) nests per edge in 2004 to 53.8 (±31.2) in 2005 and 47.5 (±26.3) in 2006. There was a significant year effect on hoopoe nesting success ($F = 5.21$, d.f. = 2, $P = 0.03$, $n = 14$), which decreased in line with moth abundance from 0.81 (±0.09) in 2004 to 0.67 (±0.17) in 2005 and 0.52 (±0.15) in 2006. In all 3 years, the log-number of hoopoes observed foraging at a pine plantation edge was significantly and positively correlated to the number of moth winter nests per edge (Fig. 4). This relationship was stronger in 2004 ($n = 30$ edges, $r^2 = 0.452$, $P < 0.0001$) than in 2005 ($n = 39$, $r^2 = 0.273$, $P < 0.001$) and 2006 ($n = 40$, $r^2 = 0.218$, $P < 0.002$). Results of ANCOVA showed a significant prey abundance effect ($F = 43.19$, d.f. = 1, $P < 0.0001$) on

Table 3 Mean ± SD values of micro-habitat attributes measured in 1-m² quadrats centred on 40 hoopoe feeding locations and 40 randomly distributed plots

Micro-habitat variables	Foraging sites	Random sites	*t*-test	*P*
Distance to the nearest nest (m)	271.5 ± 143.0	254.4 ± 133.7	−0.881	ns
Vegetation height (cm)	7.4 ± 7.1	23.8 ± 15.7	4.271	***
Bare ground (%)	25.7 ± 26.7	12.1 ± 29.3	−6.139	***
Woody debris (%)	1.8 ± 3.9	2.5 ± 2.9	1.828	ns
Litter (%)	8.1 ± 12.2	20.5 ± 27.3	1.966	ns
Bryophyte cover (%)	17.1 ± 18.2	5.3 ± 13.2	−3.756	***
Grass cover (%)	34.0 ± 24.8	41.8 ± 30.9	−0.270	ns
Shrub cover (%)	11.3 ± 14.3	17.2 ± 22.4	1.123	ns
Pteridium aquilinum (%)	0.3 ± 0.5	6.3 ± 10.3	4.392	***
Molinia caerulea (%)	3.5 ± 12.2	6.2 ± 16.6	1.159	ns
Pseudarrhenatherum longifolium (%)	7.9 ± 14.4	11.3 ± 11.8	1.751	ns
Short annual graminoids (%)	11.0 ± 15.6	6.4 ± 15.9	−2.204	*
Dicots (%)	2.0 ± 3.2	0.6 ± 2.2	−3.156	**
Calluna vulgaris (%)	6.2 ± 14.1	4.5 ± 12.8	−1.281	ns
Erica cinerea (%)	2.7 ± 5.3	4.0 ± 10.3	−0.924	ns
Erica scoparia (%)	0.7 ± 3.2	1.6 ± 3.6	1.784	ns
Ulex minor (%)	2.2 ± 3.4	1.2 ± 5.2	−2.962	**
Ulex europaeus (%)	0.3 ± 1.6	4.9 ± 12.1	4.177	***
Deciduous shrubs (%)	0.4 ± 0.9	5.1 ± 8.1	3.607	***

paired *t*-tests, d.f. = 39, $^{***}P < 0.001$, $^{**}P < 0.01$, $^{*}P < 0.05$, ns = not significant

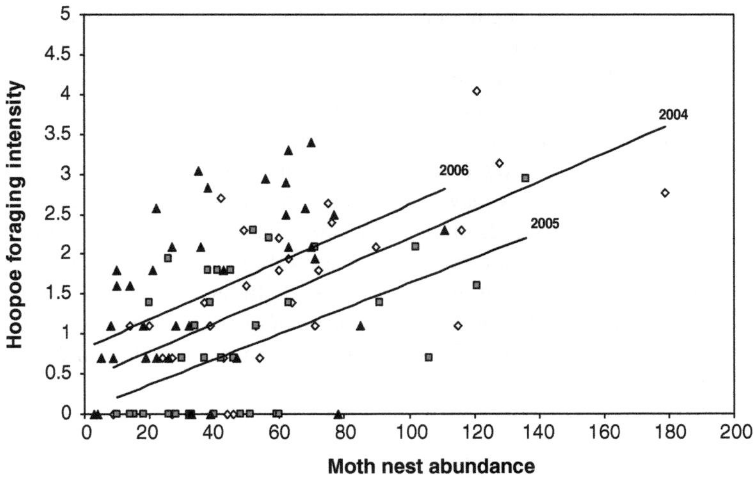

Fig. 4 Relationship between the log-number of foraging hoopoes and moth nest abundance per edge from 2004 to 2006 (black triangles: 2006, $r^2 = 0.22$, $n = 40$, $P < 0.002$; grey squares: 2005, $r^2 = 0.27$, $n = 39$, $P < 0.001$; white diamonds: 2004, $r^2 = 0.45$, $n = 30$, $P < 0.0001$)

hoopoe foraging intensity when taking into account the covariation between year and prey abundance. However, the interaction effect between year and prey abundance was not significant ($F = 0.06$, d.f. = 2, $P = 0.94$), i.e., the slopes of the three regression models were not significantly different. The overall year effect was however significant ($F = 10.91$, d.f. = 2, $P < 0.0001$), indicating that the intercepts differed according to year, in parallel to the variations in hoopoe density (Fig. 4).

Discussion

Habitat use and landscape complementation

The present study demonstrated that resource complementation between habitats at the landscape-scale was an important mechanism of habitat selection for this breeding population of hoopoes. Birds showed a preference for landscape mosaics with high habitat diversity. They selected particularly deciduous woodlands and hedgerows for the availability of deep nesting cavities in old oaks. At the home-range scale, hoopoes likewise showed a preference for habitat mosaics combining mature pine plantations, deciduous woods, hedgerows, grasslands and sand tracks, but only sand tracks and deciduous woods and hedgerows were selected more than expected from their availability. Deciduous woods and hedgerows were typical breeding sites, while foraging birds occurred mostly on sand track edges, and sometimes on grasslands. At a finer scale, foraging hoopoes selected microsites with short and sparse vegetation dominated by bryophytes, annual graminoids, dicots and dwarf gorse.

Habitat selection in birds is known to be a hierarchical process acting at multiple scales (Wiens 1995). For instance, owls choose their habitats according to trophic resources at a large scale, and according to breeding and roosting requirements at a smaller scale (Martinez and Zuberogoitia 2004). Similarly, choughs (*Pyrrhocorax pyrrhocorax*) use grazed habitats at a coarse scale and, at a finer scale, areas with the shortest swards for foraging (Whitehead et al. 2005). Our results suggest multi-scale habitat selection at three nested spatial scales:

(i) at the landscape scale, breeding hoopoes select oak woodlands embedded in a matrix of pine plantations and open habitats, (ii) at the home-range scale they prefer areas with breeding and foraging habitats in close vicinity, and (iii) at the micro-habitat scale, foraging birds select short and sparse swards along plantation-track edges. Thus, landscape mosaics with high habitat diversity are favoured because they fulfil both breeding and foraging requirements. As landscape complementation and supplementation are widespread mechanisms of multi-habitat use, they have important implications for bird conservation in mosaic landscapes (Wiens 1995; Brotons et al. 2004; Tubelis et al. 2004). In pine plantation forests of western Europe, several other threatened insectivorous birds would benefit from increasing habitat diversity at the landscape-scale through supplementation or complementation of resources, including nightjar (*Caprimulgus europaeus*, Sierro et al. 2001), wryneck (Freitag 2004), woodlark (Bowden 1990), or mistle thrush (*Turdus viscivorus*, Pons et al. 2003).

Foraging and prey availability

Foraging habitat selection results from an interaction between food abundance and accessibility, mediated by vegetation structure (Morris et al. 2001). As a result, the question arises if hoopoes feed on habitat edges because of higher prey abundance or higher accessibility compared to stand interiors? For example, nightjars did not use pine plantations as much as oak scrublands despite similar moth abundance in the two habitats because dense understorey in plantations prevent birds from foraging in flight (Sierro et al. 2001). Like other ground gleaners or probers, hoopoes feed preferably in short sward structures with c.25% bare ground. They generally avoid the interior of plantations stands because of dense understorey, but they can use them when mechanical cutting creates short vegetation or small gaps (Camprodon and Brotons 2006). Dense vegetation and impenetrable soils make arthropods inaccessible by probing or gleaning (McCracken and Tallowin 2004), and shorter and sparser swards are therefore preferred by most ground insectivores (Bowden 1990; Browne and Aebischer 2003; Whitehead et al. 2005). Hoopoes are able to use foraging sites located far from nesting cavities (Arlettaz et al. 2000; Kristin 2001). In our study area, the distance between the nest and suitable foraging sites did not seem to be a limiting factor since hoopoes undertook foraging trips of up to 600 m from the nest. Mean foraging distance was 272 m, larger than that observed in other ground insectivorous birds such as wryneck (115 m, Freitag 2004) or woodlark (118 m, Bowden 1990).

The hoopoe is a brood reduction strategist able to adjust clutch size to prey availability by selective starvation of the youngest chicks (Martin-Vivaldi et al. 1999). The influence of food availability and its accessibility on breeding success is therefore critical (Fournier and Arlettaz 2001). As a specialist predator, the hoopoe is likely to respond to fluctuations in prey abundance (Crawford and Jennings 1989; Sherry 1990), as suggested by nesting success decreasing in line with moth abundance. Lepidopterous pupae and larvae are the main preys of many insectivorous forest birds (Glen 2004). The distribution of pine processionary moth is the main factor for the occurrence of another specialist predator, the great spotted cuckoo (*Clamator glandarius*, Hoyas and López 1998). However, in western Europe, only the hoopoe can feed on buried moth pupae during the breeding season, because of its long curved bill and unique foraging technique among forest insectivorous birds (Kristin 2001). Although hoopoes commonly feed on pine processionary moth in Spain and Italy, the main prey in Switzerland is the molecricket (Arlettaz et al. 2000; Battisti et al. 2000). The decrease in moth abundance observed during the study coincided with an increase in grassland use and foraging time, which suggests that hoopoes may switch to alternative orthopteran preys (field crickets), in years of low moth density. However, as the nutritional value of moth pupae compared to field crickets is not documented, the relative importance of the two preys in the study area and their among-years variations remain to be investigated (Fournier and Arlettaz 2001).

Implications for conservation

The long-term conservation of the hoopoe in mosaic landscapes dominated by pine plantations depends on the maintenance of habitat diversity or its restoration by planting or regenerating oak woodland patches embedded in the pine plantation matrix. An appropriate management of the fringes between tracks and pine stands by regular cutting is also critical to allow hoopoes to access their preys (either moth pupae or field crickets) and will benefit other ground foraging birds (McCracken and Tallowin 2004), as well as plants and arthropods (Mullen et al. 2003). Edges between mature plantations and clearcuts may also provide suitable foraging sites if they are bordered by a short herbaceous strip both favourable to caterpillar burrowing and hoopoe probing. Moreover, previous studies have shown that clear-cutting in plantation forests lead to the establishment of a specific bird assemblage involving several threatened species (Barbaro et al. 2005; Paquet et al. 2006).

As nest site availability, together with prey availability, is a limiting factor for the hoopoe in plantation forests, we advocate the use of nestboxes to increase breeding density or restore populations in areas where cavities are lacking. Hoopoes generally respond to the establishment of nestboxes within a few years (Arlettaz et al. 2000; Kristin 2001). Moreover, breeding pairs tend to aggregate in the study area (Barbaro et al. 2007), although spatial aggregation may be caused by potentially confounding factors such as environmental heterogeneity (Cornulier and Bretagnolle 2006) or intra-specific social interactions (Martinez and Zuberogoitia 2004). In the hoopoe, displaying males tend to aggregate spontaneously where they expect to find females, and non-paired males frequently help to feed incubating females and chicks of other males (Martin-Vivaldi et al. 2002). We therefore suggest that the establishment of a dense network of nestboxes will allow the clumping of breeders in loose colonies and would increase social interactions in hoopoe populations.

In addition, the use of nestboxes may also be a tool for promoting biological control of pest insects in pine plantation forests. Predation of pine processionary moth by insectivorous birds may maintain moth populations at low densities, despite interactions with other causes of mortality such as parasitoid insects (Crawford and Jennings 1989; Battisti et al. 2000; Glen 2004). The increase of pine processionary moth populations with climate warming and the consequent potential threats to forest health and biodiversity (Hodar and Zamora 2004; Battisti et al. 2005) may be therefore mitigated by an increase in the density of functional insectivores such as the hoopoe (Jones et al. 2005). Conservation management in production forests should aim at maintaining or restoring native vegetation patches and corridors within a complex landscape matrix to enhance the functional diversity of species (Fischer et al. 2006). The hoopoe is an emblematic example of a threatened keystone species that may be favoured by such management recommendations in plantation forests.

Acknowledgements We are particularly indebted to S. Blache for initiating this study, F. Jiguet and O. Dehorter (CRBPO, Muséum National d'Histoire Naturelle, Paris) for the ringing authorization, G. Mays and P. Zeddam for additional ringing and F. Lagane for nestboxes. I. van Halder, H. Jactel, M. Deconchat, S. Saïd and J. C. Samalens helped with GIS or statistical analyses and A. Hampe, F. Burel and two anonymous referees improved the previous drafts of the paper. Many thanks to all the people involved in the field work, P. Menassieu, G. and D. Piou, R. Burlett, F. Sin, V. Dupin, A. Plichon, P. Boyer, M. Dupuich, M. Lagarde, V. Varlet and F. Jouandoudet.

References

Aebischer NJ, Robertson PA, Kenward RE (1993) Compositional analysis of habitat use from animal radio-tracking data. Ecology 74:1313–1325

Arlettaz R, Fournier J, Zbinden N (2000) Evolution démographique (1979–1998) d'une population témoin de Huppe fasciée *Upupa epops* en Valais et stratégie de conservation ciblée. Nos Oiseaux 47:19–27

Barbaro L, Pontcharraud L, Vetillard F, Guyon D, Jactel H (2005) Comparative responses of bird, carabid, and spider assemblages to stand and landscape diversity in maritime pine plantation forests. Ecoscience 12:110–121

Barbaro L, Rossi JP, Vetillard F, Nezan J, Jactel H (2007) The spatial distribution of birds and carabid beetles in pine plantation forests: the role of landscape composition and structure. J Biogeogr 34:652–664

Battisti A, Bernardi M, Ghiraldo C (2000) Predation by the hoopoe on pupae of *Thaumetopoea pityocampa* and the likely influence on other natural enemies. Biocontrol 45:311–323

Battisti A, Stastny M, Netherer S, Robinet C, Schopf A, Roques A, Larsson S (2005) Expansion of geographic range in the pine processionary moth caused by increased winter temperatures. Ecology 15:2084–2096

Bibby CJ, Burgess ND, Hill DA, Mustoe S (2000) Bird census techniques. Academic Press, London

Bowden CGR (1990) Selection of foraging habitats by woodlarks (*Lullula arborea*) nesting in pine plantations. J Appl Ecol 27:410–419

Brotons L, Herrando S, Martin JL (2004) Bird assemblages in forest fragments within Mediterranean mosaics created by wild fires. Landscape Ecol 19:663–675

Browne SJ, Aebischer NJ (2003) Habitat use, foraging ecology and diet of Turtle Doves *Streptopelia turtur* in Britain. Landscape Ecol 145:572–582

Burfield I, van Bommel F (2004) Birds in Europe: population estimates, trends and conservation status. BirdLifeInternational, Cambridge

Calenge C (2006) The package "adehabitat" for the R software: a tool for the analysis of space and habitat use by animals. Ecol Model 197:516–519

Camprodon J, Brotons L (2006) Effects of undergrowth clearing on the bird communities of the Northwestern Mediterranean Coppice Holm oak forests. Forest Ecol Manag 221:72–82

Cornulier T, Bretagnolle V (2006) Assessing the influence of environmental heterogeneity on bird spacing patterns: a case study with two raptors. Ecography 29:240–250

Crawford HS, Jennings DT (1989) Predation by birds on spruce budworm *Choristoneura fumiferana*: functional, numerical, and total responses. Ecology 70:152–163

Démolin G (1969) Comportement des adultes de *Thaumetopoea pityocampa* Schiff.: dispersion spatiale, importance écologique. Ann Forest Sci 26:81–102

Dunning JB, Danielson BJ, Pulliam HR (1992) Ecological processes that affect populations in complex landscapes. Oikos 65:169–175

Eybert MC, Constant P, Lefeuvre JC (1995) Effects of changes in agricultural landscape on a breeding population of linnets *Acanthis cannabina* L. living in adjacent heathland. Biol Conserv 74:195–202

Fischer J, Lindenmayer DB, Manning AD (2006) Biodiversity, ecosystem function, and resilience: ten guiding principles for commodity production landscapes. Front Ecol Environ 4:80–86

Fournier J, Arlettaz R (2001) Food provision to nestlings in the Hoopoe *Upupa epops*: implications for the conservation of a small endangered population in the Swiss Alps. Front Ecol Environ 143:2–10

Freitag A (2004) Measuring food availability for an insectivorous bird: the case study of wrynecks and ants. In: van Hemden H, Rothschild M (eds) Insect and bird interactions. Intercept, Andover

Glen DM (2004) Birds as predators of lepidopterous larvae. In: van Hemden H, Rothschild M (eds) Insect and bird interactions. Intercept, Andover

Hodar JA, Zamora R (2004) Herbivory and climatic warming: a Mediterranean outbreaking caterpillar attacks a relict, boreal pine species. Biodivers Conserv 13:493–500

Hoyas J, López F (1998) Distribución del críalo según la abundancia de procesionaria del pino. Quercus 149:20–22

Jones GA, Sieving KE, Jacobson SK (2005) Avian diversity and functional insectivory on north-central Florida farmlands. Conserv Biol 19:1234–1245

Julliard R, Jiguet F (2005) Statut de conservation en 2003 des oiseaux communs nicheurs en France selon 15 ans de programme STOC. Alauda 73:345–356

Kristin A (2001) Family Upupidae (Hoopoe). In: del Hoyo J, Elliott A, Sargatal J (eds) Handbook of the birds of the world, vol. 6, Mousebirds to Hornbills. Lynx Edicions, Barcelona

Kupfer JA, Malanson GP, Franklin SB (2006) Not seeing the ocean for the islands: the mediating influence of matrix-based processes on forest fragmentation effects. Global Ecol Biogeogr 15:8–20

Maizeret C (2005) Les Landes de Gascogne. Delachaux et Niestlé, Paris

Manly BJF, Miller P, Cook LM (1972) Analysis of a selective predation experiment. Am Nat 106:719–736

Manning AD, Lindenmayer DB, Nix HA (2004) Continua and Umwelt: novel perspectives on viewing landscapes. Oikos 104:621–628

Martinez JA, Zuberogoitia I (2004) Habitat preferences for long-eared owls *Asio otus* and little owls *Athene noctua* in semi-arid environments at three spatial scales. Bird Study 51:163–169

Martin-Vivaldi M, Palomino JJ, Soler M, Soler JJ (1999) Determinants of reproductive success in the Hoopoe *Upupa epops*, a hole-nesting non-passerine bird with asynchronous hatching. Bird Study 46:205–216

Martin-Vivaldi M, Martinez JG, Palomino JJ, Soler M (2002) Extrapair paternity in the Hoopoe *Upupa epops*: an exploration of the influence of interactions between breeding pairs, non-pair males and strophe length. Bird Study 144:236–247

McCollin D (1998) Forest edges and habitat selection in birds: a functional approach. Ecography 21:247–260

McCracken DI, Tallowin JR (2004) Swards and structure: the interactions between farming practices and bird resources in lowland grasslands. Ecography 146:108–114

McGarigal K, Cushman SA, Neel MC, Ene E (2002) Fragstats: spatial pattern analysis program for categorical maps. Computer software program, University of Massachusetts, Amherst

Morris AJ, Whittingham MJ, Bradbury RB, Wilson JD, Kyrkos A, Buckingham DL, Evans AD (2001) Foraging habitat selection by yellowhammers (*Emberiza citrinella*) nesting in agriculturally contrasting regions in lowland England. Biol Conserv 101:197–210

Mullen K, Fahy O, Gormally M (2003) Ground flora and associated arthropod communities of forest road edges in Connemara, Ireland. Biodivers Conserv 12:87–101

Ouin A, Aviron S, Dover J, Burel F (2004) Complementation/supplementation of resources for butterflies in agricultural landscapes. Agric Ecosyst Environ 103:473–479

Paquet JY, Vandevyvre X, Delahaye L, Rondeux J (2006) Bird assemblages in a mixed woodland-farmland landscape: the conservation value of silviculture-dependant open areas in plantation forest. Forest Ecol Manag 227:59–70

Pons P, Lambert B, Rigolot E, Prodon R (2003) The effects of grassland management using fire on habitat occupancy and conservation of birds in a mosaic landscape. Biodivers Conserv 12:1843–1860

Preiss E, Martin JL, Debussche M (1997) Rural depopulation and recent landscape changes in a Mediterranean region: consequences to the breeding avifauna. Landscape Ecol 12:51–61

R Development Core Team (2006) R: a language and environment for statistical computing. R Foundation for Statistical Computing, Vienna, Austria. ISBN 3-900051-070, URL http://www.R-project.org

Robinet O, Bretagnolle V, Clout M (2003) Activity patterns, habitat use, foraging behaviour and food selection of the Ouvéa Parakeet (*Eunymphicus cornutus uvaeensis*). Emu 103:71–80

Sherry TW (1990) When are birds dietary specialized? Distinguishing ecological from evolutionary approaches. Stud Avian Biol 13:337–352

Sierro A, Arlettaz R, Naef-Daenzer B, Strebel S, Zbinden N (2001) Habitat use and foraging ecology of the nightjar *Caprimulgus europaeus* in the Swiss Alps: towards a conservation scheme. Biol Conserv 98:325–331

Sutherland WJ, Newton I, Green RE (2004) Bird ecology and conservation: a handbook of techniques. Oxford University Press, Oxford

Tubelis DP, Cowling A, Donnelly C (2004) Landscape supplementation in adjacent savannas and its implications for the design of corridors for forest birds in the central Cerrado, Brazil. Biol Conserv 118:353–364

Virkkala R, Luoto M, Rainio K (2004) Effects of landscape composition on farmland and red-listed birds in boreal agricultural-forest mosaics. Ecography 27:273–284

Wethered R, Lawes MJ (2003) Matrix effects on bird assemblages in fragmented Afromontane forest in South Africa. Biol Conserv 114:327–340

Whitehead S, Johnstone I, Wilson J (2005) Choughs *Pyrrhocorax pyrrhocorax* breeding in Wales select foraging habitat at different spatial scales. Bird Study 52:193–203

Wiens JA (1995) Habitat fragmentation: island vs. landscape perspectives on bird conservation. Bird Study 137:97–104

Wolda H (1990) Food availability for an insectivore and how to measure it. Stud Avian Biol 13:38–43

Wolff A, Dieuleveut T, Martin JL, Bretagnolle V (2002) Landscape context and little bustard abundance in a fragmented steppe: implications for reserve management in mosaic landscapes. Biol Conserv 107:211–220

Worton BJ (1989) Kernel methods for estimating the utilization distribution in home-range studies. Ecology 70:164–168

Diversity and composition of fruit-feeding butterflies in tropical *Eucalyptus* plantations

Jos Barlow · Ivanei S. Araujo · William L. Overal · Toby A. Gardner · Fernanda da Silva Mendes · Iain R. Lake · Carlos A. Peres

Originally published in the journal Biodiversity and Conservation, Volume 17, No 5, 1089–1104.
DOI: 10.1007/s10531-007-9240-0

Abstract Production landscapes are rarely considered as priority areas for biodiversity conservation in the tropics. Tree plantations have the potential to provide a conservation service in much of the humid tropics since they are rapidly increasing in extent and present less of a structural contrast with native vegetation than many more intensive agricultural land-uses. We used hierarchical partitioning to examine the factors that influence the value of large-scale *Eucalyptus* plantations for tropical fruit-feeding butterflies (Lepidoptera: Nymphalidae) in the Brazilian Amazon. We focused on evaluating the importance of landscape versus stand-level factors in determining the diversity and composition of butterfly assemblages, and how butterfly-environment relationships vary within and between subfamilies of Nymphalidae. Native understorey vegetation richness had the strongest independent effect on the richness, abundance and composition of all fruit-feeding butterflies, as well as a subset of species that had been recorded in nearby primary forests. However, overall patterns were strongly influenced by the most abundant subfamily (Satyrinae), and vegetation richness was not related to the abundance of any other subfamily, or non-Satyrinae species, highlighting the importance of disaggregating the fruit-feeding Nymphalidae when examining butterfly-environment relationships. Our results suggest that plantations can help conserve a limited number of forest species, and serve to highlight the research that is necessary to understand better the relationship between fruit-feeding butterflies and environmental variables that are amenable to management.

J. Barlow · I. S. Araujo · W. L. Overal
Museu Paraense Emílio Goeldi, Avenida Magalhães Barata 376, Belem, Para 66040-170, Brazil

Present Address:
J. Barlow (✉)
School of Biological Sciences, Lancaster University, Lancaster LA1 4YW, UK
e-mail: jos.barlow@lancaster.ac.uk

J. Barlow · T. A. Gardner · I. R. Lake · C. A. Peres
Centre for Ecology, Evolution and Conservation, School of Environmental Sciences, University of East Anglia, Norwich NR4 7TJ, UK

F. da Silva Mendes
Universidade Federal Rural da Amazônia, Avenida Presidente Tancredo Neves 2501, Belem, Para 66077-530, Brazil

E.G. Brockerhoff et al. (eds.), *Plantation Forests and Biodiversity: Oxymoron or Opportunity?*
DOI: 10.1007/978-90-481-2807-5_9

Keywords Amazon · Biodiversity · Brazil · Conservation · Hierarchical partitioning · Species-environment relationships

Introduction

Biodiversity conservation in production landscapes is becoming increasingly recognized as a major conservation priority worldwide (Daily 2001; Lindenmayer and Franklin 2002; Fischer et al. 2006). The inadequacy of the current protected areas network (Rodrigues et al. 2004) highlights the potential importance of production areas in providing a vital complementary conservation service (Lindenmayer and Franklin 2002; Fischer et al. 2006). Tree plantations have the potential to provide a valued conservation service in much of the humid tropics as they (1) are rapidly increasing in extent, and (2) present less of a structural contrast with native vegetation than many alternative yet more intensive agricultural land-uses that are biologically impoverished (e.g. cattle ranches, soybean, cotton croplands). The coverage of plantation forestry in the tropics increased from c.17.8 million hectares in 1980 to c.70 million in 2000 (Brown 2000; FAO 2005), while it is estimated that around 1 million hectares of tropical forest are converted to tree plantations each year (FAO 2005). This coverage is likely to increase further (especially in areas of tropical forest, e.g. Fearnside 1998), in part due to both private and national investment in carbon-sequestration projects (Yu 2004) and a growing interest in biofuels and timber products (Pacala and Socolow 2004). In their latest Global Forest Resource Assessment the Food and Agriculture Organization of the UN has predicted a 50% increase in the production of industrial wood from plantations in the next 40 years (FAO 2005).

Plantation forests may support biodiversity conservation by buffering fragments of native forest, facilitating the movement of animals across the landscape matrix, and providing suitable habitat for some forest dependent species (Lindenmayer and Franklin 2002; Carnus et al. 2006; Fisher et al. 2006; Lindenmayer et al. 2006). However, despite their potential importance, the conservation value of extensive monocultures are very poorly understood (Kanowski et al. 2005), with most of the work restricted to a few well-studied taxa (birds and mammals) in temperate and subtropical regions (see Hartley 2002; Lindenmayer and Hobbs 2004). Furthermore, management recommendations for many temperate plantations focus on maximizing the amount of young "non-woodland" habitat suitable for the open-habitat specialists of greatest conservation concern (e.g. Humphrey et al. 1999; Eycott et al. 2006), and are of limited relevance for the humid tropics where most species of conservation concern are forest species.

We examined fruit-feeding butterflies (Lepidoptera: Nymphalidae) in *Eucalyptus* plantations in the north-east Brazilian Amazon. *Eucalyptus* currently accounts for around 50% of all tropical tree plantations (Evans and Turnbull 2004). Butterflies have frequently been used as indicators of the conservation value of tropical habitats and the consequences of disturbance and land-use change (e.g. Brown 1997; Hamer and Hill 2000; Koh et al. 2007), and in a previous study undertaken at the same study site fruit-feeding butterflies explained almost 57% of the variance in the responses of 14 other taxa to land-use change (J.Barlow et al. unpubl. data). However, despite their popularity as ecological and biodiversity indicators, we are only aware of three studies that have explicitly examined the value of tropical plantations for butterflies (Ramos 2000; Stork et al. 2003; Barlow et al. 2007). Moreover, we are not aware of any study that has evaluated how stand-level and landscape-level features interact to determine the conservation value of plantations for butterflies. We addressed this information deficit by examining how landscape configuration, and plantation age, productivity, and understorey structure and composition all combine to influence the species richness, diversity and composition of fruit-feeding butterflies. Observations of butterflies in plantations from five sites in the same region (Barlow et al. 2007) led us to hypothesize that (1) the presence of fruit feeding butterflies would be strongly linked to local patterns of richness and structure of the native of understorey vegetation in individual sites (and less so to landscape context), and (2) that butterfly-environment relationships would be highly specific to subfamilies.

Methods

Study site

The project was conducted within the 1.7 million hectare Jari landholding located on the border between the States of Pará and Amapá in north-eastern Brazilian Amazonia (00°27′00″–01°30′00″ S, 51°40′00″–5°20′00″ W). The area was purchased in 1968 for cellulose pulp production, and held c. 53,000 ha of *Eucalyptus* plantations at the time of study. These plantations are embedded in a largely undisturbed primary forest matrix (>1 million hectares). All sample sites had similar stocking densities (c. 900–1100 trees ha^{-1}) and no thinning had occurred. However, the understorey native vegetation is periodically suppressed, typically at 0, 1 and 3 years of age, either by labour-intensive manual removal or herbicidal treatment (Glyphosate and Isoxaflutole). The species composition of the native understorey was variable despite this clearing: Species-poor sites tended to contain only one or two species of annuals (typically from the families Asteraceae, Rubiaceae, Piperaceae, Poaceae or Cyperaceae), while species-rich sites contained many species of lianas (including *Davilla* spp., Dilleniaceae) and small pioneer trees such as *Vismia* spp. (Clusiacaeae), *Cecropia* spp. (Cecropiaceae), *Mabea taquari* and *Aparisthmium cordatum* (Euphorbiaceae). The plantations are managed in short-cycles and stands are clear-felled every 5–7 years.

Butterfly sampling

Fruit-feeding butterflies were trapped at 30 spatially independent *Eucalyptus* plantation sites) using cylindrical VanSomeren-Rydon traps (Rydon 1964; DeVries et al. 1997), baited with a standard mixture of mashed and fermented banana. The mosquito netting capture cylinder was 90 cm in height, minimising the risk of escape once butterflies had entered. Four traps were placed in the understorey of each site, spaced 100 m apart along a 300 m long transect. The baited trays were suspended 50 cm from the ground. Traps were placed in the forest in the morning (0900–1000 h) and checked the following day between 1500 and 1600 h. All sampling was undertaken during mostly-dry days in May 2005. Butterfly identification was carried out at the Museu Paraense Emílio Goeldi (MPEG) in Belém, Brazil, using reference collections and the plates and descriptions in D'Abrera (1988) and Neild (1996). Nomenclature follows Lamas (2004).

We classified butterflies as primary forest species if they had been recorded within five independent neighbouring primary forest sites spread out across the wider landscape. Each of these sites was sampled four times during the previous year, maximizing our knowledge of occupancy across the year (see Barlow et al. 2007). Each seasonal replicate ran eight understorey and eight canopy traps for five days at each site, totaling 1600 traps days of effort. All traps were located >500 m from the forest edge.

Environmental data

The structure and species richness of the native understorey vegetation was quantified in 20 m diameter circular plots at each trap location (n = 4 per site) using six variables that were recorded concurrently with butterfly sampling. Values from the four replicate plots per site were averaged to create a single site score. The variables measured were the percentage of vegetation cover on the ground, the mean vegetation height, maximum vegetation height, a score of understorey structure complexity (from 0–5, zero being *Eucalyptus* only, five being a highly developed understorey), a score of liana load on trees (from 0–5, with 0 = no lianas, and five = all *Eucalyptus* stems had lianas up to the sub-canopy), and a score of understorey vegetation richness (0–5), with 0 representing just one species, and 5 representing >25 species. We analysed the landscape composition and configuration of the study area with a vegetation and land-cover classification developed from a combination of land-use data from the landholding company and a supervised classification of a 2003 LandSat 7 (30 m pixel) satellite image. We calculated the percentage of primary forest within 1 km and 3 km buffer using a GIS (Arc-Info, Environmental Research Systems 1998).

We used the dominant height (the average total height of the 100 largest diameter, non-deformed trees per hectare) as our measure of stand productivity. Measurements of dominant height were made by Jari Celulose S.A. in the last 5 years, and all measurements were for *Eucalyptus urograndis* stands aged between 4 and 7.5 years. The relationship between *Eucalyptus* age at time of measurement and dominant height was weak (r = 0.3, P = 0.15), and most of the variance in dominant height (range 17–35 m) can be assumed to reflect local productivity rather than stand age.

Data reduction

We used Principal Components Analysis (PCA) to compose two composite variables for highly collinear independent variables (see Table 1). First, we constructed a composite variable to describe understorey structural complexity, composed of mean vegetation height, maximum vegetation height and the score of habitat structure complexity. The PCA loadings on the first factor were high for all these variables (0.96, 0.96 and 0.94, respectively), and the first factor explained 90.4% of the total variance. Second, we used a PCA to combine the percentage of primary forest within 1 km and 3 km buffers to create a score that helps describes the landscape composition around each sample site. The first factor explained 90.8% of the variance. For the purposes of analyses, the PCA score of understorey structure (+2 to remove negative numbers) and the liana load scores were $\log_{10}$ transformed to achieve approximate normality. Vegetation cover scores were Arcsine transformed.

Data analysis

All trap data was pooled within sites to maximize site-level representation. Species richness and the completeness of overall sampling were examined using sample-based

Table 1 Correlation matrix between environmental variables, with correlation coefficient (r_s) shown in bottom left, and *P*-values in top right. Variables used to create composite variables through PCA are shown by superscript characters

	Understorey Ht (Max)[a]	Understorey Ht (Ave)[a]	Habitat complexity[a]	Liana load	Vegetation cover (%)	Vegetation richness score	Stand age	Dominant height	Primary forest in 1 km[b]	Primary forest in 3 km[b]
Understorey Ht (Max)[a]		<0.00	<0.001	<0.001	0.003	<0.001	0.001	ns	ns	ns
Understorey Ht (Ave)[a]	0.92		<0.00	<0.001	0.001	<0.001	<0.001	ns	ns	ns
Habitat complexity[a]	0.93	0.88		<0.001	0.005	<0.001	<0.001	ns	ns	ns
Liana load	0.66	0.64	0.75		0.003	0.001	0.017	ns	ns	ns
Vegetation cover (%)	0.53	0.59	0.50	0.52		0.010	<0.001	ns	ns	ns
Vegetation richness score	0.72	0.75	0.81	0.59	0.46		0.005	ns	ns	ns
Stand age	0.65	0.59	0.60	0.43	0.78	0.50		ns	ns	ns
Dominant height	0.18	0.17	0.17	0.00	0.23	0.17	0.22		ns	ns
Primary forest in 1 km[b]	0.04	–0.05	0.07	0.01	0.06	0.26	0.10	0.25		<0.001
Primary forest in 3 km[b]	0.15	0.11	0.20	0.12	0.14	0.22	0.25	0.13	0.78	

[a] Combined by PCA to create understorey structure variable

[b] Combined by PCA to create a distance to primary forest variable

rarefaction curves constructed using EstimateS v.7 (Colwell 2004). Sample-based rarefaction curves were also repeated for the four most abundant subfamilies (Satyrinae, Biblidinae, Nymphinae and Charaxinae). Total estimated species richness was calculated for all sites together and for each individual site using the mean of the four commonly employed abundance-based estimators (ACE, Chao 1, Jack 1 & Bootstrap, see Colwell 2004). The mean of these estimators was used in order to minimize any bias from any particular estimator, the performance of which often varies according to differences in richness, sampling effort, and community evenness (O'Hara 2005). The similarity of butterfly species composition in a given stand in relation to that in neighbouring primary forest (using data from Barlow et al. 2007) was examined using the Jaccard similarity index to avoid the influence of sample effort in biasing species relative abundance distributions.

Like other faunal groups, the observed species richness of fruit-feeding butterflies is highly sensitive to sampling effort (e.g. DeVries et al. 1997, Molleman et al., 2006, Koh et al. 2007). We tested the validity of our results by using rarefaction to analyse data from 20 additional samples of butterfly fauna in *Eucalyptus* (4 seasonal replicates conducted at 5 sites) which had used 10 times the sample effort employed in the current study (40 trap days per sample; see Barlow et al. 2007). Although four trap days only captured 37% of the number of species sampled over a longer period, the results from four trap days were a very good predictor of the pattern of observed richness across sites when sampled with the full 40 trap days of effort ($F_{1,18} = 63$, $r^2 = 0.78$, $P < 0.001$). We are therefore confident that the results of this short-term study are a realistic representation of the patterns of fruit-feeding butterfly richness and composition within these plantations at the time of year we sampled.

Linking butterfly data with environmental variables

We were interested in revealing the most likely causal factors from within our candidate set of explanatory variables. Traditional model selection techniques often fail in this task due to high levels of multicollinearity among explanatory variables (Graham 2003), resulting in a spurious understanding of the nature of particular species-environment relationships (Mac Nally 2000). To minimize the influence of multi-collinearity among related explanatory variables we used hierarchical partitioning (Chevan and Sutherland 1991) to examine the independent effects of the seven key environmental variables (understorey vegetation richness, understorey vegetation structure, liana load on trees, vegetation cover on ground, landscape configuration, dominant height [plantation productivity], and *Eucalyptus* age) on six dependent variables of interest; observed butterfly richness, abundance, diversity (Simpson's diversity), the similarity of species composition to primary forest of all fruit-feeding butterflies (Jaccard's index), and the richness and abundance of known primary forest species. In addition, we examined the influence of the same explanatory variables on the abundance of the four dominant subfamilies, and the 13 most abundant species (i.e. all species with more captures than sample sites, $n = 30$). Hierarchical partitioning is a regression technique in which all possible linear models are jointly considered in an attempt to identify the most likely causal factors, providing a measure of the effect of each variable that is largely independent from that of other variables (Chevan and Sutherland 1991, Mac Nally 2000). Patterns of species abundance were modelled using Poisson errors and a goodness of fit based on r-square. The significance of independent effects was calculated using Mac Nally's (2002) randomization test

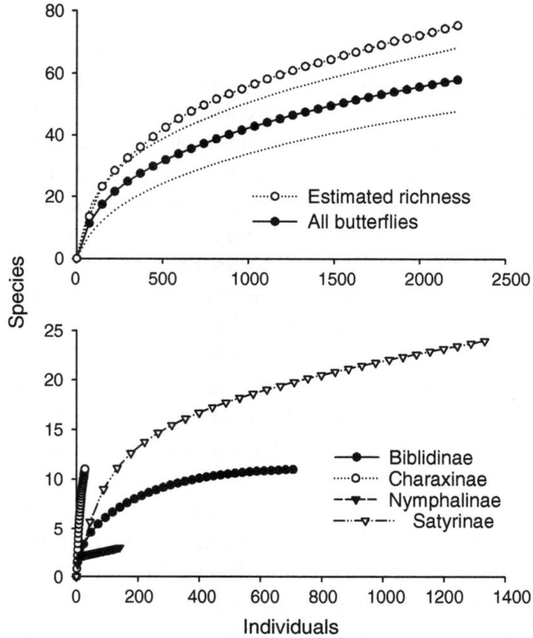

Fig. 1 (**a**) Sample-based rarefaction curves for fruit-feeding butterflies in *Eucalyptus* plantations (Black symbols). Dotted lines denote 95% confidence intervals. Clear symbols lines show the mean estimated richness from four different richness estimators (ACE, Chao 1, Jacknife & Bootstrap). (**b**) Sample-based rarefaction curves for the four most abundant subfamilies

Table 2 Correlations between the abundance of the four most abundant butterfly subfamilies, with correlation coefficient (r_s) shown in bottom left, and *P*-values in top right. $n = 30$ for all correlations

	Biblidinae	Charaxinae	Nympalinae	Satyrinaae
Biblidinae		0.34	0.36	0.35
Charaxinae	–0.18		0.19	0.61
Nympalinae	0.18	0.25		<0.001
Satyrinae	0.18	0.10	0.67	

with 1000 iterations. Hierarchical partitioning and associated randomization tests was implemented using the hier.part package freely available in the R statistical program (http://www.r-project.org). Finally, because hierarchical partitioning only partitions the variance explained by selected predictor variables, we also calculated a measure of overall model fit for each species, based on the explained deviance (R^2_{dev}) of a General Linear Model (Mac Nally 2002).

Results

We captured a total of 2200 butterflies and 56 species at the 30 sampled *Eucalyptus* plantation sites (Fig. 1). Capture success was highly variable across sites, ranging from as few as six to as many as 350 individual butterflies. Capture success was also unevenly distributed across species: the five and 10 most abundant species accounted for 62% and 85% of all captures respectively. *Hamadryas feronia* (Biblidinae) was the only species present at all sites and the second most abundant overall (14% of total captures). Sample rarefaction curves for individual subfamilies were highly variable, increasing very rapidly for the Charaxinae, while the Biblidinae appeared to have reached their asymptote (Fig. 1). Of six pairwise correlations between the abundance of individual subfamilies, only that between the Nymphalinae and Satyrinae was significant ($P = 0.05$, Table 2).

Hierarchical partitioning revealed a strong independent effect of understorey vegetation richness on the abundance, richness and species composition of fruit-feeding butterflies captured in plantations, and on the richness and abundance of the primary forest species (Fig. 2). Liana load was the only other significant environmental variable (Fig. 2), and it had a significant effect on butterfly species richness. None of the environmental variables had a significant effect on species diversity. Analysing results separately by subfamily revealed that the results for all fruit-feeding butterflies were strongly influenced by the Satyrinae, which was the most abundant subfamily (Fig. 1) and the only subfamily that was significantly influenced by vegetation richness (Fig. 2). No other subfamily appeared to be significantly influenced by any of the environmental variables we recorded, and although the dominant height and the landscape configuration appeared to have much stronger effects on the abundance of the Charaxinae than the other variables, the overall model fit was rather low (Fig. 2).

The analysis of the 13 most abundant species (with >30 captures in total) revealed a fairly consistent influence of vegetation richness on species from within the Satyrinae (vegetation richness had a significant effect for six of the eight species; Table 3), although the percentage ground cover had a stronger effect than vegetation richness for *Yphthimoides renata*, plantation age and liana load had significant effects on *Cissia terrestris*, while none of the variables had significant effects on *Magneuptychia libye* (Table 3). None of the measured environmental variables had a strong effect on the abundance of the three most abundant species of *Hamadryas* examined, reflecting the pattern for the Biblidinae as a whole. In addition, none of the measured variables had strong effects on the most abundant species of Nymphalinae (*Colobura dirce*), but plantation productivity (dominant height) had a strong and negative effect on the abundance of *Historis odius* (Nymphalinae).

Discussion

This short-term study of fruit-feeding butterflies in *Eucalyptus* plantations in the Brazilian Amazon supported our *a priori* hypotheses that (1) local stand-level vegetation structure and compositional factors would be more important than landscape context for fruit-feeding butterflies, and (2) butterfly–environment relationships would be highly specific to subfamilies. We examine the strength and validity of these butterfly-environment relationships focusing on vegetation richness, highlight future research priorities, and discuss the wider conservation implications of this study.

Butterfly-environment relationships

Overall, we found very few species-environment relationships that did not involve vegetation richness (Fig. 2, Table 3). Although butterfly species richness has been related to vegetation richness at large spatial scales (Thomas and Mallorie 1985; Kerr et al. 2001), more detailed statistical analyses suggest that such positive relationships may only be correlative, with both groups responding to similar environmental factors (Hawkins and Porter 2003). The link between vegetation richness and butterfly richness at local scales has stronger support: Gilbert and Smiley (1978) found a positive relationship between the

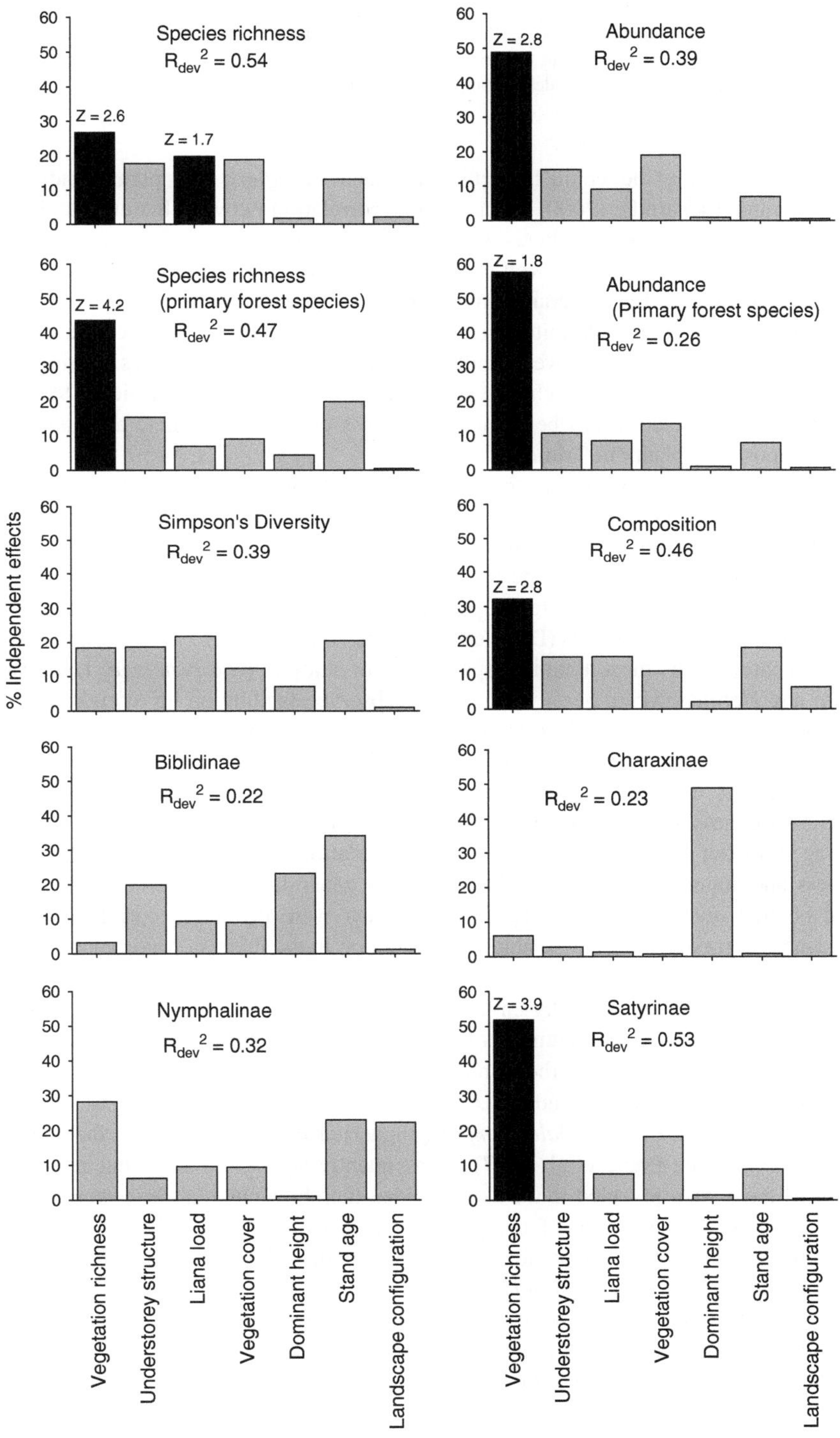
% Independent effects
Species richness
R_{dev}^2 = 0.54
Z = 2.6
Z = 1.7
Abundance
R_{dev}^2 = 0.39
Z = 2.8
Species richness
(primary forest species)
R_{dev}^2 = 0.47
Z = 4.2
Abundance
(Primary forest species)
R_{dev}^2 = 0.26
Z = 1.8
Simpson's Diversity
R_{dev}^2 = 0.39
Composition
R_{dev}^2 = 0.46
Z = 2.8
Biblidinae
R_{dev}^2 = 0.22
Charaxinae
R_{dev}^2 = 0.23
Nymphalinae
R_{dev}^2 = 0.32
Satyrinae
R_{dev}^2 = 0.53
Z = 3.9
Vegetation richness
Understorey structure
Liana load
Vegetation cover
Dominant height
Stand age
Landscape configuration

◀**Fig. 2** Distribution of percentage independent effects of measured environmental variables on fruit-feeding butterflies in *Eucalyptus* plantation forests. Black bars represent significant effects ($P < 0.05$) as determined by randomization tests. All significant effects described positive relationships. R^2_{dev} is the total deviance explained by a generalized linear model encompassing all measured variables

number of species of heliconid butterflies and their *Passiflora* host plants, and Steffan-Dewenter and Tscharntke (2000) report a close correlation between butterfly and vegetation richness in European grasslands. Furthermore, host-plant specificity is a key correlate of extinction risk in butterflies (Koh et al. 2004), and the presence or absence of a small number of specific host plants could have a large influence on butterfly diversity. However, these relationships are not ubiquitous, and Schulze et al. (2004) and Veddeler et al. (2005) failed to find a relationship between understorey richness and the richness of fruit-feeding butterfly in a study of land-use change in Indonesia, and Singer and Ehrlich (1991) found no evidence of a relationship between the richness of forest Satyrinae and their monocotyledonous host plants in Trinidad.

Although our results appear to lend strong support for the relationship between vegetation richness and butterfly richness at local scales, there are some reasons to interpret these butterfly-environment relationships with caution. The positive effect of vegetation richness was only found within the Satyrinae, which is a subfamily composed of generalists whose larvae are able to feed upon many different species of grasses, sedges and other monocotyledonous plants (DeVries 1997, Singer and Ehrlich 1991). For example, the larvae of three of the most abundant species in our study (*Cissia penelope*, *Yphthimoides renata* and *Hermeuptychia hermes*) have been observed feeding on up to eight species of grass or sedge in fragmented forests of Trinidad, and consumed all grass or sedge species offered to then in the laboratory (Singer and Ehrlich 1991). With such low levels of host specificity, it is unsurprising that the same study failed to find a relationship between Satyrinae richness and the number of species of available host plants, and it is difficult to envisage how our coarse scale observations of vegetation richness would have affected the richness and abundance of a group dominated by generalists. An alternative explanation could be that our estimates of vegetation richness were positively correlated with the abundance of grasses and sedges (that were common in the understorey in many sites, but not recorded specifically).

The lack of significant butterfly-environment relationships for the other subfamilies may also relate to our coarse measurement of environmental variables in general. For example, the most abundant genera of the Biblidindae (*Hamadryas*) are common in studies of *cerrado* vegetation (Pinheiro and Ortiz 1992) and disturbed tropical forests, where they track the abundance of their *Dalechampia* (Euphorbiaceae) hostplants (Shahabuddin and Ponte 2005; Uehara-Prado et al. 2007). Consequently species-environment relationships may well have been revealed if we had recorded the abundance of *Dalechampia* (which was present in at least some of the plots), and possibly some of the other species of Euphorbiaceae that were often abundant in the understorey (such as *Mabea taquari*, *Aparisthmium cordatum*, and *Manihot* spp.).

The Nymphalinae were dominated by just two species (Appendix), both of which are known to feed upon the pioneer tree *Cecropia* spp. Although a previous study did not find a relationship between *Historis odius* and their *Cecropia* food plants (Shahabuddin and Ponte 2005), the strong negative effect of plantation productivity on *H. odius* suggests that the relationship between productivity and *Cecropia* abundance warrants further investigation.

Table 3 Distribution of percentage independent effects of measured environmental variables on patterns of abundance for 13 species of fruit-feeding butterflies in *Eucalyptus* plantation forests

Subfamily	Species	R^2_{dev}	Vegetation richness	Understorey structure	Liana load	Vegetation cover	Dominant height	Stand age	Landscape configuration
Satyrinae	*Cissia penelope*	0.46	**37.0**	11.0	14.4	17.6	1.7	17.9	0.4
Biblidinae	*Hamadryas feronia*	0.18	9.4	12.1	2.9	6.8	42.4	16.5	10.0
Satyrinae	*Paryphthimoides argulus*	0.38	**61.1**	7.1	10.4	12.6	1.8	4.8	2.1
Biblidinae	*Hamadryas februa*	0.30	2.8	21.9	4.8	10.9	16.0	37.8	5.8
Satyrinae	*Taygetis laches*	0.64	**31.5**	9.4	6.8	20.6	6.8	**21.5**	3.3
Biblidinae	*Hamadryas amphinome*	0.35	5.8	13.5	11.5	17.6	7.7	43.4	0.4
Satyrinae	*Paryphthimoides vestigiata*	0.69	**53.9**	6.2	12.1	4.0	2.8	13.7	7.1
Nymphalinae	*Colobura dirce*	0.42	19.2	5.3	4.7	8.5	10.5	34.8	17.1
Satyrinae	*Yphthimoides renata*	0.73	**25.1**	13.1	12.6	**31.7**	3.0	11.3	3.2
Satyrinae	*Hermeuptychia hermes*	0.59	**48.1**	7.5	6.1	8.6	6.1	20.9	2.6
Satyrinae	*Cissia terrestris*	0.83	6.6	7.1	**26.2**	13.5	6.0	**39.8**	0.7
Nymphalinae	*Historis odius*	0.64	3.5	6.0	8.4	9.0	**56.1**(–)	14.8	2.2
Satyrinae	*Magneuptychia libye*	0.18	7.8	2.5	10.4	52.2	6.2	9.9	11.0

Values in bold represent significant effects ($P < 0.05$) determined by randomization tests. The only significant effect that depicts a negative relationship is shown in parentheses (*Historis odius*). R^2_{dev} is the total deviance explained by a generalized linear model encompassing all variables

Finally, the ability of butterflies to utilize plantations may be linked to factors other than host-plant availability, include their flight morphology and ability to avoid avian predators (Chai and Srygley 1990). Our data provided limited support for the influence of landscape structure on the Charaxinae as the availability of primary forest in neighbouring buffer areas was strongly linked to their abundance (Fig. 2). The Charaxinae are typically composed of powerfully flying forest species (Neild 1996), and it is possible that sites closer to primary forest were more likely to capture transient butterflies moving across the matrix, or individuals that perceive plantation edges as natural forests gaps.

Disaggregating the responses of fruit-feeding Nymphalidae

Aggregating species' responses can mask patterns of change if groups or species respond in contrasting ways (Manning et al. 2004; Lindenmayer et al. 2005). Our results show that butterfly-environment relationships can be strongly affected by the aggregation of subfamilies that may exhibit distinct ecological responses to patterns of habitat change, supporting previous assessments on responses of butterfly genera (Uehara-Prado et al. 2007) and studies of change across larger spatial scales (Brown and Freitas 2000). They also suggest that studies that failed to reveal any significant relationships between patterns of Nymphalidae richness and local vegetation (e.g. Schulze et al. 2004; Veddeler et al. 2005) may have analysed their data at an inappropriate taxonomic level.

Conservation implications

Although commercial plantations are in no way a replacement for native primary forests (e.g. Barlow et al. 2007), the *Eucalyptus* plantations that we examined were far from the "biological deserts" they are often portrayed as in the literature (see Kanowski et al. 2005). This study highlights the potential importance of the native understorey vegetation for the abundance, richness and diversity of some species of fruit-feeding butterflies, supporting similar findings from other taxa in tree plantations elsewhere in the world (e.g. Curry 1991; Chey et al. 1997; Humphrey et al. 1999; Lindenmayer and Hobbs 2004).

Whilst discussions about patterns of species richness within the wider countryside (Daily 2001) are of limited relevance for conservation unless the species are of conservation value (e.g. Petit and Petit 2003), we found similar results whether we consider all species, or only those known to occur in neighbouring areas of primary forest. We show that the conservation value of these forests for some subfamilies of fruit-feeding butterflies can be maximized if plantation managers tolerate a species-rich native understorey, but acknowledge that a comprehensive understanding of wider butterfly-environment relationships requires a more detailed examination of vegetation structure and composition as well as species richness. Finally, this short-term study provides just a snapshot of the patterns of butterfly diversity within plantations, and much longer-term work is required to examine different taxa, and across different seasons, years, and successive silvicultural rotations. The rapid loss of primary forest habitats and the growth of plantations in many areas of the world underline the urgency with which this work needs to be undertaken.

Acknowledgements We thank the Brazilian Ministério de Ciências e Tecnologia (CNPq) and Ministério do Meio Ambiente (MMA-IBAMA) for permissions to conduct this research. We are very grateful to Grupo Orsa and the staff of Orsa Florestal and Jari Celulose S.A. in Monte Dourado, Brazil, for permission to work in their landholding, as well as logistical support throughout the duration of the project and for making the plantation dominant height data available to us. The project was funded by the UK Government Darwin Initiative, National Geographic Society, Conservation Food and Health Foundation and Conservation International. This is publication number 11 of the Land-Use Change and Amazonian Biodiversity project.

Appendix

Summary of captures. Determination of primary forest species is based on indepedent data from Barlow et al. (2007). The subfamilies Coliadinae (Pieridae), Ithomiinae, and Liminitidinae (Nymphalidae) are not fruit-feeding

Subfamily	Species	Present in primary forest	Occupancy (out of 30 sites)	Abundance
Biblidinae	*Biblis hyperia*	No	3	3
Biblidinae	*Catonephele acontius*	Yes	9	10
Biblidinae	*Dynamine arene*	No	3	4
Biblidinae	*Ectima thecla*	Yes	1	1
Biblidinae	*Hamadryas amphinome*	Yes	18	133
Biblidinae	*Hamadryas arinome*	Yes	2	2
Biblidinae	*Hamadryas februa*	Yes	24	227
Biblidinae	*Hamadryas feronia*	Yes	30	310
Biblidinae	*Hamadryas iphthime*	No	3	13
Biblidinae	*Nessaea obrina*	Yes	3	4
Brassolinae	*Catoblepia generosa*	Yes	1	1
Brassolinae	*Eryphanis automedon*	Yes	1	1
Brassolinae	*Opsiphanes invirae*	Yes	2	2
Charaxinae	*Archaeoprepona demophon*	Yes	3	4
Charaxinae	*Archaeoprepona demophoon*	Yes	2	2
Charaxinae	*Fountainea ryphea*	Yes	3	3
Charaxinae	*Hypna clytemnestra*	Yes	1	1
Charaxinae	*Memphis acidalia*	Yes	5	6
Charaxinae	*Memphis moruus*	Yes	1	1
Charaxinae	*Memphis oenomais*	Yes	1	1
Charaxinae	*Memphis vicinia*	No	3	5
Charaxinae	*Memphis xenocles*	No	1	1
Charaxinae	*Siderone galanthis*	Yes	1	2
Charaxinae	*Zaretis itys*	Yes	2	2
Coliadinae	*Eurema albula*	No	1	2
Coliadinae	*Eurema nise*	No	1	1
Coliadinae	*Phoebis sennae*	No	1	1
Ithomiinae	*Hypothyris euclea*	No	1	1
Limenitidinae	*Adelpha pollina*	No	1	2
Morphinae	*Morpho helenor*	Yes	1	1
Nymphalinae	*Anartia jatrophae*	No	1	1
Nymphalinae	*Colobura dirce*	Yes	18	97

Appendix continued

Subfamily	Species	Present in primary forest	Occupancy (out of 30 sites)	Abundance
Nymphalinae	*Historis odius*	Yes	17	43
Satyrinae	*Caenoptychia boulleti*	Yes	1	1
Satyrinae	*Caeruleuptychia scopulata*	No	1	1
Satyrinae	*Chloreuptychia agatha*	Yes	1	1
Satyrinae	*Cissia myncea*	Yes	5	8
Satyrinae	*Cissia penelope*	Yes	20	423
Satyrinae	*Cissia terrestris*	No	9	67
Satyrinae	*Erichthodes erichtho*	No	1	1
Satyrinae	*Hermeuptychia hermes*	Yes	17	91
Satyrinae	*Magneuptychia antonoe*	Yes	1	1
Satyrinae	*Magneuptychia libye*	Yes	14	36
Satyrinae	*Magneuptychia newtoni*	No	10	11
Satyrinae	*Magneuptychia tricolor*	Yes	1	1
Satyrinae	*Pareuptychia binocula*	Yes	1	1
Satyrinae	*Pareuptychia hesionides*	Yes	3	5
Satyrinae	*Paryphthimoides argulus*	No	12	288
Satyrinae	*Paryphthimoides numeria*	No	8	14
Satyrinae	*Paryphthimoides vestigiata*	Yes	10	102
Satyrinae	*Taygetis cleopatra*	Yes	3	11
Satyrinae	*Taygetis echo*	Yes	2	4
Satyrinae	*Taygetis kerea*	Yes	3	4
Satyrinae	*Taygetis laches*	Yes	21	140
Satyrinae	*Taygetis virgilia*	Yes	8	29
Satyrinae	*Yphthimoides renata*	Yes	18	92
Total				2200

References

Barlow J, Overal WL, Araujo IS, Gardner TA, Peres CA (2007) The value of primary, secondary and plantation forests for fruit-feeding butterflies in the Brazilian Amazon. J Appl Ecol 44(5):1001–1012

Brown C (2000) The Global Outlook for Present and Future Wood Supply from Forest Plantation. Working Paper No GFPOS/WP/03, FAO, Rome, Italy

Brown KS Jr. (1997) Diversity, disturbance, and sustainable use of Neotropical forests: insects as indicators for conservation monitoring. J Insect Conserv 1:25–42

Brown KS Jr., Freitas AVL (2000) Atlantic forest butterflies: Indicators for landscape conservation. Biotropica 32:934–956

Carnus JM, Parrotta J, Brockerhoff E, Arbez M, Jactel H, KremerA, Lamb D, O'Hara K, Walters B (2006) Planted forests and biodiversity. J Forestry 104:65–77

Chai P, Srygley RB (1990) Predation and the flight, morphology, and temperature of Neotropical rain-forest butterflies. Am Nat 135:748–765

Chey VK, Holloway JD, Speight MR (1997) Diversity of moths in forest plantations and natural forests in Sabah. Bull Entomol Res 87:371–385

Colwell RK (2004) Estimates: statistical estimation of species richness and shared species from samples. Version 7. User's guide and application published at: http://viceroy.eeb.uconn.edu/estimates (accessed in February 2007)

Curry GN (1991) The influence of proximity to plantation of diversity and abundance of bird species in an exotic pine plantation in north-eastern New South Wales. Wildl Res 18:299–314
D'Abrera BL (1988) Butterflies of the neotropical region, part V. Nymphalidae (conc) & Satyridae. Hill House, Victoria
Daily GC (2001) Ecological forecasts. Nature 411:245
DeVries PJ, Murray D, Lande R (1997) Species diversity in vertical, horizontal, and temporal dimensions of a fruit-feeding butterfly community in an Ecuadorian rainforest. Biol J Linn Soc 62:343–364
Evans J, Turnbull J (2004) Plantation forestry in the tropics. Oxford University Press, Oxford
Eycott AE, Watkinson AR, Dolman PM (2006) Ecological patterns of plant diversity in a plantation forest managed by clearfelling. J Appl Ecol 43:1160–1171
FAO (2005) Global Forest Resources Assessment 2005, FAO, Rome, Italy. Available at http://www.fao.org/forestry (accessed January 2007)
Fearnside PM (1998) Plantation forestry in Brazil: Projections to 2050. Biomass & Bioenergy 15:437–450
Fischer J, Lindenmayer DB, Manning AD (2006) Biodiversity, ecosystem function, and resilience: ten guiding principles for commodity production landscapes. Front Ecol Environ 4:80–86
Gilbert LE, Smiley JT (1978) Determinants of local diversity in phytophagous insects: host specialists in tropical environments. In: Diversity of Insect Faunas. Symposium of the Royal Entomological Society of London 9:89–104
Graham MH (2003) Confronting multicollinearity in ecological multiple regression. Ecology 84:2809–2815
Hamer KC, Hill JK (2000) Scale-dependent effects of habitat disturbance on species richness in tropical forests. Conserv Biol 14:1435–1440
Hartley M (2002) Rationale and methods for conserving biodiversity in plantation forests. For Ecol Manage 155:81–95
Hawkins BA, Porter EE (2003) Does herbivore diversity depend on plant diversity? The case of California butterflies. Am Nat 161:40–49
Humphrey JW, Hawes C, Peace AJ, Ferris-Kaan R, Jukes MR (1999) Relationships between insect diversity and habitat characteristics in plantation forests. For Ecol Manage 113:11–21
Kanowski J, Catterall CP, Wardell-Johnson GW (2005) Consequences of broadscale timber plantations for biodiversity in cleared rainforest landscapes of tropical and subtropical Australia. For Ecol Manage 208:359–372
Kerr JT (2001) Butterfly species richness patterns in Canada: Energy, heterogeneity, and the potential consequences of climate change. Conservation Ecology **5**(1): 10. Available at: http://www.consecol.org/vol5/iss1/art10/
Koh LP (2007) Impacts of land use change on South-east Asian forest butterflies: a review. J Appl Ecol 44(4):703–713
Lamas G (2004) Checklist: Part 4 A. Hesperioidea-Papilionoidea. Association for Tropical Lepidoptera. Scientific publishers, Gainsville, USA
Lindenmayer DB, Franklin JF (2002) Conserving biodiversity: a comprehensive multiscaled approach. Island Press, Washington
Lindenmayer DB, Cunningham RB, Fischer J (2005) Vegetation cover thresholds and species responses. Biol Conserv 124:311–316
Lindenmayer DB, Franklin JF, Fischer J (2006) General management principles and a checklist of strategies to guide forest biodiversity conservation. Biol Conserv 131:433–445
Lindenmayer DB, Hobbs RJ (2004) Fauna conservation in Australian plantation forests-a review. Biol Conserv 119:151–168
Mac Nally R (2000) Regression and model-building in conservation biology, biogeography and ecology: The distinction between and reconciliation of 'predictive' and 'explanatory' models. Biodivers Conserv 9:655–671
Mac Nally R (2002) Multiple regression and inference in ecology and conservation biology: further comments on identifying important predictor variables. Biodivers Conserv 11:1397–1401
Manning AD, Lindenmayer DB, Nix HA (2004) Continua and Umwelt: novel perspectives on viewing landscapes. Oikos 104:621–628
Molleman F, Kop A, Brakefield PM, DeVries PJ, Zwaan BS (2006) Vertical and temporal patterns of biodiversity of fruit-feeding butterflies in a tropical forest in Uganda. Biodivers Conserv 15:107–121
Neild AFE (1996) The butterflies of Venezuala. Part 1: Nymphalidae I (Limentidinae, Apaturinae, Charaxinae). Meridian Publications, London
O'Hara RB (2005) Species richness estimators: how many species can dance on the head of a pin. J Anim Ecol 74:375–386
Pacala S, Socolow R (2004) Stabilization wedges: Solving the climate problem for the next 50 years with current technologies. Science 305:968

Petit LJ, Petit DR (2003) Evaluating the importance of human-modified lands for Neotropical bird conservation. Conserv Biol 17:687–694

Ramos FA (2000) Nymphalid butterfly communities in an Amazonian forest fragment. J Res Lepidoptera 35:29–41

Rydon A (1964) Notes on the use of butterfly traps in East Africa. J Lepidopterists' Soc 18:51–58

Schulze CH, Waltert M, Kessler PJA, Pitopang R, Shahabuddin, D. Veddeler, M. Muhlenberg, S. R. Gradstein, C. Leuschner, I. Steffan-Dewenter, and T. Tscharntke. (2004) Biodiversity indicator groups of tropical land-use systems: comparing plants, birds, and insects. Ecol Appl 14:1321–1333

Shahabuddin G, Ponte CA (2005) Frugivorous butterfly species in tropical forest fragments: correlates of vulnerability to extinction. Biodiv Conserv 14:1137–1152

Singer MC, PR Ehrlich (1991) Host specialization of satyrine butterflies, and their responses to habitat fragmentation in Trinidad. J Res Lepidoptera 30:248–256

Steffan-Dewenter I, Tscharntke T (2000) Butterfly community structure in fragmented habitats. Ecol Lett 3:449–456

Stork NE, Srivastava DS, Watt AD, Larsen TB (2003) Butterfly diversity and silvicultural practice in lowland rainforests of Cameroon. Biodiv Conserv 12:387–410

Thomas CD, Mallorie HC (1985) Rarity, species richness and conservation - butterflies of the Atlas Mountains in Morocco. Biol Conserv 33:95–117

Uehara-Prado M, Brown KS, Freitas AVL (2007) Species richness, composition and abundance of fruit-feeding butterflies in the Brazilian Atlantic Forest: omparison between a fragmented and a continuous landscape. Glob Ecol Biogr 16:43–54

Veddeler D, Schulze CH, Steffan-Dewenter I, Buchori D, Tscharntke T (2005) The contribution of tropical secondary forest fragments to the conservation of fruit-feeding butterflies: effects of isolation and age. Biodiv Conserv 14:3577–3592

Yu CM (2004) Seqüestro florestal de carbono no Brasil: dimensões políticas, socioeconômicas e ecológicas. Annablume, Sao Paulo

Impact of landscape and corridor design on primates in a large-scale industrial tropical plantation landscape

Robert Nasi · Piia Koponen · John G. Poulsen · Melanie Buitenzorgy · W. Rusmantoro

Originally published in the journal Biodiversity and Conservation, Volume 17, No 5, 1105–1126.
DOI: 10.1007/s10531-007-9237-8

Abstract Tropical plantations are rapidly expanding as a source of industrial wood. In Indonesia, such large-scale industrial plantations are generally made of large mono-specific blocks interspersed with natural forest remnants. The extent and biodiversity value of these remnants vary as laws and regulations on their design and management are either unclear, without solid scientific basis or left to the interpretation of private companies responsible for the plantations. Our study area comprises of three *Acacia mangium* plantations, which have on average 18% of their total area set aside from production and conserved as natural forests. These remnant natural forests may, if appropriately designed and managed, be used to mitigate the negative impact of plantations on biodiversity by providing some degree of connectivity with and between remaining natural forest patches (such as the Tesso Nilo conservation area). We sampled natural vegetation in one and primate diversity in all three plantation sector and examined patterns of primate species richness and abundance with relation to spatial arrangement and dimensions of conservation area, which has been set aside from plantation production. We demonstrate unambiguously the critical importance of a well-connected network of natural forest corridors in the plantation landscape to maintain primates and discuss the potential biodiversity value of natural forest remnants in broad-scale industrial landscapes.

Keywords Tropical plantations · Landscape structure · Landscape level management · Biodiversity · Connectivity · Fragmentation · Primates · *Acacia mangium*

R. Nasi · P. Koponen (✉) · J. G. Poulsen
CIFOR (Centre for International Forestry Research), Bogor Barat, 16680, Indonesia
e-mail: p.koponen@cgiar.org

R. Nasi
CIRAD, Montpellier, France

M. Buitenzorgy
University of Wageningen, Wageningen, The Netherlands

W. Rusmantoro
Lantana, Bogor, Indonesia

Introduction

Between 1985 and 1997, the rate of deforestation in Indonesia doubled and 20 million ha of natural forest was lost, including 61% of lowland forests in Sumatra (FWI/GFW 2002 and references cited therein). Experts have predicted that if these rates continue, intact tropical lowland forests in most of Indonesia may be completely lost by 2012 (Holmes 2002). Estimations of the extent of forest cover change in Indonesia vary due to the difficulties in measurements and inconsistent definitions (FWI/GFW 2002; Sunderlin 1999). Nevertheless fast-growing tropical wood plantations are rapidly expanding as a source of industrial wood for fibre and pulp (Cossalter and Pye-Smith 2003). This trend towards plantation forestry over the past two decades has occurred largely through the conversion of natural forests (Barr 2001). Existing guidelines for industrial plantations regarding the ecological sustainability have not adequately responded to this rapid development and should be improved to cover all key factors of biodiversity (Marjokorpi and Salo 2006). According to Marjokorpi and Salo (2006) threats due to fragmentation are considered in a general manner in most of the guidelines and wildlife corridors are perceived as a measure to mitigate their effects. The Center for International Forestry Research (CIFOR) has linked criteria and indicators to the code of practice for industrial tropical tree plantations (Poulsen et al. 2001), but broad implementation is still lacking.

At present more than 2 million ha across Indonesia are under fast growing plantations and in total 9 million ha are targeted for plantation development (FWI/GFW 2002). In Indonesia, plantations are generally made of large mono-specific blocks interspersed with natural forest remnants. The extent and value of these forest remnants vary as laws and regulations are unclear, lack solid scientific basis or are left to the interpretation of the plantation company (Cossalter and Pye-Smith 2003).

Central Sumatra belongs to the Sundaland biodiversity hotspot and third of its mammal species are endemic (Whitten et al. 2000), but less than 8% of the primary vegetation remains (Myers et al. 2000) and fragmentation has dramatically changed the habitats (FWI/GFW 2002). In Riau, the Tesso Nilo natural forest complex covers almost 200,000 ha, representing one of the largest intact rain forests remaining on the island of Sumatra and one of the most biologically diverse forests on the earth (Gillison 2001). The lowland tropical rain forest complex is surrounded by a plantation landscape, which consists of a mosaic of forest patches under various intensities of management and patches subject to diverse human activities. *Plantation stands* are managed for maximum productivity and typically have little scope for change to alternative within-stand management regimes and techniques. *Conservation areas* within plantation concessions are typically either riparian forests left primarily to protect water courses or are larger areas called Kawasan pelestarian plasma nutfah (KPPN), which are particularly left for biodiversity conservation. Tesso Nilo forest complex borders many Riau Andalan Pulp and Paper's (RAPP) concessions (only the Eastern part of Tesso Nilo area has national park status). Therefore, the conservation area inside the plantation, may be of high importance for biodiversity. In this context we assume that one sustainable landscape management aim, additional to the maximum production of industrial wood, is the persistence of fragmentation sensitive plants and animals. Fragmentation of forest cover has profound ecological significance and is the subject of considerable technical literature. In brief, small populations in fragmented or heavily harvested landscapes run much greater risks of reduced reproduction, genetic deterioration and extinction (Nason and Hamrick 1997). Furthermore, forest fragments are especially vulnerable to fire (Buechner and Dawkins 1961; Nepstad et al. 1999) invasion by weedy species and other processes of habitat erosion (Gascon et al. 2000; Laurance et al. 1997; Laurance and Williamson 2001). However, depending on species' ecology and behaviour, the impact of fragmentation on any given assemblage usually remains hard to assess. Some species are edge specialists, or benefit from an increased diversity of habitats while others may not even cross open ground or approach a forest edge (Chapman et al. 2006; Costa et al. 2005; Gonzalez-Solis et al. 2001; Meijaard et al. 2005; Newmark 1991). Primates were chosen for this study because although their ecology and response to forest disturbance (such as selective logging) is fairly well known, there is relatively poor literature on impacts of habitat fragmentation and loss at landscape level (Baranga 2004; Chapman and Onderdonk 1998; Laidlaw 2000; Meijaard et al. 2005).

Natural forest remnants inside plantations may, if appropriately designed, be used to mitigate the negative impact of large-scale industrial plantations on biodiversity by linking remaining natural forest patches in structurally diverse ways and at various scales. However, this will depend on the organism considered (Forman 1995; MacDonald 2003 and references in there). Thus, natural forest areas set aside from production may contribute as corridors to maintain ecological integrity and resilience of the plantation landscape (Laidlaw 2000; Turner and Corlett 1996). In general, corridors may function as habitat and provide shelter, nesting sites, refuge for biota, but in the strictest sense particularly act as connections for organism movement between habitat patches (Beier and Noss 1998). Therefore, they may allow a species in a single habitat to be saved from, or to re-colonize after, local extinction. Corridors may also permit migratory species to move between seasonal habitats in areas where fragmentation of habitat has jeopardized their movement (Forman 1995; Lindenmayer and Franklin 2002).

To assess the impact of large-scale industrial plantations, which are dominated by fast growing species on primates, we examined patterns of primate species richness and abundance in relation to spatial arrangement and dimensions, respectively, of areas set aside from plantation production (particularly riparian forests, potentially called corridors), in three large-scale industrial *Acacia mangium* plantation landscapes in Riau, Central Sumatra. We tested the null hypotheses that presence of primates are independent of patch connectivity, width, distance to roads, crown closure, age of neighbouring plantation stands and height, respectively, for patches set aside from plantation.

This study was part of a project called Biodiversity in plantations, which was managed jointly by CIFOR and Riau Andalan Pulp and Paper (RAPP). The project was designed to monitor and evaluate the ecological, environmental and social conditions of tropical large-scale industrial plantations, and to explore opportunities for influencing their overall landscape management. One of the main aims was to find ways to balance the goals of sustainable production with conservation of biodiversity and maintenance of environmental and social services in the landscape. In particular, this study explored the role and potential of riparian forests as wildlife corridors.

Materials and methods

Study area and management practices

The study was conducted within the RAPP concession area in Riau province (Central-East Sumatra, Indonesia) (Fig. 1a). Three large *Acacia mangium* plantation sectors Baserah (27,580 ha), Teso East (18,496 ha), and Teso West (2,0391 ha) were selected for the study (Fig. 1), because they have a range of conservation areas set aside from production (natural riparian forests and various sized protected areas called KPPN) as well as plantation stands

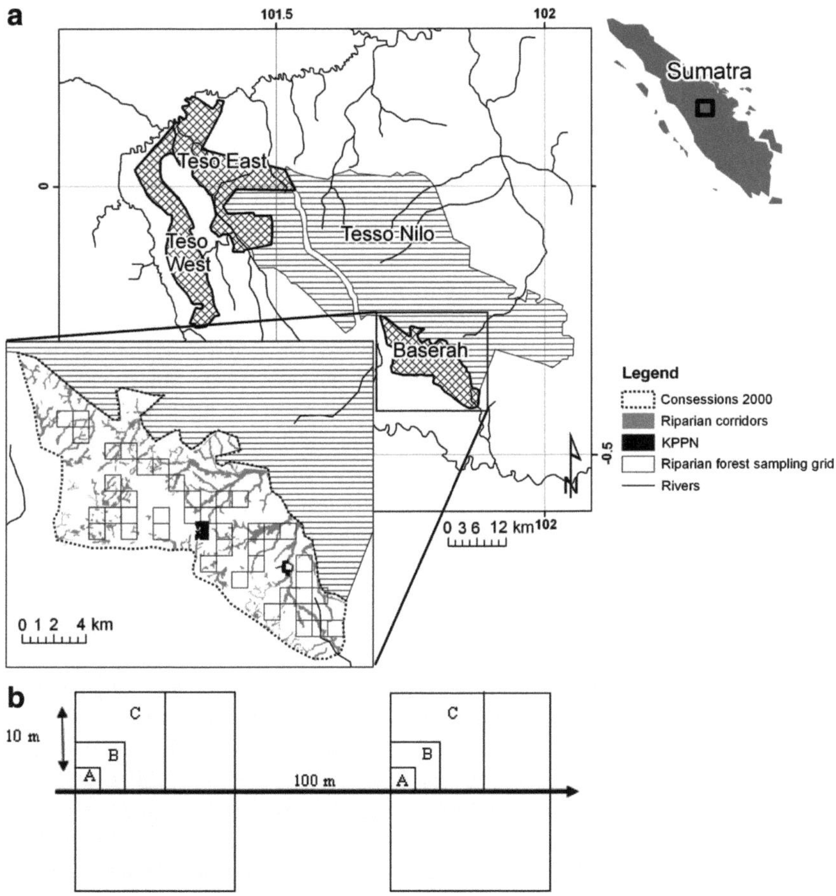

Fig. 1 (**a**) Map of three surveyed plantation sectors in Riau, Sumatra; (**b**) Floristic inventory of the riparian forests in Baserah sector followed a transect on one side of the river bank. Along the transect plots were established for trees (largest plot) and A: seedlings, B: saplings, C: poles

of various age (Table 1). Two of the sectors (Baserah and Teso East) are located adjacent to the Tesso Nilo forest complex making it important to consider landscape connectivity and biodiversity persistence in conservation areas outside Tesso Nilo. All three sectors have been established at a similar time and separate compartments inside the sectors were planted between 1993 and 1999 (thus, the age of those compartments varied from 1 to 77 months in 2000).

Legislation for sustaining ecology in the establishment of plantations is fragmented and not well connected with the land use planning process (M. Stuewe, A. Nawir personal communication). The area studied was previously called HTI (Hutan tanaman industri) concession, but the term HTI has changed to simpler "*hutan tanaman*" or plantation forest. Presently, Indonesian law allows 70% of the concession area to be planted with the main timber or pulp species. (H. Witono personal communication). The remaining 30% of non-planted area is divided between conservation, local use and infrastructure, 10% of the concession area should be allocated for conservation either being natural riparian forests or as KPPN (protected forest). In addition 10 % should be allocated for "*tanaman unggulan*",

Table 1 Land use within the three sectors (Baserah, Teso East and Teso West) of RAPP concession area in Riau, Central Sumatra in 2001

Sector	Conserved natural forest[a]	Planted (indigenous plywood)	Unplanted[b]	Planted (*Acacia mangium*)	Inoperable	Infrastructure	Total
Baserah (ha)	3,281	0	1,562	12,030	10,274	433	27,580
%	12	0	6	44	37	3	100
Teso East (ha)	4,845	0	984	10,549	2,249	319	18,946
%	26	0	5	56	12	2	100
Teso West (ha)	3,357	3,054	112	9,578	3,958	332	20,391
%	16	15	1	47	19	2	100

[a] Conserved natural forest: riparian forests and protected areas called Kawasan pelestarian plasma nutfah (KPPN)

[b] Unplanted area, may be lowland rain forest of variable quality (over logged); Inoperable: local rubber or oil palm plantations but mostly degraded natural vegetation (often shrubland) , which is claimed by local people

which are natural forests enriched with local high value timber species. A further 5% should be allocated for both infrastructure and "*tanaman kehidupan*", (plants supporting livelihoods) such as timber, fruit trees or food crops of local community interest (H. Witono personal communication Ministerial degree 70, 1995). Requirements for KPPN areas are slightly unclear. There is an absolute legal requirement to conserve all plants and animals with protected status, but the minimum sufficient area is not mentioned. Cultural sites (burial grounds, places of worship) and special feature sites (caves, karst landscape and hot springs) must also be protected. Rules for the protected riparian forests within plantation have a basis in UU 41, 1999 article 50, which stipulates the dimensions for set aside forests on rivers, dams and lake banks. Before this law rules varied among districts and the current law's role has not been well established or consistently promoted. Thus, timber concessionaires have been left without rigorous guidance, clear translation of rules into day-to day management practices or strong incentives (A. Nawir personal comminication). For instance in Kampar district, the required riparian forest buffer width previously depended on river width (as required by Presidential degree 32 of 1990, article 16); in Indra Giri Hulu region it depended on river branching (from main river until tertiary river branch, Regional decree, KPTS/368/XI/1998). When RAPP established plantations in 1993, its interpretation was that rivers less than 3 m wide needed no riparian forest buffer; those more than 3 m wide needed at least 25 m on each bank (nearly all riparian buffers in RAPP three study sectors fall into this category); any 50 m wide rivers needed 250 m buffer zone including the actual river. The original vegetation before planting or conversion was mostly low land rain forest, which was in some areas selectively logged, particularly for commercial Dipterocarp species. In southern parts of Baserah sector there were also over-logged or degraded. Areas intended for plantation stands were subsequently clear-cut leaving primarily riparian forests as buffer strips and some larger KPPN areas untouched. All individuals of tree species of special concern, to biodiversity conservation or use by local communities were theoretically left standing, even if in otherwise clear-cut areas. Consequently the extent of conserved natural forest is the result of the interpretations by plantation managers of various regulations and represents between 12% and 26% of individual sector areas. According to Indonesian regulations, vegetation on slopes over 40% must be conserved but slope length is not specified and RAPP uses 500 m without objection from the Forestry Department (C. Munoz personal comminication).

Table 2 The number of 100 ha cells available and selected, respectively, for primate surveys in each three sector at RAPP, Riau Province, Sumatra

Sector	Percentage of riparian forest in the cell (%)	Number of all available cells			Number of selected cells		
		Simple[a]	Complex[b]	Total	Simple[a]	Complex[b]	Total
Baserah	0	–	–	26	–	–	8
	10	24	6	30	6	2	8
	20	24	10	34	6	2	8
	30	5	27	32	3	5	8
	40	1	13	14	–	8	8
	Total cells	54	56	136	15	17	40
Teso East	0	–	–	4	–	–	4
	10	4	–	4	4	–	4
	20	6	11	17	4	–	4
	30	11	8	19	3	1	4
	40	3	8	11	–	4	4
	Total cells	24	27	55	11	5	20
Teso West	0	–	–	54	–	–	8
	10	13	6	19	5	3	8
	20	6	11	17	6	2	8
	30	9	12	21	2	6	8
	40	4	11	15	1	7	8
	Total cells	32	40	132	14	18	40
KPPN[c], Baserah	100	–	–	2			2

[a] Simple, Riparian forest is cutting through or present as one undivided line within the cell

[b] Complex, Riparian forest is present as a dividing or branching line or as multiple patches within the cell

[c] KPPN, Protected forest mainly for biodiversity conservation purposes, sampling in this cell only covers natural forest

Sampling design

Three sectors (Baserah, Teso East and Teso West) were each divided into a 100 ha (1 × 1 km) grid, based on maps provided by RAPP. Each grid cell (100 ha) was originally classified according to two main variables: (i) the proportion of land set aside from production (to the nearest 10%: 0%, 10%, 20%, 30% and 40%) and (ii) the spatial complexity of the landscape within a cell (simple or complex). A cell with a *simple* landscape contained a single, linear undivided riparian forest strip or patch. A *complex* landscape comprised of either several isolated patches and/or divided riparian forests (following more than one main river course). This resulted in a 4 × 2 landscape matrix of available cells and an additional control set consisting of cells with 100% of the area as plantation (Table 2). Within each plantation sector, cells were randomly selected (by using a random number generator) for each landscape pattern combination including cells with no conservation areas. A total of 100 cells were selected representing all possible combinations; 40 from Baserah and Teso West, 20 from Teso East. Amongst the eighty cells with riparian forests (10%, 20%, 30%, 40%), the number of cells with complex and simple landscape configurations were almost even (52.5% complex and 47.5% simple). This approach was used to maximize the sampled range of other landscape characteristics. Two existing KPPN areas inside Baserah sector were surveyed separately. Primates were surveyed twice at the sites described previously and in the second survey we added individual cells next to the Baserah sector, one to rubber plantation and one to Tesso Nilo natural forest.

We gathered 18 variables from each cell (listed in Table 3, except age and landscape complexity). Connectivity was recorded for each cell by noting from maps if riparian forests were connected to Tesso Nilo natural forest complex or KPPN (there were three separate KPPN areas in Teso East, one in Teso West and two in Baserah). If riparian forests were connected, the distance from each constituent cell to Tesso Nilo and KPPN were recorded both as (a) the most direct distance and (b) the distance through contiguous riparian forests. Riparian forest width (<50, 50–150, >150 m) was measured with tape in the centre of the riparian strip at five evenly spaced locations per kilometre and average crown closure (<10, 10–20, 20–30, 30–40, >40%) at 10 evenly spaced locations per kilometre. At the same time siltation (expressed as 'silted or not silted'), river width (<2, 2–5, >5 m), recent signs of disturbance (particularly illegal logging activities (indicated as present or not) were recorded. A subjective measure for forest quality was allocated on the basis of crown closure, presence of fruit trees and perceived general condition of the forest using a three point qualitative scale (3 = good forest condition with closed canopy, presence of fruit trees for feeding). Riparian forest length in total was identified on a six-point scale divided into 100 m intervals (six if length >600 m) and the length (to the closest 50 m) and number of roads (all types, irrespective of width) per cell and their intersections with riparian forests (intersected or not) was measured from the maps. The age of plantation stands (months) and their percentage cover of each cell was determined by superimposed a point grid on maps provided by RAPP. Landsat image (1992) was used to estimate which cells in Baserah sector had suffered before the plantation establishment from over-logging and had thus more an open canopy (open canopy or patchy forest; closed canopy).

Vegetation surveys

Due to the limited time and funding, vegetation was surveyed only in Baserah sector in each of the 32 cells, which contained riparian forests and in addition in two KPPN areas. Baserah seemed to be most promising in inclusion of variation in riparian forest connectivity, width and quality. One transect in the same direction with the river in each cell was made in the middle part of the riparian forest strip on one side of the river. Transects consisted of plots (20 × 20 m), which were 100 m apart, and each transect in a cell consisted of a minimum of nine plots (Fig. 1b).

A total of 347 plots were sampled in Baserah. In each plot trees with dbh (diameter at breast height; 1.3 m) >20 cm were identified in the 20 × 20 m area; poles (dbh 10–20 cm) within one systematically placed 10 × 10 m^2; saplings (>1.5 m high and dbh $\leq$ 10 cm) from one systematically placed 5 × 5 m^2 and seedlings ($\leq$1.5 m high and dbh < 10 cm) from one systematically placed 2 × 2 m^2 (Fig. 1b). Experts from the Bogor Herbarium were consulted to ensure correct species identification. Species nomenclature follows the Tree flora of Malaya (Whitmore 1972). The height of trees and poles was estimated and dbh was measured with a tape.

Primate surveys

Primates were surveyed with equal search effort throughout 100 cells (40 cells in Baserah, 20 in Teso East and 40 in Teso West) from the beginning of February until the end of April 2000. Surveys were conducted only during dry and calm weather, in the early morning (between 6 and 9 am) and late afternoon (between 4 and 7 pm) while walking along eight pre-defined and parallel, evenly spaced 1 km long transects (observation width using binoculars 50 m on both sides of the transect) laid across each cell. Surveys were conducted

Table 3 Overview of the differences between sampling cell (1 km^2) variables in surveyed three *Acacia mangium* plantation sectors (Baserah, Teso East and Teso West) in Riau, Sumatra based on Kruskall–Wallis (K–W) test or Phi-test

Variable	Measurement	Baserah	Teso East	Teso West	Test	Value	Significance[a]
Crown closure (1–5; 5: very open, >40 %)	Ordinal	2.6 (16% >30%)	3.8 (88% >30%)	4.5 (100% >30%)	K–W	46.43	0.00*
Riparian forest width (1–3; 3: widest, >150 m)	Ordinal	1.3 (81% <50 m)	1.6 (56% <50 m)	1.7 (44% <50 m)	K–W	6.95	0.03*
River width (1–3; 3 widest, >5 m)	Ordinal	2.3 (50% 2–5 m)	1.8 (46% 2–5 m)	1.9 (50% 2–5 m)	K–W	4.94	0.08
Riparian forest quality (1–3; 3: good)	Ordinal	1.5 (16% good)	2.3 (50% good)	2.3 (50% good)	K–W	13.02	0.00*
Riparian forest length (1–6; 6 longest)	Ordinal	2.9 (22% >400 m)	3.8 (75% >400 m)	4.5 (75% >400 m)	K–W	21.40	0.00*
Distance from KPPN through riparian forest (mean km)[b]	Ratio	2.3	2.07	2.03	K–W	2.65	0.27
Distance from KPPN (mean km, most direct)	Ratio	1.62	1.31	1.69	K–W	1.86	0.40
Distance from Tesso Nilo through riparian forest (mean km)[c]	Ratio	4.28	4.48	8.35	K–W	17.19	0.00*
Distance from Tesso Nilo (mean km, most direct)	Ratio	3.05	4.48	8.35	K–W	27.17	0.00*
Number of roads	Ratio	2.42	3.35	3.45	K–W	6.16	0.00
Length of roads (km)	Ratio	2.30	1.9	2.3	K–W	3.86	0.14
Riparian forests intersected by roads	Nominal	44%	88%	53%	Phi	0.14	0.45
Visual disturbance, illegal logging	Nominal	75%	88%	100%	Phi	0.34	0.01*
Siltation	Nominal	5%	10%	7.5%	Phi	0.08	0.76
Direct linkage with Tesso Nilo	Nominal	53%	75%	0%	Phi	0.54	0.00*
Direct linkage with KPPN	Nominal	84%	94%	100%	Phi	0.27	0.06

[a] Level of significance ($\alpha = 0.01$)

[b] KPPN = protected forest mainly for conserving biodiversity inside plantations

[c] Natural forest complex next to the plantation sectors

*The difference between sectors is significant

twice during the study period, and the average values were used for analyses. All surveys were done by the same two observers and any between two observer variability was continuously monitored and consistently found to be negligible (less than 2.5% on both species identification and count of number of individuals). Each observed primate was identified, its height above the ground was recorded and location noted on a detailed (1:1,000 scale) map. Surveys were conducted twice at each site during the study period, and the average values were used for analyses. An additional (below called second) survey on a sub sample of four common primate species (*Hylobates agillis*, *Presbytis melalophos*, *Macaca fascicularis* and *Macaca nemestrina*) was carried out following the same protocol between 10 and 26 February 2000. This second survey was conducted only in the Baserah sector where additional identification of feeding trees was made. In this second survey two additional cells were surveyed, one in a rubber plantation and one in north of Baserah, inside Tesso Nilo natural forest.

Data analyses

To analyse the difference between the three sectors with respect to all gathered independent variables, two kinds of tests were used. For ordinal or ratio variables, we used Kruskal–Wallis test, a nonparametric approximation for one-way ANOVA. For nominal variables, we used Phi-test. Phi is a χ^2 based association that involves dividing the χ^2 statistic by the sample size and taking the square root of the result (Ranta et al. 1989). We used Spearman ρ (Ranta et al. 1989) to analyse the differences between all cells using 18 variables (variables listed in Table 3 and we added landscape complexity and age). Primate analyses are based on number of primate individuals per species and species richness per cell and relate to the impacts of land cover and landscape structure on patterns of primate species distribution and composition. We applied Kruskal-Wallis test and Spearman ρ matrix to analyse correlation between landscape variables and presence of primates. Analyses of primate data were conducted using non-parametric procedures due to lack of normality, and because transformation to normality was not feasible. Due to the large size of the tables, significant results of Spearman ρ are mainly described in the text.

Importance value index (IVI; Mueller-Dombois and Ellenberg 1974) was calculated for each tree and pole species as a sum of relative frequency, relative density and relative basal area and for saplings and seedlings as a sum of relative frequency and relative density. IVI value for one species is a sum of tree, pole, sapling and seedling IVI values. A species with a high IVI occurs with a higher density; occupies more space and is distributed relatively more uniformly than a species with a low IVI. Species with high IVI are considered as structurally important.

Results

Overall landscape structure

Riparian forests (corridors)

Most riparian forests in the three sectors may be considered as corridors, which are connected through other riparian forests to either KPPN conservation area inside the plantation or remaining natural forest area outside sector boundaries (Tesso Nilo forest complex) or

Table 4 Correlation (Spearman's ρ) between some characteristics of riparian forests inside three *Acacia mangium* concession sectors in Riau, Sumatra

Characteristics of riparian forests in the cell	Quality	Canopy closure	Length	Covered area	Width
Quality	1	0.532*	0.331*	-0.096^{ns}	-0.216^{ns}
Canopy closure		1	0.522*	0.533*	0.062^{ns}
Length			1	0.497*	0.240**
Covered area				1	0.342*
Width					1

*Significant at the 0.001 level (2-tailed)

**Significant at the 0.01 level (2-tailed)

^{ns} Not significant

both (Table 3). The overall landscape connectivity seems theoretically ensured as only 16% of the cells with riparian forests in Baserah are totally unconnected and isolated and in other sectors connectivity is even better. On the other hand KPPN conservation areas are not designed specifically to be well connected to natural forests (in Baserah 40% of the cells with riparian forests were linked to both KPPN and Tesso Nilo, in Teso East only 10%), although the majority are connected to Tesso Nilo through riparian forests. The size of KPPN areas on land use maps is very variable. For instance, in Baserah one KPPN area is 132 ha, the other 32 ha. The latter is comparable in size with riparian forest patches. The average distances from riparian forests to KPPN are similar in all three sectors. A large proportion of corridors are less than 50 m wide with the Baserah sector having the highest proportion (81%) of these narrower corridors. Riparian forests of the Baserah sector are significantly shorter and more open (20–30%) and generally of lower quality than the ones in two Teso sectors (Table 3) reflecting the fact that the southernmost areas in Baserah were degraded or over-logged during the plantation establishment. *Acacia mangium* is typically harvested in Riau after it has reached 6–7 years, and the average age for all three sectors was 31 months. We did not find any differences in the impact of the age of surrounding plantation compartments to the riparian forests.

Because the analyses at the riparian forest level are related to the quality of forests, which is affected mainly by felling operations in the border of riparian forests, siltation and design of the landscape (dimensions of width, covered area, length) of individual riparian forest, we show only relevant parts of the Spearman ρ matrix (18 variables, data from 100 cells, total of 22 significant correlations from 171 at $\alpha = 0.01$ level) results. Riparian forest width, length, quality, crown closure and proportion in cell are positively correlated (Table 4) and thus the longest continuous riparian forests are generally the widest, have the highest crown closure and the best overall quality. Based on the full Spearman ρ matrix we concluded that landscape complexity and the average width of riparian forest within cells were consistently correlated with proportion of land set aside. Thus, the width of riparian forest increased with the proportion of riparian forest ($r = 0.349$, $P = 0.002$) and cells with complex landscape were characterized by larger proportions of land set aside ($r = 0.582$, $P < 0.001$).

Disturbance in the riparian forests

More than 60% of the riparian forests have been subject to some sort of disturbance, generally illegal logging activities (Table 3). Most disturbed riparian forests were isolated strips, i.e. not connected to Tesso Nilo or to KPPN areas ($r = -0.462$, $P < 0.001$), and occurred predominantly in older plantations ($r = 0.416$, $P < 0.001$). Mean canopy closure in Baserah sector was lower (0–50%) in previously unlogged or only moderately disturbed riparian forests than in previously over logged riparian forests (75–100%). This counter-intuitive result can be explained by two main factors (i) some set aside areas were not logged because originally they were very poorly stocked and with a degraded canopy, (ii) logging gaps are quickly colonized by low, dense canopy pioneer species like *Macaranga* species, increasing the canopy closure.

Roads

The length of roads per cell is similar for both Baserah and Teso West sectors and lower in Teso East (Table 3). Due to the cell selection (Table 2), the three sectors are not different with respect to landscape complexity, but when all cells with riparian forests ($n = 80$) are pooled, the length of roads is similar in complex and simple landscape cells (Kruskall–Wallis tests, $r = 0.176$, $P = 0.119$) irrespective of both of the proportion of land set aside per cell ($r = -0.272$, $P = 0.015$) and whether cells are connected to KPPN areas or Tesso Nilo (respectively $r = -0.175$, $P = 0.121$; $r = -0.051$, $P = 0.651$). The total length of roads was correlated with increasing age of plantation stands ($r = 0.220$, $P = 0.05$). This may either reflect that (a) cells with older plantation stands have invariably been under management (by the plantation company) for more years than those cells with younger plantation stands, or, (b) that road management has changed (i.e. has become more efficient) in recent years.

Primates

Species diversity and abundance

Eight species of primates were recorded during the two surveys, three gibbon species: *Hylobates agilis*, *H. lar*, *H. syndactilus*, two macaques: *Macaca fascicularis*, *M. nemestrina* and two langurs: *Presbytis femoralis*, *P. melalophos* and one unidentified species each of *Presbytis* and *Hylobates*. During the first survey, which covered all three sectors, 84 individuals belonging to seven species were recorded (*H. agilis*, *H. lar*, *H.* sp., *H. syndactilus*, *M. nemestrina*, *P. femoralis*, *P. melalophos* and unidentified *P.* sp.). Species-specific abundance and presence differed between sectors (Table 5) and the highest occurrence of individuals (38) was recorded from Baserah while far fewer were observed from Teso West (18). During the second survey which focused on four species (*H. agilis*, *P. melalophos*, *M. fascicularis*, *M. nemestrina*) and aimed mainly to better understand the behaviour of these species, 62 individuals were censused in corridors and an additional 56 individuals were counted in two KPPNs, natural forest (one site, nine individuals) and rubber plantation (one site, 28). The biggest groups (more than 20 individuals) were found for the two *Macaca* species though some solitary individuals were also recorded, *Presbytis* species were found either solitary or in small groups (3–9 individuals) and *Hylobates* were solitary or in pairs. *M. fascicularis* was mainly found in riparian forests while *M. nemestrina* was only recorded from KPPN areas (Table 6). Not more than three species were recorded at the same time in the same location and generally (62% of the observations) only one species was recorded per time in one location.

Table 5 Distribution of primate species across the three sectors (first survey) in Riau, Central Sumatra. Species conservation status according to IUCN is indicated below

Species	Number/ percentage	Baserah	Teso East	Teso West	All sectors
Hylobates agilis	N	–	2	4	6
(LR/nt 1994 IUCN)[a]	%	–	33.3	66.7	100
Hylobates sp.	N	3	–	–	3
(i.e. unidentified at sp. level)	%	100	–	–	100
Hylobates lar	N	8	–	4	12
(LR/nt 1994 IUCN)	%	66.3	–	33.3	100
Hylobates syndactilus	N	5	–	–	5
	%	100	–	–	100
Macaca nemestrina	N	4	1	–	5
(VU 1994 IUCN)[b]	%	80	20	–	100
Presbytis femoralis	N	18	16	7	41
(LR/nt 1994 IUCN)	%	43.9	39	17.1	100
Presbytis melalophos	N	–	9	–	9
(LR/nt 1994 IUCN)	%	–	100	–	100
Presbytis sp.	N	–	–	3	3
(i.e. unidentified at sp. level)	%			100	100
Number of individuals (species)		**38 (5)**	**28 (4)**	**18 (4)**	**84 (8)**
%		**45.2**	**33.3**	**21.4**	**100**

[a] LR/nt = Lower risk or near threatened

[b] VU = Vulnerable

Table 6 Distribution of four primate species (second survey) in Baserah sector and in adjacent rubber plantation and natural forest of Tesso Nilo in Riau, Central Sumatra

Habitat	Number/ percentage	*Hylobates agilis*	*Macaca fascicularis*	*Macaca nemestrina*	*Presbytis melalophos*	All species
Riparian forests	N	23	23	–	16	62
	%	37.1	37.1	–	25.8	100
KPPN	N	4	8	25	19	56
	%	7.1	14.3	44.6	33.9	100
Tesso Nilo	N	–	–	–	9	9
	%	–	–	–	100	100
Rubber plantation	N	–	28	–	–	28
	%	–	100	–	–	100
All habitats	N	**27**	**59**	**25**	**44**	**155**
	%	**17.4**	**38.1**	**16.1**	**28.4**	**100**

Habitat of primates

Primates were detected in 30 out of the 100 sampled cells (two surveys pooled): Baserah (16 cells, 40% of those surveyed in Baserah), Teso East (6, 30%), Teso West (8, 20%). Primates were never found in planted areas except for *M. fascicularis*, which was also found in rubber and *Acacia* plantations (the latter observed during the first survey). Primates were only detected in riparian forests linked to either KPPN or Tesso Nilo. However a large proportion (45%) of cells with connected riparian forests had no primates at the time of survey. Based on K–W test, presence of riparian forest, proportion of set aside area inside the cell, river width, complexity of the structure of riparian forest and level of disturbance (includes number and length of roads and riparian forest quality) have an significant impact on the richness and

Table 7 Impact of different factors (all cells, $n = 100$, pooled) on primates' species richness and number of individuals (density) in Riau, Sumatra based on Kruskal–Wallis test

Cell factors	Number of primate individuals			Primate species richness		
	χ^2	df[a]	Level of significance	χ^2	df[a]	Level of significance
Riparian forest present	8.04	1	0.01*	8.11	1	0.01*
Proportion of riparian forests	6.34	3	0.1*	6.65	3	0.08*
Landscape complexity	2.22	1	0.14*	3.07	1	0.08*
Quality of riparian forests	2.7	2	0.26	2.32	2	0.31
Riparian forest width	3.45	2	0.18	3.22	2	0.20
River width	5.74	2	0.06	5.22	2	0.07
Crown closure	9.65	4	0.05*	10.64	4	0.03*
Level of disturbance	3.27	1	0.07*	3.79	1	0.05*
Existing siltation	0.72	1	0.4	0.67	1	0.42
Riparian forests intersected by roads	0.09	1	0.77	0.17	1	0.68
Linkage with natural forest	2.93	3	0.40	2.94	3	0.40

[a] df = Degrees of freedom

*Significant at the level $\alpha = 0.1$

abundance of primates when all species are pooled (Table 7). Primates were most frequently observed in cells with 20–30% of area as corridors, riparian forests along wider rivers with less disturbance but still being more common in open than closed riparian forests.

Vegetation

Diversity and abundance

A total of 347 plots were measured giving a sampled area of 0.14, 0.9, 3.5 and 13.9 ha for seedlings, saplings, poles and trees respectively (Table 8). More than 17,900 individuals belonging to around 250 species (all size classes in KPPN and riparian forest areas pooled) were recorded during the surveys, ranging from the relatively ubiquitous *Knema laurina* (Myristicaceae), *Syzygium fastigiatum* (Myrtaceae), and *Helicia excelsa* (Proteaceae) (Table 8) to the rare and local *Aquilaria malaccensis* (Thymelaeaceae), *Artocarpus dadah* (Moraceae), and *Milettia atropurpurea* (Leguminosae). The average stand basal area for poles and trees was 14.3 m^2 ha^{-1} (Table 8), which is a low but not unusual for Dipterocarp lowland tropical forests. From the 30 most important plant species in Baserah (importance index, IVI, trees, poles, saplings and seedlings pooled in Table 9) only six species from riparian forests (Table 9a) were shared with the 30 most important species from the KPPN area (Table 9b).

Food availability for primates

The four species from second survey were all recorded eating fruits and leaves. Most of the plant species with highest IVI were species, that primates were observed to eat (leaves and fruits are not separated here), although in KPPN they were less commonly eaten (Table 9). In riparian forests only *Calophyllum lowii* was not observed as a food source for any of the four primate species. Primates observed in the area are mostly arboreal diurnal species and seemed to favour feeding on poles and trees, in the highest canopy level available (*P. melalophos*. mainly in average at 15 m, *M. nemestrina* at 18 m, *H. agilis* at 28 m and

Table 8 Overall vegetation inventory results from conservation areas (results are pooled from riparian forests and two protected KPPN forest areas) in Baserah concession sector in Riau, Sumatra

Size class	Species (number)	Individuals (number)	Mean density ($number^{-1}$ ha)	Mean basal area (m^{-2} ha)
Seedlings	144	7,855	55,326	–
Saplings	181	6,420	7,240	–
Poles	127	1,830	536	8.1
Trees	127	1,840	131	14.3
All size classes	238	17,932	–	22.4

M. fascicularis 12 m) but for individuals feeding was recorded at all levels (e.g. *M. fascicularis* from 0 to 25 m, *M. nemestrina* from 0 to 40 m). Besides the listed most important plant species in Table 9, the diet of the primates was very varied, e.g. *Macaca fascicularis* was observed to feed at least from 36 other tree species and *Hylobates agilis* from 22 other tree species.

Discussion

Conservation value of the plantation landscape

Even areas under intensive production may be valuable from a conservation biology perspective, particularly in the light of continuing discussion on "segregate or integrate" patterns at the landscape level (Bennet et al. 2006; Hartley 2002; Noordwijk et al. 1997). Overall, the real potential for maximising the biological value of plantation landscapes is by provision of incentives and data for better landscape design and management of areas set aside for conservation. To achieve maximum value from a biodiversity and conservation biology standpoint, plantation landscapes should be designed so that they are penetrable and permeable for those biodiversity components, which are of conservation concern. On the other hand they should be impenetrable and impermeable for pests, weeds, and other invasive organisms (Beier and Noss 1998; Laidlaw 2000; Lindenmayer and Franklin 2002). Second, from a social standpoint, the priority must be to design plantation landscapes in a way that minimizes the adverse impacts and maximises benefits for the local people and communities living in and around these areas (Maturana et al. 2005; Mitchell and Craig 2000; Nawir et al. 2003). Even degraded lands prior to their conversion to pulp wood plantations have been valuable to local people by providing a wide range of products and services. These are often not compensated for on land conversion due to the lack of existing markets. (Maturana et al. 2005; Mitchell and Craig 2000; Nawir et al. 2003). Conservation areas may still be important for livelihoods of local people, but little work has been done in Sumatra to research the potential in combining the conservation and livelihood aims in terms of natural forest corridors in plantation landscapes.

In a survey of Tesso Nilo forest complex carried out by Gillison (2001) a high total of more than 900 species, including all vascular plants was recorded, with an average of 140 species from a 40 × 5 m plot. The present survey considered only woody vegetation and found over 200 tree species from riparian forests, a level comparable to the richness of the large preserved forest complex or other tropical rainforests of the region. However, the composition of the riparian forest is different to KPPN areas (Table 9), and this should be taken into account when designing the conservation areas within plantation landscape.

Table 9 Most common, abundant and largest tree species from (a) riparian forests and (b) KPPN area are arranged by the total importance value and with observations of being food sources (either fruits or leaves) for four primates studied in Baserah sector (second survey), in Riau, Sumatra

(a) Species arranged by importance values, riparian areas				(b) Species arranged by Importance values, KPPN areas			
Scientific name	Importance index			Scientific name	Importance index		
	Total in KPPN	Total, riparian forest	Food source[a]		Total in KPPN	Total, riparian forest	Food source[b]
Syzygium claviflorum	0.2	45	P	Not id42	51.2	0.8	–
Knema laurina	0.5	44.1	All	*Artocarpus anisophyllus*	42.1	0.6	–
Syzygium fastigiatum	0.3	26.3	P	*Cratoxylum glaucum*	27.6	2.4	Mf
Santiria oblongifolia	0.9	24.5	H, P, Mn	Not id54	25.0	0	–
Helicia excelsa	0.2	15.6	H, P, Mn	*Ardisia sumatrana*	19.3	0.2	Mf
Shorea leprosula	1	9.1	P, Mn, Mf	Not id24	10.5	0	–
Calophyllum lowii	0.7	8.3	–	Not id49	10.2	0	–
Dillenia sp1	0.5	8.3	All	*Scapium maropodum*	10.1	0.4	H
Horsfieldia subglobosa	0.7	7.9	H, P, Mn	*Dyera costulata*	8.4	0.4	–
Nephelium lappaceum	4.8	7.2	All	***Quercus lucida***	8.2	6.4	H, Mn
Lithocarpus sp1	3	7.1	All	*Vatica resak*	7.6	0	–
Diospyros cf. *coriacea*	0.8	6.6	H, P	*Palaquium obovatum*	7.3	2	H
Palaquium sp1	0.9	6.5	P	*Syzygium antisepticum*	7.1	1.7	Mn
Quercus lucida	8.2	6.4	P	Not id35	7.0	3.4	–
Ochanostachys amentacea	2.1	6.1	H, P, Mn	*Ficus sumatrana*	6.8	0.3	–
Griffithianthus merillii	0.2	5.8	P, Mn	*Sindora sumatrana*	6.7	0.4	–
Pometia pinnata	6.1	5.8	P	***Xylopia ferruginea***	6.1	4.3	–
Not id41	2.3	5.6	P	***Pometia pinnata***	6.1	5.8	–
Scapium linearicarpum	3.3	5.6	All	*Baccaurea bracteata*	5.4	2.4	H
Artocarpus kemando	0.4	5.6	H, P, Mn	*Castanopsis inermis*	5.4	0.3	–
Dialium indum	4.7	5.3	P	***Nephelium lappaceum***	4.8	7.2	Mf
Antidesma trunciflorum	0.6	5.3	P, Mf	***Dialium indum***	4.7	5.3	Mf
Koompasia malaccensis	3.3	5.1	P, Mn	*Aporosa dioica*	4.3	2.2	H
Anisophyllea disticha	1.2	5	P	*Barringtonia sacrostachys*	4.2	1.1	–
Shorea seminis	2.8	4.9	H, P, Mn	*Avicennia alba*	4.1	2.2	–

Table 9 continued

(a) Species arranged by importance values, riparian areas				(b) Species arranged by Importance values, KPPN areas			
Scientific name	Importance index			Scientific name	Importance index		
	Total in KPPN	Total, riparian forest	Food source[a]		Total in KPPN	Total, riparian forest	Food source[b]
Trycalisia malaccensis	1.4	4.4	H, P, Mn	not id55	3.9	0	–
Xylopia ferruginea	6.1	4.3	All	*Myristica argenta*	3.7	0.3	–
Endospernum peltatum	0.2	4.2	P	*Polyalthia lateriflora*	3.7	0.2	–
Shorea sp1	0.5	4	P	***Koompasia malaccensis***	3.3	5.1	–
Hopea sp1	0.4	3.7	H, P, Mn	Not id33	3.3	0	–
All species (not shown until total of 400)	**58.3**	**303.3**		**All species (not shown until total of 400)**	**318**	**55.3**	

Only the 30 most important (with highest IVI values) species are shown. Species common to both sites are in bold

[a] H, *Hylobates agilis*; P, *Presbytis melalophos*; Mf, *Macaca fascicularis*; Mn, *Macaca nemestrina*

[b] Not recorded for *Presbytis melalophos*

There is a lack of inventory data for primates in Tesso Nilo area, but existing previous surveys are in line with our results: according to Gillison (2001) the Riau Province Forest Department survey in June 1992 recorded the same two *Macaca* species and *Hylobates syndactilus* in theTesso Nilo Forest Complex. In addition, they reported *Nycticebus coucang* (slow loris). Gillison (2001) observed in the vicinity of Tesso Nilo in selectively logged low land forests *M. fascicularis* and *M. nemestrina*, and in a 30-year-old rubber plantation and swamp forest *H. agilis.* A later survey conducted by LIPI (Indonesian Institute of Sciences) using similar methods to ours reported, *H. agilis*, *P. femoralis* and *M. nemestrina* from Tesso Nilo (Suyanto et al. 2003). Thus it seems that we observed the most common species of the area in the conservation area inside the plantations. Nevertheless, since these species differ in their ecology, we expected to see greater differences in their abundance as a reaction to habitat disturbance. Our survey does not conclude whether the observed primates are using the riparian forests and/or KPPN areas as part of their permanent home range or merely as movement conduits. This distinction would be important in order to better understand the value of the conservation areas that have been set aside from production. Groups or species, such as *Presbytis melalophos* with smaller home ranges are generally more territorial, though this behaviour is affected by resource supply. Some studies have noticed that in a corridor situation there is not only an increased overlap of home ranges, but also increased territoriality, which may constrain animal movements (Marsh and Wilson 1981 and cited references in there). *Macaca fascicularis*, which has a feeding strategy that is more opportunistic than that of gibbons and langurs, usually has a strong preference for riparian and swamp forests, where their small size may be beneficial in lower, denser vegetation (Marsh and Wilson 1981). In our study, this species was only observed in Baserah sector and there only in KPPN areas in stead of in riparian forests. However, *M. fascicularis* has been observed frequently also from disturbed areas and forest edges (Marsh and Wilson 1981; Meijaard et al. 2005). Langurs may be more severely affected by selective logging than gibbons, since the plant species that often provide substantial amounts of food for them (e.g. several species of Leguminosae) are also commonly selectively logged (Marsh and Wilson 1981). Due to possibility for folivorous diet most langurs survive better than gibbons in situations with a scarcity of fruit trees (Marsh and Wilson 1981). The latter were more abundant in both surveys potentially reflecting this difference in their ecology.

Effect of landscape connectivity on biodiversity

When habitat of primates becomes fragmented or disturbed, primates normally have two ways of responding to the spatial and temporal change in their resources; either they travel longer distances in order to get the needed resources (Johns 1986 and cited references in there) or they become less selective (Meijaard et al. 2005). To evaluate the role of small sized riparian remnant patches over a longer time period and better understand their contribution to the dynamic landscapes of primates, monitoring is needed (Mbora and Meikle 2004; Onderdonk and Chapman 2000; Turner 1996; Tutin et al. 1997). We observed unambiguously the critical importance of a well-connected network of natural forest corridors and other natural forest patches in the plantation landscape to maintain primates. First, primates were not abundant in any of the *A. mangium* plantation stands, irrespective of stand age, size and location, suggesting that the plantation stand element of the landscape is of little value to primates. Second, primates were only observed in riparian forests that were connected to either KPPN area or Tesso Nilo, showing that primates are very sensitive to the connectivity of natural forest patches in the landscape. This implies that any areas set aside from production should be connected to the larger conservation area or natural forest, to maintain its value for primates. Further we found that primate diversity is better conserved if relatively large areas within a given production area were set aside. Occurrence of primates in riparian forest is dependent on both the connectivity to large areas of natural forest or for conservation, and on changes in forest structure, particularly crown closure. However, if less than 20–30% of the total area was set aside from production, impact to primates was observed.

An important result of the present study is that extent of the concession area which is compulsorily set aside from plantation production might be sufficient to maintain primates in the area provided this 20–30% of the land area is appropriately distributed so as to balance (a) overall connectivity of corridors and/with larger conservation areas, (b) habitat quality in terms of crown closure and disturbance of corridors and (c) there is still a substantial amount of natural forest left in the landscape. A wider implication for plantation concessions is that corridors should be located so that connectivity to conservation areas or other large areas of natural forest is maintained at the regional scale, thereby increasing the long term regional presence of primates. Corridors can potentially serve as habitat for annual migrations and for daily movements. Maintaining the possibility for dispersal movements is a major concern for conservation biologists because this movement is vital to keeping the primates of a reserve "connected" with conspecifics living in other reserves, or outside the reserves (Beier and Noss 1998).

Integrated planning

The need to allow for faunal movement implies that it may not be appropriate that corridors are simply distributed randomly in the landscape, rather, priority should be given to areas within the concession that will ensure that reserves and other large natural forest areas outside the concession are kept well-connected to these set aside areas. The fundamental problem is that often landscape level planning of roads, KPPN and corridors, respectively, are conducted as virtually separate and independent activities, with very little or no coordination between managers of relevant departments (Gunarso and Davie 2000). Our results emphasize the need for integrated landscape level planning at the concession level. This must explicitly consider the three main landscape elements—plantation stands, riparian forests as corridors, and conservation areas (KPPN). It must also consider the tradeoffs between them in land allocation decisions and also how road infrastructure and distribution may affect the conservation value of natural forest corridors and conservation areas.

Impact of roads

We found that primate density was higher at low road density (in terms of both number and length), but we were not able to ascertain the extent of causality. During both planning and implementation of plantation establishment the first major landscape level intervention is the construction of road infrastructure (or canals in peat lands). Usually, there is flexibility in response to topography and settlement locations with respect to design of road infrastructure at the landscape level in lowland areas in Sumatra. Furthermore, current regulations often do not provide specific guidelines on the design of roads with relation to corridors (H. Witono personal communication). Most companies aim for minimum compliance with existing regulations, yet changes in road infrastructure plans can often be done at little increased economic costs.

KPPN

Often, regulations on design and management of conservation areas (such as KPPN) in tropical plantation landscapes, are not clear and lack theoretical and scientific underpinning. The impact of area and dimensions of conservation areas upon the biodiversity conservation value of these areas, was not assessed in the present study, but it is an important issue, which deserves more attention including research (Pfund et al. 2006).

Effect of illegal logging on primates

We found that logging in riparian forests occurred predominantly in older plantation stand areas ($r = 0.416$; $\alpha = 0.01$). This may be because riparian forests in these older plantation areas had been under active management for a longer time than more recently developed areas, and hence may have accumulated impacts (i.e. had been exposed for a longer period). Second, the occurrence of logging in riparian forests was strongly related to road infrastructure. Logging had occurred in all those riparian forests that were intersected by roads. This would imply that riparian forests are less likely to be logged if located at greater distance from roads.

Food availability for primates

The food preference differed between primate species. Based on the range of food sources, *Presbytis melalophos*, feeding on all the most important tree species in riparian areas, is considered to be a generalist compared to *Macaca nemestrina* with more restricted range of food source species. Perhaps consequently, *P. melalophos* was found more widespread compared to *M. nemestrina*, which had a very restricted distribution comprising only those areas in which its food source species were abundant, i.e. in wider riparian areas or KPPN conservation area. Some earlier studies have considered that both *M. fascicularis* and *M. nemestrina* were largely opportunistic and thus to a higher degree able to survive in disturbed land use mosaics than more specialized frugivores (Johns and Skorupa 1987). Both *H. agilis* and *Presbytis* species were common in riparian forests, which is in line with other studies (Johns and Skorupa 1987). Generally speaking, wide range of food sources has been considered to be an important factor for the survival of primates in disturbed South East Asian forests (Johns 1986) and our results show that riparian forests may offer such conditions. Consequently to ensure that the riparian corridors remain a suitable habitat for primates, further studies and monitoring are needed to investigate the influence of corridor management (such as the effect of felling operations on corridors) and landscape level characteristics (including corridor design) on food availability in natural forest corridors.

Conclusions

The fact that conservation areas are required and that they occur in Indonesia within large-scale industrial plantation concessions with potentially 30% of the land being set aside without any provision for biodiversity conservation, is very positive. In very fragmented landscape, such as that of Riau province, remaining patches of natural forest become critical in retaining native biodiversity. In this study we have described the plantation landscape and evaluated how its design seemed to affect in its year 2000 form the occurrence of primates. But to be able to base decisions on a solid understanding of the processes in this landscape, our data are not yet sufficient to evaluate the functionality of the landscape for the survival of primates in a longer run. Therefore, further comparative studies and monitoring are needed to better understand the dynamics of primate subpopulations in a large-scale plantation landscapes to improve land use planning taking the conservation of primates into account. Specifically, this should involve comparison of areas at different distances from natural forest, monitoring of the availability and regeneration of food species in corridors and analysis of the population or demographic structure of primates in natural forest corridors under different landscape designs as well as under varying adjacent plantation management.

Acknowledgements We express our gratitude to the reviewers for their insightful comments that have greatly improved the manuscript. We are grateful to RAPP for permitting access to their lands, and for providing skilled field assistants and extensive logistical support. John Bathgate and Jean-Laurent Pfund provided critical comments on earlier drafts and Trudy O'Connor greatly helped to improve the language. We also wish to acknowledge European Union and Finland, Ministry of Foreign Affairs for funding part of this study.

References

Baranga D (2004) Forest fragmentation and primates' survival status in non-reserved forests of the 'Kampala area', Uganda. Afr J Ecol 42:70–77

Barr C (2001) Political economy of fiber & finance in Indonesia's pulp & paper industries—banking on sustainability: structural adjustment and forestry reform in post-suharto Indonesia. CIFOR and WWF's Macroecnomics Program office, Washington DC

Beier P, Noss RF (1998) Do habitat corridors provide connectivity. Conserv biol 12:1241–1252

Bennet AF, Radford JQ, Haslem A (2006) Properties of land mosaics: implications for nature conservation in agricultural environments. Biol Conserv 133:250–264

Buechner HK, Dawkins HC (1961) Vegetation change induced by elephants and fire in Murchison Falls National Park, Uganda. Ecology 42:752–766

Chapman CA, Onderdonk DA (1998) Forests without primates: primate/plant codependency. Am J Primatol 45:127–141

Chapman CA, Speirs ML, Gillespie TR et al (2006) Life on the edge: gastrointestinal parasites from the forest edge and interior primate groups. Am J Primatol 68:397–409

Cossalter C, Pye-Smith C (2003) Fast-wood forestry, myths and realities. CIFOR, Bogor

Costa LP, Leite YLR, Mendes SL et al (2005) Mammal conservation in Brazil. Conserv Biol 19:672–679

Forman RTT (1995) Land mosaics. The ecology of landscapes and regions. Cambridge University press, Cambridge

FWI/GFW (2002) The state of the forest: Indonesia. Forest Watch Indonesia and Washington DC, Global Forest Watch, Bogor, Indonesia

Gascon C, Williamson GB, da Fonseca GAB (2000) Receding forest edges and vanishing reserves. Science 288:1356–1358

Gillison AN (2001) Vegetation survey and habitat assessment of the Tesso Nilo forest complex; Pekanbaru, Riau Province, Sumatra, Indonesia. WWF-US, pp 1–68

Gonzalez-Solis J, Guix JC, Mateos E et al (2001) Population density of primates in a large fragment of the Brazilian Atlantic rainforest. Biodivers Conserv 10:1267–1282

Gunarso G, Davie J (2000) Biodiversity conservation in Indonesia: policy and politics. In: Tacconi L (ed) Biodiversity and ecological economics. Participation, values and resource management. Earthscan, London, 254 pp

Hartley MJ (2002) Rationale and methods for conserving biodiversity in plantation forests. For Ecol Manage 155:81–95

Holmes D (2002) Where have all the forests gone? Environment and social development East Asia and pacific region discussion paper, vol 24462. World Bank, 38 pp
Johns AD (1986) Effects of selective logging on the behavioral ecology of West Malaysian primates. Ecology 67:684–694
Johns AD, Skorupa JP (1987) Responses of rain-forest primates to habitat disturbanceÖ a review. Int J Primatol 8:157–191
Laidlaw RK (2000) Effects of habitat disturbance and protected areas on mammals of peninsular malaysia. Conserv biol 14:1639–1648
Laurance WF, Laurance SG, Ferreira LV et al (1997) Biomass collapse in Amazonian forest fragments. Science 278:1117–1118
Laurance WF, Williamson B (2001) Positive feedbacks among forest fragmentation, drought and climate change in the amazon. Conserv biol 15:1539–1535
Lindenmayer DB, Franklin JF (2002) Conserving forest biodiversity: a comprehensive multiscaled approach. Island press, Washington
MacDonald MA (2003) The role of corridors in biodiversity conservation in production forest landscapes: a literature review. Tasforests 14:41–50
Marjokorpi A, Salo J (2006) Operational standards and guidelines for biodiversity management in tropical and subtropical forest plantations—how widely do they cover the ecological framework? Biodiversity management in fast-growing tree plantations: a case study from West Kalimantan, Indonesia. Ph.D. Thesis. Painosalama Oy, pp 45–68
Marsh CW, Wilson WL (1981) A survey of primates in Peninsular Malaysian forests. Final report for the Malaysian primates research programme. Universiti Kebangsaan Malaysia and University of Cambridge, UK
Maturana J, Hosgood N, Suhartanto AA (2005) Moving towards company-community partnerships elements to take into account for fast-wood plantation companies in Indonesia CIFOR. Working paper no. 29, CIFOR, Bogor, 49 pp
Mbora DNM, Meikle DB (2004) Forest fragmentation and the distribution, abundance and conservation of the Tana River red colobus (Procolobus rufomitratus). Biol Conserv 118:67–77
Meijaard E, Sheil D, Nasi R et al (2005) Life after logging: reconciling wildlife conservation and production forestry in Indonesian Borneo. CIFOR, Bogor
Mitchell ND, Craig JL (2000) Managing the matrix: realigning paradigms toward sustainability. In: Craig JL, Mitchell N, Saunders DA (eds) Conservation in production environments, managing the matrix. Surrey Beaty & Sons Pty limited, Chipping Norton, pp 26–34
Mueller-Dombois D, Ellenberg H (1974) Aims and methods of vegetation ecology. John Wiley and Sons, New York
Myers N, Mittermeier RA, Mittermeier CG et al (2000) Biodiversity hotspots for conservation priorities. Nature 403:853–858
Nason JD, Hamrick JL (1997) Reproductive and genetic consequences of forest fragmentation: two case studies of Neotropical canopy trees. J Hered 88:264–276
Nawir AA, Santoso L, Mudhofar I (2003) Towards mutually-beneficial company-community partnerships in timber plantation: lessons learnt from Indonesia, 1st edn, vol 26. CIFOR, Bogor
Nepstad DC, Verssimo A, Alencar A et al (1999) Large-scale impoverishment of Amazonian forests by logging and fire. Nature 398:505–508
Newmark WD (1991) Tropical forest fragmentation and the local extinction of understory birds in the Eastern Usambara Mountains, Tanzania. Conserv Biol 5:67–78
Noordwijk MV, Tomich TP, Foresta HD et al (1997) To segregate—or to integrate? Agroforest Today 9(1):6–9
Onderdonk DA, Chapman CA (2000) Coping with forest fragmentation: the primates of Kibale National Park, Uganda. Int J Primatol 21:587–611
Pfund J-L, O´Connor T, Koponen P et al (2006) Transdisciplinary research to promote biodiversity conservation and enhanced management of tropical landscape mosaics Patterns and processes in forest landscapes, consequences of human management. IUFRO Landscape ecology working party, Locorotondo, Bari, Italy
Poulsen J, Applegate G, Raymond D (2001) Linking criteria and indicators to a code of practice for industrial tropical tree plantations. CIFOR, Bogor
Ranta E, Rita H, Kouki J (1989) Biometria, tilastotiedettä ekologeille. Yliopistopaino, University press, Helsinki
Sunderlin W (1999) Development of the forestry sector in Indonesia. In: Palo M, Uusivuori J (eds) World forests, society and environment. Kluwer academic publishers, London, pp 214–221

Suyanto A, Sinaga MH, Saim A (2003) Biodiversitas mammalia di Tesso Nilo. Bidang zoologi, Pusat Penelitian Biologi—LIPI, WWF, Bogor, 12 pp

Turner IM, Corlett R.T. (1996) The conservation value of small, isolated fragments of lowland tropical rain forest. Tree II:330–333

Tutin CEG, White LJT, Mackanga MA (1997) The use by rain forest mammals of natural forest fragments in an equatorial African savanna. Conserv Biol 11:1190–1203

Whitmore TC (ed) (1972) Tree flora of Malaya, a manual for foresters, vol 1–4. Longman Malaysia, Kuala Lumpur

Whitten T, Damanik SJ, Anwar J et al (2000) The ecology of Sumatra, vol 1. Periplus Editions (HK) Ltd., Singapore

Non-native plantation forests as alternative habitat for native forest beetles in a heavily modified landscape

Stephen M. Pawson · Eckehard G. Brockerhoff · Esther D. Meenken · Raphael K. Didham

Originally published in the journal Biodiversity and Conservation, Volume 17, No 5, 1127–1148.
DOI: 10.1007/s10531-008-9363-y

Abstract The once extensive native forests of New Zealand's central North Island are heavily fragmented, and the scattered remnants are now surrounded by a matrix of exotic pastoral grasslands and *Pinus radiata* plantation forests. The importance of these exotic habitats for native biodiversity is poorly understood. This study examines the utilisation of exotic plantation forests by native beetles in a heavily modified landscape. The diversity of selected beetle taxa was compared at multiple distances across edge gradients between each of the six possible combinations of adjacent pastoral, plantation, clearfell and native forest land-use types. Estimated species richness (Michaelis–Menten) was greater in production habitats than native forest; however this was largely due to the absence of exotic species in native forest. Beetle relative abundance was highest in clearfell-harvested areas, mainly due to colonisation by open-habitat, disturbance-adapted species. More importantly, though, of all the non-native habitats sampled, beetle species composition in mature *P. radiata* was most similar to native forest. Understanding the influence of key environmental factors and stand level management is important for enhancing biodiversity values within the landscape. Native habitat proximity was the most significant environmental correlate of beetle community composition, highlighting the importance of retaining native remnants within plantation landscapes. The proportion of exotic beetles was consistently low in mature plantation stands, however it increased in pasture sites at increasing distances from native forest. These results suggest that exotic plantation forests may provide important alternative habitat for native forest beetles in landscapes with a low proportion of native forest cover.

S. M. Pawson · R. K. Didham
School of Biological Sciences, University of Canterbury, Private Bag 4800, Christchurch, New Zealand

S. M. Pawson (✉) · E. G. Brockerhoff
Scion, P.O. Box 29237, Fendalton, Christchurch, New Zealand
e-mail: Steve.Pawson@scionresearch.com

E. D. Meenken
Crop and Food Research, Private Bag 4704, Christchurch, New Zealand

E.G. Brockerhoff et al. (eds.), *Plantation Forests and Biodiversity: Oxymoron or Opportunity?*
DOI: 10.1007/978-90-481-2807-5_11

Keywords Coleoptera · Edge effects · Habitat fragmentation · Habitat isolation · Landscape ecology · Matrix habitat · Natural forest · Pasture · Pitfall trapping · Plantations

Introduction

Habitat loss and fragmentation are recognised as critical agents of species decline (Tilman et al. 1994; Sala et al. 2000; Brooks et al. 2002; Fahrig 2003; Reed 2004; Ewers and Didham 2006a). In New Zealand, the impacts of habitat loss on biodiversity have been most severe in the fertile lowland forest environments that were best suited for conversion to pastoral agriculture (Norton 2001). Over 90% of the original forest cover has been removed in some regions and significant changes in land use still occur (Ewers et al. 2006; Walker et al. 2006, but Brockerhoff et al. 2008). Afforestation of 1.8 million hectares of exotic (primarily *Pinus radiata*) forest (ca. 25% of New Zealand's current total forest cover) in the form of intensively-managed plantation forests has established large areas of forest habitat that have been missing from these landscapes for many decades (Anon 2005). In a landscape matrix otherwise dominated by pastoral farming, matrix habitats such as plantation forests that have similar characteristics to native forest (hereafter referred to as a 'low contrast' matrix habitat), are increasingly recognised for their potential contribution to forest biodiversity preservation (Humphrey et al. 1999; Anon 2000; Carnus et al. 2006).

Because plantation forests are intensively managed for the commercial production of timber and other forest products, they are typically composed of just one or a few tree species (predominantly *Pinus radiata* in New Zealand) grown in even-aged stands that are repeatedly harvested by clearfelling. As a consequence, plantation forests are often assumed to support a low abundance and diversity of indigenous species, and have been referred to as 'biological deserts' (see Brockerhoff et al. 2001). Contrary to such perceptions, research has shown that managed plantation forests can support a diverse array of native understorey plants (Allen et al. 1995; Geldenhuys 1997; Ogden et al. 1997; Brockerhoff et al. 2003), birds (Ryder 1948; Weeks 1949; Clout 1984; Clout and Gaze 1984), and invertebrates (Humphrey et al. 1999; Hutcheson and Jones 1999; Bonham et al. 2002; Woodcock et al. 2003; Humphrey 2005; Mesibov 2005; Oxbrough et al. 2005; Carnus et al. 2006). Furthermore, plantation forests can contribute to the maintenance of ecosystem integrity by buffering the microclimate of native forest remnants from external influence (Norton 1998; Brockerhoff et al. 2001; Hartley 2002; Denyer et al. 2006), and by providing a low contrast forest environment suitable for many species dispersing between remnant native habitats in the landscape (Norton 1998; Hale et al. 2001).

Internationally, there is increasing pressure to reduce the real and perceived negative environmental effects of plantation forestry, and to enhance sustainable timber production (Hock and Hay 2003). An array of stand-level initiatives including the management of harvest debris, legacy management and the manipulation of stand composition, vertical structure and age has been evaluated as a means to enhance biodiversity in managed forests (Kerr 1999; Franklin et al. 2000; Bonham et al. 2002; Mazurek and Zielinski 2004). However, external influences at larger spatial scales, such as landscape composition and connectivity, are now recognised as critical determinants of biodiversity within plantations, and the influence of stand level management should be considered within a landscape context (Humphrey et al. 2004; Lindenmayer and Hobbs 2004; Barbaro et al. 2005). Unfortunately, advances in this vein are hampered because most landscape ecological research continues to focus on natural ecosystems (Norton 2001; Fazey et al. 2005), and there is little understanding of the contribution that managed exotic habitats, such as plantation forests, can make to regional biodiversity conservation relative to unmanaged native remnants.

The objective of this study was to determine the role of modified habitats, particularly non-native plantation forests as alternative habitat for native species in heavily fragmented landscapes. We concentrate on plantation forests as they now dominate the matrix habitat surrounding remnant native habitat in some regions of the central North Island and the eastern South Island. Beetle biodiversity in native forest habitat and three human-modified ecosystems (pasture, mature production forest stands, and recently clearfell-harvested production stands) in the highly fragmented landscape of the central North Island was compared. We focused in particular on beetle diversity within three families; ground beetles (Carabidae), chafer beetles (Scarabaeidae) and bark beetles (Curculionidae: Scolytinae) as these taxa provided greater trophic breadth than a reliance on single invertebrate groups. By sampling across gradients between habitats we were also able to investigate the presence of edge-mediated changes in the abundance of beetles.

Methods

Study sites and collection of beetles

The study was conducted in the central North Island of New Zealand (Fig. 1), a region historically subject to infrequent catastrophic disturbances, predominantly from the Taupo Volcanic Zone (Froggatt and Lowe 1990; Wilmshurst and McGlone 1996). Before European colonisation the vegetation of this volcanic plateau was a mosaic of seral shrub-heaths and frost flats at higher altitudes, and lush mixed podocarp–broadleaved forests on lowland terraces (Wardle 1991). Current patterns of indigenous vegetation are a reflection of substantial changes in land use over the last 200 years (McGlone 1989; Roche 1990). Exotic pasture (dominated by ryegrass, *Lolium perenne*, and clover, *Trifolium repens*) and plantation forests of *P. radiata* now surround highly fragmented, isolated patches of native habitat. Plantations now account for 64% of the forest cover in the political region where the study occurred (Ewers et al. 2006), and most were established on land previously cleared for pastoral farming that subsequently proved to be economically unsustainable. Native vegetation is now limited to a few large intact areas of forest (total area 344,000 ha in the region, with the study site also bordering the largest (691,587 ha) remaining native forest fragment in the North Island, but of the 497 native forest patches in the Whakatane region <5% of them are >100 ha in size (Ewers et al. 2006)), many small privately-owned forest remnants (including a substantial network throughout plantation forests along riparian margins and steep gullies), pockets of indigenous shrubland, areas of fire-induced shrubland, and the significant but often unrecognised native plant component within the understorey of plantation forests (Allen et al. 1995; Ogden et al. 1997; Brockerhoff et al. 2003).

Beetles were collected along three independent replicate edge gradients in each of the six possible comparisons between pairs of the following habitats: mature 26-year-old *P. radiata*, recently clearfelled *P. radiata* stands, native forest and pasture (Fig. 1). Along each of the 18 edge gradients, individual pitfall traps were placed at seven distances from the habitat boundary, at −125, −25, −5, 0, +5, +25 and +125 m perpendicular to the edge (negative distances arbitrarily assigned to one of the habitats), giving 126 pitfall traps in total. The logarithmic scale applied to the sampling design reflects the a priori assumption

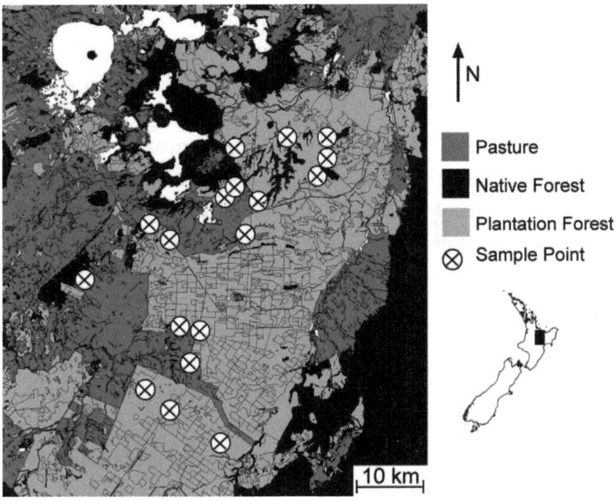

Fig. 1 Plantation forest (predominantly *P. radiata*) and pasture dominate the study area in the central North Island, New Zealand. Native forest remnants are scattered throughout, with a large native forest remnant to the east

that changes in species abundances would occur most rapidly near habitat edges (Didham et al. 1998; Ewers et al. 2007). A standard pitfall trap design consisted of a circular, 680 ml polypropylene plastic container of 100 mm diameter buried to ground level. Two white plastic guide panels 1.2 m long and 0.10 m high were placed at ground level in a cross-design, in an attempt to increase trap catch by channelling ground-dispersing arthropods towards the central collecting cup. A 70% monoethylene glycol (antifreeze) solution was used as a preservative and changed at approximately monthly intervals. Samples were subsequently transferred into 70% alcohol for storage prior to analysis. Insects were sorted using a 6–50 × Zeiss stereomicroscope. Carabidae, Scarabaeidae and Scolytinae (Curculionidae) were identified to species level from the pitfall samples. These three taxonomic groups were chosen to provide a balance between identifying all Coleoptera (which was not technically feasible due to resource limitations) and the other extreme of relying on a single taxonomic 'indicator' group. The families were selected to provide a range of trophic groups: Scarabaeidae are herbivores, Scolytinae are subcortical feeders in wood, and Carabidae are generally predators. Little is currently known about the diet of New Zealand carabids, but it is assumed that they have similar trophic roles to those found in other countries (Larochelle and Larivière 2001). Further reference to beetle abundance, diversity or composition here refers to these three selected families only.

It is widely recognised that pitfall trapping has inherent biases (Ward et al. 2001). As with all passive trapping systems, such as malaise or flight interception trapping, capture rates in pitfall traps represent an integrated measure of local density and wider-scale activity rates, and cannot necessarily be used as a direct measure of local population density (Southwood 1994; Lang 2000). As such, references in this paper to 'abundance' refer to activity-density, i.e., relative number of individuals present in a pitfall trap in a given habitat. Furthermore, conclusions regarding changes in relative abundance of species refer to changes in the relative activity-density of different species, rather than to identifiable changes in local population density. However, treatment effects are still meaningful in terms of altered habitat utilisation patterns, as shown by Nittérus and Gunnarsson (2006).

Pitfall trap monitoring was undertaken six times at monthly intervals between November 2002 and February 2003 and between December 2003 and February 2004. The number of individuals of each species captured (within traps) was used in estimated species richness analyses, but for other analyses the capture rates of individual species were unit-standardised by converting to catch per 100 trap-days to account for minor differences in sampling duration between traps that resulted from the schedule of sample collection.

Collection of environmental variables

Landscape-scale environmental variables were derived for each site from a Geographic Information System (GIS) of 15 data layers underlying the Land Environments of New Zealand (LENZ) classification system (Leathwick et al. 2002), using ArcView v.3.2. The underlying LENZ data layers included seven climate variables, seven soil variables, and a slope variable (Leathwick et al. 2002). Individual landscape-level LENZ attributes were incorporated into composite variables using a principal components analysis (PCA) of non-collinear variables (see Gates and Donald (2000)), and the resulting axis scores used as environmental variables (PCA-LENZ 1, PCA-LENZ 2 and PCA-LENZ 3) in ordination analyses of beetle data (Table 1).

Understorey vegetation surveys were conducted within a 2.5 × 2.5 m quadrat centred on the pitfall trap. The relative abundance of individual plant species was quantified in four vertical strata (ground: 0.0–0.3 m, shrub: 0.3–2.0 m, sub canopy: 2.0–10.0 m and canopy:

Table 1 Description and units of measurement of environmental variables included in constrained CCA ordination (Fig. 4)

Abbreviation	Description	Units of measurement
Dist	Distance along transect	Metres
Long	Longitude (also expresses collinear effects of $Long^2$, $Long^3$, $Long^2 * Lat$, and $Long * Lat^2$)	NZMG Longitude/ 1,000,000
Lat	Latitude (also expresses collinear effects of Lat^2 and Lat^3)	NZMG Latitude/ 1,000,000
500 m-nat	Proportion of native vegetation within 500 m radius	Proportion
500 m-exo	Proportion of exotic vegetation within 500 m radius	Proportion
1000 m-nat	Proportion of native vegetation within 1000 m radius	Proportion
1000 m-exo	Proportion of exotic vegetation within 1000 m radius	Proportion
5000 m-nat	Proportion of native vegetation within 5000 m radius	Proportion
5000 m-exo	Proportion of exotic vegetation within 5000 m radius	Proportion
Adj-N	Adjacent stand to site is native	Categorical
Adj-P	Adjacent stand to site is pasture	Categorical
Adj-M	Adjacent stand to site is *Pinus radiata* 26 yr	Categorical
Adj-C	Adjacent stand to site is clearfell	Categorical
DW-1-5	Dead wood, categorical scale 1–5, i.e., 4 levels	Categorical
D-1-5	Drainage, categorical scale 1–5, i.e., 4 levels	Categorical
L-1-5	Leaf litter, categorical scale 1–5, i.e., 4 levels	Categorical
PCA-Veg1	PCA axis 1 scores of understorey vegetation surveys	Ordination scores
PCA-Veg2	PCA axis 2 scores of understorey vegetation surveys	Ordination scores
PCA-Veg3	PCA axis 3 scores of understorey vegetation surveys	Ordination scores
PCA-LENZ 1	PCA axis 1 scores of LENZ environmental information	Ordination scores
PCA-LENZ 2	PCA axis 2 scores of LENZ environmental information	Ordination scores
PCA-LENZ 3	PCA axis 3 scores of LENZ environmental information	Ordination scores

>10.0 m). Identified species were assigned to one of seven cover classes (<1, 1–5, 6–10, 11–25, 26–50, 51–75 and 76–100%). For subsequent analysis, stratum data for each species were converted to a single weighted value that reflected the percentage of cover in each of the four vertical stratigraphic units (ground, shrub, sub-canopy and canopy), using the following formula:

$$\text{Vegetation cover} = \sum_{N-tiers}^{i=1} \text{midpoint of \% cover} * \log_{10} (\text{tier depth} + 1)$$

Tier depths for ground and shrub layers were at 0.3 and 2.0 m, respectively, in all habitats, but tier depths for sub-canopy and canopy layers (when present) varied from 4.0 to 10.0 m depending on habitat type. Single weighted cover values for each species were $\log_e(x + 1)$ transformed and a principal components analysis (PCA) was performed using Canoco V. 4.01 (ter Braak and Smilauer 1999). The resulting principal component axes scores provided a measure of the understorey vegetation community associated with each pitfall trap and were included as environmental variables in subsequent ordination analyses of beetle relative abundance data (PCA-Vege1, PCA-Vege2 and PCA-Vege3; Table 1). Canopy cover was also calculated from an analysis of hemispherical photographs using the software package Hemiview, Version 2.1, and was expressed as a percentage.

Estimates of ground cover complexity around each pitfall trap were made concurrently with the vegetation surveys using a five point qualitative scale: leaf litter cover (1 = 0–5% cover, 2 = 6–30%, 3 = 31–50%, 4 = 51–70%, 5 = 71–100%); drainage (1 = poor, surface water present within the plot even during prolonged dry spells, 2 = low–med, 3 = medium, at least one place within the plot that "squelches" when walked on, 4 = med–high, 5 = well-drained, no surface water even during prolonged rain); and dead-wood cover (1 = none or very little dead wood, at most a few twigs, 2 = low–med, 3 = medium amount of deadwood, at least one log >10 cm diameter, 4 = med–high, 5 = much dead wood, plot is difficult to move in due to the amount of deadwood). Categorical variables for each measure of ground-cover complexity were then incorporated into the ordination analysis using binary dummy variables ($n - 1$ categories for each variable).

The proportion of native and exotic plantation forests within a 500, 1000 and 5000 m radius of the centre of each pitfall trap gradient was calculated using data from the Land Cover Database V2 (LCDB2) (Terralink 2004). Proportional cover values for land use types within the landscape were then included as explanatory variables in multivariate ordination analyses (Table 1).

Assessment of species richness and community composition

Species accumulation curves were calculated using the sample-based rarefaction index (Mao Tau, in Estimate-S v7.5.0), rescaled and expressed in terms of number of individuals (Colwell 2004). Associated confidence intervals were calculated in Estimate-S using a general binomial mixture model with 100 randomisations (Colwell et al. 2004). The expected asymptote of the species accumulation curve was calculated by extrapolation beyond the sampled data range using the Michaelis–Menten richness estimator (Colwell and Coddington 1994; Colwell 2004).

Variation in beetle species composition between habitats was analysed using multivariate ordination techniques. An unconstrained correspondence analysis (CA) was conducted on log-transformed relative species abundances. A total of 39 variables characterising spatial attributes, vegetation structure and local environmental factors were collected for each of the sampling sites (Table 1). After removing collinear variables, the remaining factors were included in a canonical correspondence analysis (CCA), and a forward selection procedure was used to identify potential confounding variables (such as spatial autocorrelation among trap locations) that explained significant variation in beetle community composition (Ter Braak 1995). Both longitude and latitude were significant and thus considered to represent spatial autocorrelation in the data, and were subsequently added to the model as covariables (Borcard et al. 1992). A final partial CCA (pCCA) was then conducted on 29 environmental variables and the two covariables, using inter-sample distances and biplot scaling. All ordination analyses were conducted in Canoco V. 4.01 (ter Braak and Smilauer 1999). A multi response permutation test (MRPP) was conducted in PC-Ord for Windows (Version 4.01) to quantify the significance of differences in community composition between site groupings based on habitat type in the pCCA axes 1 and 2 biplot.

Assessment of community and individual responses between different habitat types

The relative abundance of native and exotic beetles in different habitat types was analysed with a split-plot repeated measures ANOVA (Genstat Version 9). The exotic status of carabids was determined using Larochelle and Larivière (2001), and the exotic status of scarabaeids and scolytines was determined by the authors. Habitat comparisons (e.g., pasture versus clearfell) were assigned as the main plot and individual habitat types as sub-plots. A factor was created called 'compartment group' which provided an identifier for each replicate pairing of habitat types. By nesting habitat type within 'compartment group' it is possible to assess the effect of adjacent habitat on beetle relative abundance (measured as average catch per 100 trap days). For this particular analysis, the 5, 25 and 125 m pitfall traps were assigned as repeated measures within the sub-plot. Treatments were beetle origin (native or exotic), habitat type (native, mature *P. radiata*, clearfell and pasture) and distance from boundary between the habitats. For the ANOVA analyses, examination of residual plots indicated that a $\log_e(x + 0.01)$ transformation was required to meet the assumptions of homogeneity of variances and normality of residuals. The degrees of freedom in the analysis were adjusted to account for the lack of independence between repeated measures, and the least significant difference between treatment means (i.e., minimum difference between means above which they are significantly different) was calculated with a confidence of 95%.

Species characteristic of particular habitats were identified by the Indval procedure of Dufrêne and Legendre (1997). Indval sample groupings were assigned a priori on the basis of habitat type (clearfell, pasture, native forest and mature *P. radiata*). Habitat types were constructed by grouping pitfall traps along transects, whereby traps at 125 m and 25 m in each habitat were considered representative of their particular habitats. This was a conservative approach to avoid the most severe edge-effects that are present at the habitat boundaries and at pitfall traps 5 m either side.

Differences in the relative abundance of native and exotic species were then further analysed across entire habitat gradients to determine whether the shape of the edge response function varied between habitat types. Variation in the proportional representation of exotic beetles across habitat edges was modelled by testing the fit of five continuous response functions of increasing complexity: null, linear, power, logistic and unimodal (Ewers and Didham 2006b). Response functions were calculated in R v2.4.0 (R-Team 2006) using a single average proportional abundance of exotic species for each trap, which was the average value pooled across the six trapping periods (so as to avoid pseudoreplication). The best-fit model, out of the five response functions tested, was selected using

Akaike weights (calculated from Akaike information criterion, AIC, values), which give the probability that a particular model is the best fit to the data from the set of models that are evaluated.

Results

Estimated species richness

The target species groups were very well characterised by the level of sampling effort employed, as illustrated by species accumulation curves approaching an asymptote (Fig. 2a). Pooling across all habitats, the actual species richness was equivalent to 97.7% of the estimated species richness asymptote, as calculated by Michaelis–Menten running means.

There was no significant variation in beetle species accumulation curves as a function of habitat type, either for all species combined (Fig. 2a), or for native beetle species considered separately (Fig. 2b). In both cases, clearfell habitats had the highest Michaelis–Menten estimated species richness, whereas native forest had a lower estimated species richness than production habitats. The principal cause of this was the low exotic beetle species richness in native forest (Appendix 1), which was further illustrated by an increase

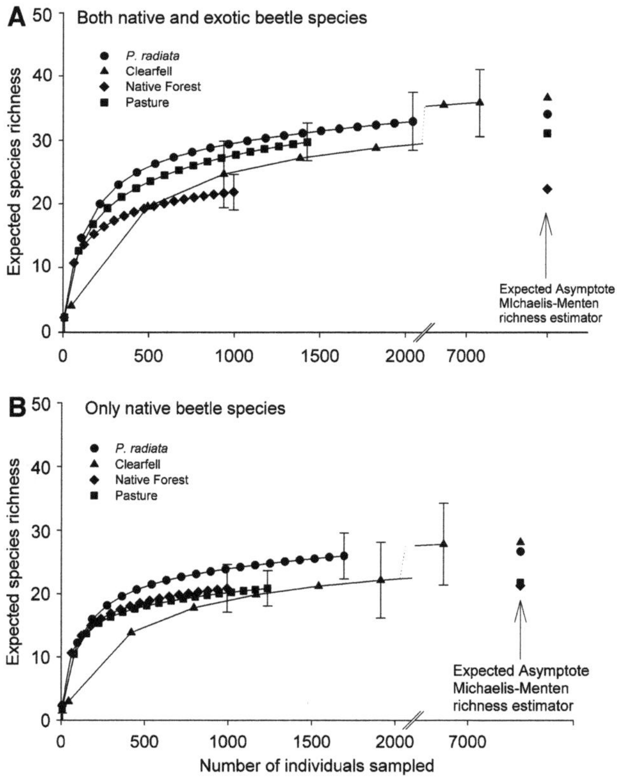

Fig. 2 Species accumulation curves for (**a**) all beetle species and (**b**) only native species, calculated for different habitat types by sample-based random re-sampling. The x-axis is rescaled to the number of individuals and error bars denote 95% confidence intervals. The Michaelis–Menten method was used to estimate species richness. All analyses were conducted using Estimate-S (Colwell 2004)

in the slope and estimated species richness of the native forest species accumulation curve relative to other habitat types when comparing native beetles only (Fig. 2b), as opposed to all beetles (Fig. 2a).

Relative beetle abundance in different habitat types

A combined total abundance of 9,974 Carabidae (28 species), 1,433 Scarabaeidae (11 species) and 633 Scolytinae (3 species) were collected from all habitat types. The majority of species (75%) were native, however eight carabid species, one scarabaeid species and two scolytine species were exotic (Appendix 1). There was strong evidence of an interaction between the origin of beetle species (exotic versus native) and their relative abundance in different habitat types (Habitat · Group Effect, $F_{3,78} = 25.11$, $P < 0.001$, Table 2). Recent clearfells had the greatest pooled mean beetle relative abundance of the four habitats sampled (140.8/100 trap days), due largely to the dominance of one native species, *Cicindela tuberculata* (78% of individuals). Exotic beetle relative abundance in

Table 2 Results of repeated measures ANOVA of beetle species abundances with respect to habitat comparison type, habitat type, species origin and distance from habitat edge

Source of variation	d.f.	Sums of squares	Mean square	*F*	*P*
Compartment stratum					
Habitat comparison	5	324.66	64.93	11.07	<0.001
Residual	12	70.41	5.87	2.79	
Compartment · Subplot stratum					
Habitat type	3	114.47	38.16	18.16	<0.001
Habitat comparison · Habitat type	3	32.49	10.83	5.16	0.016
Residual	12	25.21	2.10	0.97	
Compartment · Subplot · Distance stratum					
Distance	2	6.30	3.15	1.45	0.244
Distance · Habitat comparison	10	12.40	1.24	0.57	0.828
Distance · Habitat group	6	7.03	1.17	0.54	0.774
Distance · Habitat Comparison · Habitat group	6	22.15	3.69	1.70	0.140
Residual	48	103.95	2.17	1.70	
Compartment · Subplot · Distance · Replication stratum					
Origin	1	303.13	303.13	237.68	<0.001
Distance · Origin	2	0.41	0.20	0.16	0.853
Habitat comparison · Origin	5	94.28	18.86	14.79	<0.001
Habitat group · Origin	3	96.06	32.02	25.11	<0.001
Distance · Habitat comparison · Origin	10	21.03	2.10	1.68	0.101
Distance · Habitat group · Origin	6	10.58	1.76	1.41	0.233
Habitat comparison · Habitat group · Origin	3	11.28	3.76	3.00	0.038
Distance · Habitat comparison · Habitat group · Origin	6	5.91	0.99	0.77	0.594
Residual	78	97.38	1.25		
Total	215	1353.56			

A correction factor of 0.9658 was applied to the d.f. of the distance term and its interactions to adjust for potential correlation between pitfall traps sampled from the same trap gradient. Origin refers to whether the beetle was native or exotic native habitat was extremely low compared to all other habitat types (Fig. 3). There was evidence of lower exotic species relative abundance in mature *P. radiata* habitat compared to recent clearfells (Fig. 3). However, there was no significant difference in the relative abundance of native beetles between habitat types (Fig. 3).

Average beetle relative abundance did not change with distance into habitat, or distance as an interaction with other factors in the repeated measures ANOVA (Table 2). However, a number of species, including the exotic species *Hypharpax australasiae* and the native species *Scopodes* spp., *Demetrida natsuda* and *Lecanomerus sharpi* were present in open habitats, such as clearfells and pasture, but not in mature forest (Appendix 1). Four species were specific to a single habitat type, three of which, *Scopodes edwardsi* (clearfell), *Ataenius brouni* (forest) and *Notagonum submetallicum* (pasture), may be transient species given their low relative abundance, whereas the fourth species, *Acrossidius tasmaniae*, is a common exotic pasture pest (Appendix 1).

Variation in beetle community composition between habitat types

Twelve of the twenty-nine environmental variables tested in the forward selection procedure of the canonical analysis were significant predictors of variation in beetle community composition between sites (Table 3). Given their potential as confounding factors, latitude and longitude were incorporated as covariables in a pCCA. Axes 1 and 2 of the pCCA explained 6.3 and 5.6% of the total variance in species relative abundances, respectively, and 17.7 and 15.8% of the species-environment relationship, respectively

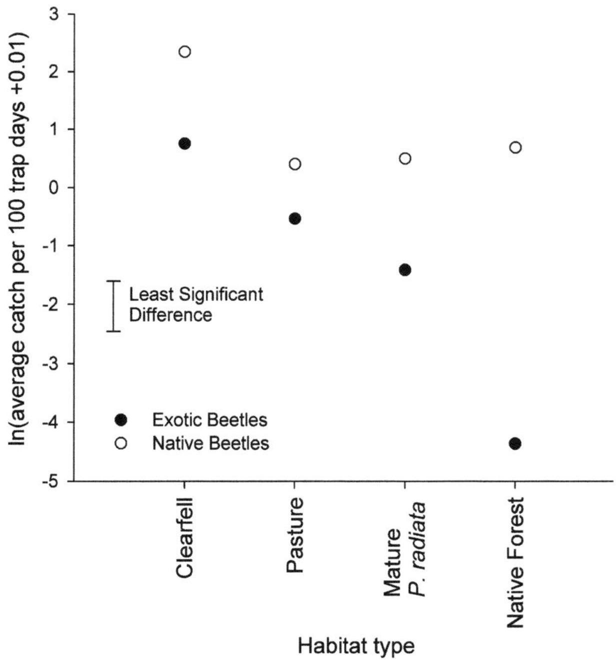

Fig. 3 The relative abundance of native and exotic beetle species in different habitat types. The least significant difference is calculated at $P = 0.05$ with 26.06 d.f. (see "Methods" section for an explanation of the statistical analysis)

Table 3 Significant environmental variables and associated intraset correlations from partial canonical correspondence analysis

Environmental variables	Eigenvalue, $\sum$ of eigenvalues = 1.49	*F*	*P*	Intraset correlations	
				Axis 1	Axis 2
500 m Native forest	0.16	4.88	0.002	**0.67**	−0.03
PCA LENZ 1	0.15	4.45	0.002	**−0.24**	**−0.51**
PCA Vege 1	0.11	3.52	0.002	**0.60**	0.04
500 m Exotic	0.10	2.96	0.002	−0.10	**−0.37**
Adjacent pasture	0.10	3.43	0.002	−0.15	−0.07
5,000 m Exotic	0.09	3.04	0.002	0.08	−0.10
PCA LENZ 3	0.09	2.81	0.002	**−0.18**	−0.09
5,000 m Native	0.08	2.89	0.002	**0.24**	0.12
Drainage 5	0.06	1.83	0.014	−0.05	**−0.24**
Litter 1	0.08	1.50	0.042	**−0.28**	−0.01

Eigenvalues, *F*-values and *P*-values are from the forward selecting regression procedure in Canoco V. 4.02 (ter Braak and Smilauer 1999). Intraset correlations in bold are significant at $P < 0.05$

(Fig. 4a). The four habitat types formed statistically distinct groupings (MRPP, A = 0.214, $P < 0.001$, Fig. 4a). Although mature *P. radiata* sites shared multivariate space with other habitat elements, comparison of the centroids for each habitat type showed that *P. radiata* stands and native forest were most similar in species composition (Fig. 4a). Overall, variation in beetle species composition was best explained by the proportion of native forest within 500 m of the sample location, with sites most strongly correlated with this environmental variable along pCCA axis 1 (Table 3). The second strongest correlation with Axis 1 was with the axis 1 scores of the PCA analysis of understorey vegetation, PCA-Veg1 (Table 3). Axis 2 was most strongly correlated with the PCA axis 1 of the LENZ data layers (Table 3).

Individual species responses

In general, Indval indicator values for individual taxa were low, but values for two species exceeded 50 (*C. tuberculata* in clearfells and *C. zealandica* in pasture) indicating a strong habitat association. A further 14 species had maximum indicator values greater than 25 (Table 4). Clearfell and native forest had the most distinctive assemblages, with eight and five species, respectively, exceeding an indicator value of 25. Native forest was dominated by native indicator species, whereas many of the clearfell species were exotic in origin. In contrast pasture and mature *P. radiata* had a predominantly generalist fauna, with two and three indicator species respectively. Partial canonical correspondence analysis axis scores of beetle species with significant indicator values greater than 25 were superimposed on the plot of significant environmental variables from the pCCA (Fig. 4b). The three species of Scolytinae were clearly associated with recent clearfells, as were *Cicindela tuberculata*, *Platynus macropterus* and *Scopodes prasinus*. *Saphobius squamulosus*, *Mecodema occiputale*, *Holcaspis mordax*, *Dichrochile maura* and *Ctenognathus bidens* were associated with native forest (Fig. 4b). As expected, the native grass grub, *Costelytra zealandica* (a common pest of improved pasture), was the species most indicative of pastoral habitat (Table 4).

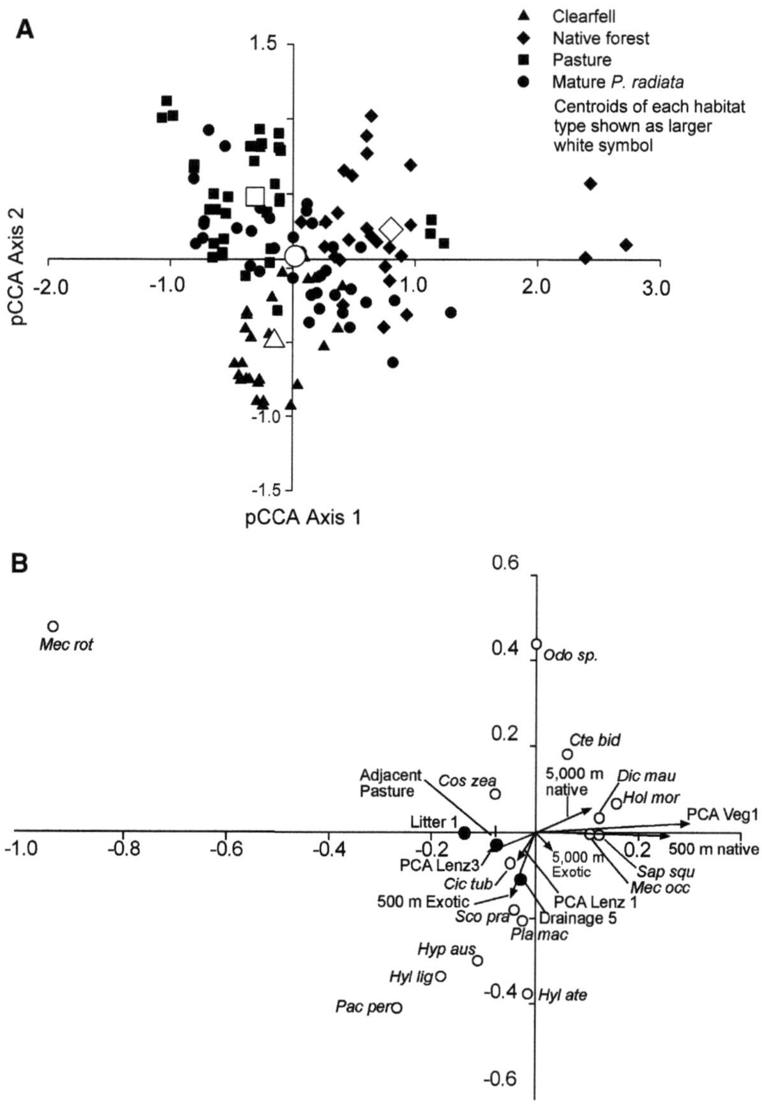

Fig. 4 (**a**) Site-based partial canonical correspondence analysis (pCCA) of beetle species relative abundance for Carabidae, Scarabaeidae and Scolytinae species in different habitat types. (**b**) Biplot of significant environmental variables (Table 3), with the most abundant taxa and those with significant indicator values for at least one habitat type overlaid (abbreviations as in Table 4)

Changes in the proportion of exotic beetles across habitat boundaries

Exotic and native beetle relative abundance varied significantly between habitat types, but this was dependent on the adjacent habitat type (Habitat Comparison · Habitat Group · Origin $F_{3,78} = 3.00$, $P < 0.038$, Table 2). There was negligible invasion by exotic beetles at all distances into native forest despite the presence of exotic species in adjacent production habitats (Fig. 5a). Proportional representation of exotic beetles in *P. radiata* was low at all distances from the native forest edge, whereas the fraction of exotic species

Table 4 Indicator values for beetle taxa in different habitats

Family	Species	Abbreviation	Clearfell	Native	Pasture	Mature *P. radiata*	Significance
Carabidae	*Cicindela tuberculata*	*Cic tub*	85	0	1	0	0.001
Scarabaeidae	*Costelytra zealandica*	*Cos zea*	2	0	53	4	0.001
Carabidae	*Platynus macropterus*	*Pla mac*	39	0	0	0	0.001
Scarabaeidae	*Odontria* sp.	*Odo* sp.	15	5	5	36	0.005
Carabidae	*Ctenognathus bidens*	*Cte bid*	2	34	2	2	0.003
Carabidae	*Mecodema occiputale*	*Mec occ*	12	34	1	9	0.004
Carabidae	*Dichrochile maura*	*Dic mau*	0	33	2	1	0.001
Carabidae	*Hypharpax australis*	*Hyp aus*	31	0	0	5	0.001
Carabidae	*Rhytisternis miser*	*Rhy mis*	27	0	31	2	0.003
Scarabaeidae	*Saphobius squamulosus*	*Sap squ*	5	31	0	4	0.002
Scolytinae	*Hylastes ater*	*Hyl ate*	31	0	0	27	0.005
Carabidae	*Mecyclothorax rotundicollis*	*Mec rot*	30	0	15	0	0.002
Carabidae	*Holcaspis mordax*	*Hol mor*	10	29	7	16	0.038
Carabidae	*Scopodes prasinus*	*Sco pra*	29	1	0	0	0.001
Scolytinae	*Hylurgus ligniperda*	*Hyl lig*	26	0	0	15	0.001
Scolytinae	*Pachycotes peregrinus*	*Pac per*	19	0	0	25	0.004

Values were calculated using the methodologies of Dufrêne and Legendre (1997) on the basis of a priori selected habitat groupings using PC-ORD V. 4.01 (McCune and Mefford 1999)

increased with distance into pasture (Fig. 5a). There was an unusual unimodal abundance pattern in recent clearfells, with exotic beetle relative abundance peaking at 5 m into the clearfell habitat before declining with increasing distance from the forest boundary (Fig. 5a). Plantation stands of *P. radiata* and recently disturbed clearfell habitat had a much lower proportional abundance of exotic beetles than intensively managed pastoral grassland adjacent to native forest (Fig. 5a). Despite the fact that both clearfells and pastoral grassland are structurally open habitats, the proportional abundance of exotic species was much higher in pasture, even when pasture was directly adjacent to clearfell habitat (Fig. 5b).

Discussion

High native beetle biodiversity in plantation forests

In New Zealand's central North Island, intensively-managed exotic plantation forests provided significant habitat for many native beetle species in the three pre-selected beetle families/subfamilies sampled. This is consistent with other recent invertebrate studies in New Zealand (Brockerhoff et al. 2005; Pawson 2006; Berndt et al. 2008) and Australia (Bonham et al. 2002; Mesibov 2005). Estimated species richness was higher in plantations (including recently clearfelled stands) than native forest (Fig. 2a). However, low beetle diversity in native forest was partially a reflection of the apparent 'resilience' of native forest to invasion by exotic species. *Hylastes ater*, a common bark-beetle pest species of *P. radiata* was the only exotic species recorded in native forest sites. In contrast, pasture

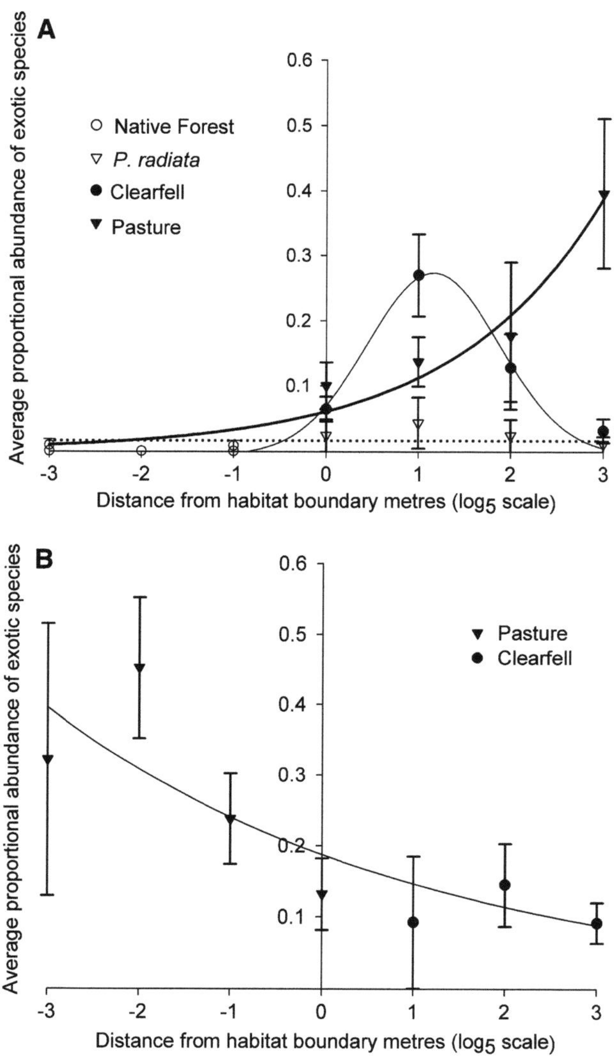

Fig. 5 Average proportional abundance of exotic beetle species across gradients between habitats: (**a**) native forest versus clearfell, *P. radiata* and pasture, and (**b**) pasture versus clearfell. Curves represent continuous response functions fitted using the methods of Ewers and Didham (2006b)

sites had nine exotic species (31% of all species), and *P. radiata* and clearfell sites had eight species each (24% of all species). Harris and Burns (2000) also observed limited exotic species invasion in native kahikatea (*Dacrycarpus dacrydioides*) forest fragments of the Waikato district (∼120 km from our study site), despite the dominance of exotic species in adjacent pasture. Harris and Burns (2000) attributed this to the difference in light levels between native forest and pasture, preventing the establishment of adventive plant species and their host-specific adventive beetles. Our results do not support this conclusion as the canopy closure of native forest and plantations was very similar (calculated using hemispherical photographs), yet *P. radiata* stands supported many more exotic beetle species. However, few of the species considered here were host-specific species, unlike many of the Malaise-trapped flying species of Harris and Burns (2000). Alternative explanations include the difference in disturbance history of the native forest compared to the managed production habitats, the origin (exotic or native) of plantation tree species, and their potential influence on beetle community composition. Disturbance is much more frequent and intense in plantation forests and pasture, and disturbance processes are known to facilitate establishment of invasive species (Hobbs and Huenneke 1992; Lozon and MacIsaac 1997).

If exotic species are excluded from the analysis the estimated species richness falls into two groups, mature plantation habitat and clearfells with higher richness, and native forest and pasture with lower richness (Fig. 2b). The lower native beetle species richness in pasture may reflect both the lack of native host plant species in managed exotic grass swards (Harris and Burns 2000; Ecroyd and Brockerhoff 2005), historical rarity of natural grasslands and their associated beetle communities in the central North Island (see appendix in Kennedy et al. 1978), or the choice of taxa that were sampled.

Relative abundance of exotic species in plantation forests

Beetle relative abundance varied significantly between habitat types, but this was dependent on the adjacent habitat type and on beetle species origin (exotic versus native) (Table 2). Exotic beetles were almost absent from native forest, irrespective of the adjacent habitat type, implying that there may be some attributes of undisturbed native forest that limit establishment of exotic species. Plantation forests are already recognised as a suitable microclimatic buffer for native remnants (Denyer et al. 2006), and the low proportional relative abundance of exotic species in mature *P. radiata* stands (Fig. 5a) suggests that they may also provide a better 'temporary' biological buffer from exotic species compared to alternative pastoral land uses. The word temporary is important to bear in mind, as pine plantations are harvested regularly and the proportional relative abundance of exotic beetles was high immediately adjacent to the native forest boundary in recently clearfelled stands (Fig. 5a). However, forest boundaries are known to act as a barrier to the dispersal of some insect species (Cant et al. 2005), and this may partially explain the unimodal relationship in exotic species dominance with distance away from the forest edge in clearfells. If biodiversity protection and the exclusion of exotic species from adjacent native forest were of critical importance, the use of a long rotation species as a 'buffer strip' could potentially be beneficial.

The proportional relative abundance of exotic beetles in pasture decreased exponentially with increasing proximity to native forest (Fig. 5a), suggesting that there may be increased spill-over of native species into the adjacent pasture habitat (Magura et al. 2001). This may partially explain the greater than anticipated total native beetle diversity in pasture sites as a whole. Whether the native beetles in the pasture samples represent resident populations is unclear, as the dispersal of insects between managed and natural ecosystems is common (see review by Rand et al. 2006). However, additional sampling of pasture sites that are more isolated from natural and plantation forests is required before a definitive statement can be made about the relative importance of dispersal versus resource utilisation in the matrix. In contrast, comparisons between mature *P. radiata* stands and clearfells in our study were conducted deep within the plantation estate, often many kilometres from the influence of alternative habitat types. As such, there is a very low probability that the rich native beetle community sampled in plantation–clearfell comparisons were the result of dispersal from adjacent non-plantation habitat.

Plantations as 'surrogates' for native forest

Of the habitat types sampled, mature *P. radiata* stands appeared to provide the best non-native habitat to augment remaining native forest fragments in this study area. Species composition in mature plantation stands was more similar to native forest than to either pasture or recently clearfelled stands for the three beetle families sampled (Fig. 4a). The environmental drivers regulating the similarity in species composition between the two habitats are unknown. However, we would expect plantation forests to provide an equivalent microclimate to native forest due to their similar canopy cover and known ability to ameliorate microclimate edge effects (Denyer et al. 2006). Furthermore, the leaf-litter and soil chemical properties in *P. radiata* stands are more likely to be analogous to native forest than are the soil properties of open pastoral habitat (Parfitt et al. 1997; Alfredsson et al. 1998). However, plantation forests are dynamic and individual stands are clearfelled about every 28 years in New Zealand. Although harvesting can be locally destructive, it is not necessarily detrimental to landscape-level species persistence if a spatial mosaic of different successional forest stages can be maintained within the landscape (Butterfield 1997; Magura et al. 2003; Pawson 2006).

The high species richness in clearfells (36 species), and their distinctive fauna (Table 4), is consistent with European studies (Niemela et al. 1993; Koivula et al. 2002), which have shown that both open-habitat species and surviving populations of forest generalist species co-exist (at least temporarily) in clearfells. In our study area, native open habitat species such as *Cicindela tuberculata* colonised recent clearfells, as did a number of exotic open-habitat species: e.g., *Anisodactylus binotatus*, *H. australis*, *H. australasiae*, *L. verticalis*, *L. vestigialis* and *R. miser* (Larochelle and Larivière 2001). However, this increase in species richness has been found to be a transient phenomenon, with the relative abundance of these species decreasing in nearby plantation stands greater than four years old (Pawson 2006).

Despite the richness of the beetle fauna in clearfells and mature plantation stands, native forest had a distinct fauna (reflected in the Indval indicator species analysis results, Table 4) characterised by a higher relative abundance of some native species. This, combined with the fact that proximity to native forest was the strongest environmental predictor (Table 3) of beetle community composition in non-native habitats, highlights the importance of retaining native forest within the plantation matrix (Humphrey et al. 2004; Lindenmayer and Hobbs 2004)

Conclusions

In New Zealand, extensive habitat loss and fragmentation have left scattered, isolated native forest remnants spread throughout a landscape matrix dominated by plantation forest and improved pastoral grassland. However, the matrix of modified production ecosystems can provide a considerable extension to the potential habitat for some native beetle species that were previously perceived to be restricted to native forest remnants. Of course, different matrix habitats are not equivalent in quality or habitat-suitability for native beetles. Mature *P. radiata* plantations support native beetle communities that are most similar in composition to those in native forest. Disturbed and open habitat areas are more prone to invasion by exotic species than native forests, where exotic beetle species were uncommon.

Further work is required to understand spatio-temporal variation in the relationship between the mosaic of different-aged plantation stands and how they interact with native habitat at a landscape scale. In particular, some of the most important questions to address are how native forest insects disperse between regenerating plantation stands, whether this is affected by proximity to native habitat, and whether stand-level plantation management techniques, such as legacy management, influence these dispersal processes. Despite the importance of proximity to native habitat, existing native remnants within many New Zealand plantations are often aggregated. Further research is required to determine the value of restoring small native remnants dispersed throughout existing plantations, and their role as source populations for beetle recolonisation of regenerating plantation stands.

Acknowledgements We would like to thank Amy Leighton, Sylvia McLaren, Jo Schaab, Cleland Wallace, Carl Wardhaugh, Michael Watson and Marijn deZwart for assistance with field work and beetle sorting, and David Norton for organising the vegetation sampling, and commenting on an earlier draft of the manuscript. Barbara Hock (Scion) kindly provided an analysis of the proportion of native and exotic habitat around each trap. Simon Grove kindly gave comments on an earlier draft of the manuscript. This work was funded by the University of Canterbury, Scion (via NZ Foundation for Research, Science and Technology contract C04X0304 and associated NSOF), and a Tertiary Education Commission Enterprise scholarship in conjunction with Fletcher Challenge Forests (Dave Lowry), with additional assistance from Kaingaroa Timberlands Ltd (Colin Maunder).

Appendix 1

Average catch per 100 trap days of individual beetle taxa sampled at different distances from the habitat edge into clearfelled plantation forest, mature *P. radiata* forest, native forest, and pasture

Species	Clearfell (1 yr)			*P. radiata* 26 yr			Native forest			Pasture		
	5 m	25 m	125 m	5 m	25 m	125 m	5 m	25 m	125 m	5 m	25 m	125 m
Exotic												
Carabidae												
Anisodactylus binotatus	0.07	0.06		0.03		0.02				0.07		
Anomotarsus illawarrae				0.02								
Hypharpax australasiae	0.12		0.25							0.05	0.05	0.19
Hypharpax australis	0.97	0.37	0.68	0.21	0.06	0.18				0.04		
Lecanomerus verticalis	0.05	0.28	0.37	0.02	0.05	0.05				0.34		0.24
Lecanomerus vestigialis	0.24	0.55	0.25	0.05	0.09	0.25					0.07	0.04
Rhytisternis miser	1.06	2.46	2.41	0.08	0.08	0.47				2.03	2.64	2.11
Scolytinae												
Hylastes ater	3.02	4.78	2.36	3.02	2.32	2.12	0.05		0.05	0.60		
Hylurgus ligniperda	0.64	1.30	0.69	0.60	0.19	0.19				0.06	0.06	

Appendix continued

Species	Clearfell (1 yr)			*P. radiata* 26 yr			Native forest			Pasture		
	5 m	25 m	125 m	5 m	25 m	125 m	5 m	25 m	125 m	5 m	25 m	125 m
Scarabaeidae												
Acrossidius tasmaniae											2.74	
Native												
Carabidae												
Allocinopus sculpticollis		0.05			0.02					0.12		
Amarotypus edwardsii	0.05	0.24	0.40	0.19	0.08	0.09	0.37	0.07		0.33	0.12	
Aulacopodus calathoides	0.52	1.16	1.26	0.38	0.34	0.42	0.99	1.51	0.29	0.73	0.41	1.09
Cicindela parryi	6.94	7.34	2.23	1.28	16.28	0.37	2.27	0.14		1.97	0.10	
Cicindela tuberculata	88.98	105.60	136.40	0.09	2.28	0.92		0.08		22.77	0.48	0.42
Ctenognathus adamsi	0.18	0.91	1.76	0.09	0.40	0.40	1.04	1.79	1.10	1.32	0.57	1.84
Ctenognathus bidens	0.13	0.45	0.18	0.18	0.19	0.13	6.18	1.46	1.39	0.50	0.64	0.26
Demetrida natsuda			0.05									
Dichrochile maura		0.06		0.06		0.06	0.23	0.71	0.24	0.16	0.16	0.10
Holcaspis mordax	1.53	1.47	1.03	2.93	1.68	1.57	1.94	3.27	2.69	2.30	1.72	2.94
Holcaspis mucronata	0.15	0.10	0.05	0.11	0.15	0.15						
Lecanomerus sharpi	0.11	0.10	0.05				0.26	0.37				
Mecodema occiputale	2.91	1.57	0.80	3.73	3.06	2.06	4.33	3.40	2.94	0.81	0.97	0.05
Mecyclothorax rotundicollis	0.40	1.19	2.26		0.05	0.23				0.51	1.42	1.39
Notagonum submetallicum												0.07
Pentagonica vittipennis		0.06		0.18	0.32	0.37						
Platynus macropterus	0.34	1.21	14.16		0.09	0.15					0.06	0.12
Scopodes edwardsi	0.06											
Scopodes multipunctatus	0.07	0.05										
Scopodes prasinus	0.32	1.14	0.10			0.05	0.06	0.12				
Syllectus anomalus		0.15		0.07	0.02				0.05			
Scolytinae												
Pachycotes peregrinus	1.12	0.66	0.38	0.54	0.73	0.49	0.04			0.04		

Appendix continued

Species	Clearfell (1 yr)			*P. radiata* 26 yr			Native forest			Pasture		
	5 m	25 m	125 m	5 m	25 m	125 m	5 m	25 m	125 m	5 m	25 m	125 m
Scarabaeidae												
Ataenius brouni						0.03						
Costelytra sp. A	0.89	0.30	0.16	0.05	1.36	0.19						
Costelytra zealandica	0.88	0.71	0.16	0.88	0.51	0.05	0.41	0.07		7.35	2.31	6.86
Odontria magnum				0.03	0.03	0.05	0.13	0.07				0.10
Odontria piciceps	2.58	1.36	0.74	2.09	4.84	1.44	1.11	0.55	0.63	1.31	0.72	1.03
Odontria sylvatica	0.75	0.30	0.56	0.14	0.02	0.02	0.86	0.52	0.12	0.27	0.24	0.10
Pyronota "red form"				0.05	0.02					0.25	0.21	1.20
Pyronota festiva	0.13	0.44	0.06	0.46	0.15	0.08	1.17	0.62	0.17	1.35	0.58	0.31
Saphobius squamulosus	0.24	1.07	0.51	0.56	0.49	0.49	1.93	3.70	2.43			
Saphobius sp.							0.42		0.05	0.15		
Stethaspis longicornis	0.12	0.13	0.06	0.03	0.02		0.06	0.05	0.11			

References

Alfredsson H, Condron LM, Clarholm M et al (1998) Changes in soil acidity and organic matter following the establishment of conifers on former grassland in New Zealand. For Ecol Manage 112:245–252

Allen RB, Platt KH, Coker REJ (1995) Understorey species composition patterns in a *Pinus radiata* plantation on the central North Island volcanic plateau, New Zealand. N Z J For Sci 25:301–317

Anon (2000) The New Zealand biodiversity strategy: our chance to turn the tide. Whakakohukihukitia Te Tai Roroku Ki Te Tai Oranga. Department of Conservation, Wellington

Anon (2005) New Zealand forest industry; Facts and figures 2004/2005. New Zealand Forest Owners Association, New Zealand Forest Industries Council and Ministry for Agriculture and Forestry, Wellington

Barbaro L, Pontcharraud L, Vetillard F et al (2005) Comparative responses of bird, carabid, and spider assemblages to stand and landscape diversity in maritime pine plantation forests. Ecoscience 12: 110–121

Berndt LA, Brockerhoff EG, Jactel H (2008) Relevance of exotic pine plantations as a surrogate habitat for ground beetles (Carabidae) where native forest is rare. Biodivers Conserv: in press

Bonham KJ, Mesibov R, Bashford R (2002) Diversity and abundance of ground-dwelling invertebrates in plantation vs. native forests in Tasmania, Australia. For Ecol Manage 158:237–247

Borcard DP, Legendre P, Drapeau P (1992) Partialling out the spatial component of ecological variation. Ecology 73:1045–1055

Brockerhoff EG, Ecroyd CE, Langer ER (2001) Biodiversity in New Zealand plantation forests: policy trends, incentives and the state of our knowledge. N Z J For 46:31–37

Brockerhoff EG, Ecroyd CE, Leckie AC et al (2003) Diversity and succession of adventive and indigenous vascular understorey plants in *Pinus radiata* plantation forests in New Zealand. For Ecol Manage 185:307–326

Brockerhoff EG, Berndt L, Jactel H (2005) Role of exotic pine forests in the conservation of the critically endangered New Zealand ground beetle *Holcaspis brevicula* (Coleoptera:Carabidae). N Z J Ecol 29:37–44

Brockerhoff EG, Shaw WB, Hock B et al (2008) Re-examination of recent loss of indigenous cover in New Zealand and the relative contributions of different land uses. N Z J Ecol: in press

Brooks TM, Mittermeier RA, Mittermeier CG et al (2002) Habitat loss and extinction in the hotspots of biodiversity. Conserv Biol 16:909–923

Butterfield J (1997) Carabid community succession during the forestry cycle in conifer plantations. Ecography 20:614–625

Cant ET, Smith AD, Reynolds DR et al (2005) Tracking butterfly flight paths across the landscape with harmonic radar. P Roy Soc Lond B 272:785–790

Carnus J-M, Parrotta JA, Brockerhoff EG et al (2006) Planted forests and biodiversity. J Forest 104:65–77

Clout MN (1984) Improving exotic forests for native birds. N Z J For 29:193–200

Clout MN, Gaze PD (1984) Effects of plantation forestry on birds in New Zealand. J Appl Ecol 21:795–815

Colwell RK (2004) Estimate S: Statistical estimation of species richness and shared species from samples. Available from persistent URL: http://purl.oclc.org/estimates

Colwell RK, Coddington JA (1994) Estimating terrestrial biodiversity through extrapolation. Philos T Roy Soc B 345:101–118

Colwell RK, Mao CX, Chang J (2004) Interpolation, extrapolating, and comparing incidence-based species accumulation curves. Ecology 85:2717–2727

Denyer K, Burns BR, Ogden J (2006) Buffering of native forest edge microclimate by adjoining tree plantations. Austral Ecol 31:478–489

Didham RK, Hammond PM, Lawton JH et al (1998) Beetle species responses to tropical forest fragmentation. Ecol Monogr 68:295–323

Dufrêne M, Legendre P (1997) Species assemblages and indicator species: The need for a flexible asymmetrical approach. Ecol Monogr 67:345–366

Ecroyd CE, Brockerhoff EB (2005) Floristic changes over 30 years in a Canterbury plains kanuka forest remnant, and comparison with adjacent vegetation types. N Z J Ecol 29:279–291

Ewers RM, Didham RK (2006a) Confounding factors in the detection of species responses to habitat fragmentation. Biol Rev (Camb) 81:117–142

Ewers RM, Didham RK (2006b) Continuous response functions for quantifying the strength of edge effects. J Appl Ecol 43:527–536

Ewers RM, Kliskey AD, Walker S et al (2006) Past and future trajectories of forest loss in New Zealand. Biol Conserv 133:312–325

Ewers RM, Thorpe S, Didham RK (2007) Synergistic interactions between edge and area effects in a heavily fragmented landscape. Ecology 88:96–106

Fahrig L (2003) Effects of habitat fragmentation on biodiversity. Annu Rev Ecol Syst 34:487–515

Fazey I, Fischer J, Lindenmayer DB (2005) What do conservation biologists publish? Biol Conserv 124: 63–73

Franklin JF, Lindenmayer BD, MacMahon JA et al (2000) Threads of continuity: ecosystem disturbances, biological legacies and ecosystem recovery. Conserv Biol Prac 1:8–16

Froggatt PC, Lowe DJ (1990) A review of late Quaternary silicic and some other tephra formations from New Zealand: the stratigraphy, nomenclature, distribution, volume and age. New Zeal J Geol Geophys 33:89–109

Gates S, Donald PF (2000) Local extinction of British farmland birds and the prediction of further loss. J Appl Ecol 37:806–820

Geldenhuys CJ (1997) Native forest regeneration in pine and eucalypt plantations in Northern Province, South Africa. For Ecol Manage 99:101–115

Hale ML, Lurz PWW, Shirley MDF et al (2001) Impact of landscape management on the genetic structure of red squirrel populations. Science 293:2246–2248

Harris RJ, Burns BR (2000) Beetle assemblages of kahikatea forest fragments in a pasture-dominated landscape. N Z J Ecol 24:57–67

Hartley MJ (2002) Rationale and methods for conserving biodiversity in plantation forests. For Ecol Manage 155:81–95

Hobbs RJ, Huenneke LF (1992) Disturbance, diversity, and invasion: implications for conservation. Conserv Biol 6:324–337

Hock B, Hay E (2003) Forest certification in New Zealand: how are we doing? N Z J For 47:17–23

Humphrey JW (2005) Benefits to biodiversity from developing old-growth conditions in British upland spruce plantations: a review and recommendations. Forestry 78:33–53

Humphrey JW, Hawes C, Peace AJ et al (1999) Relationships between insect diversity and habitat characteristics in plantation forests. For Ecol Manage 113:11–21
Humphrey JW, Peace AJ, Jukes MR et al (2004) Multiple-scale factors affecting the development of biodiversity in UK plantations. In: Honnay O, Verheyen K, Bossuyt B, Hermy M (eds) Forest biodiversity: lessons from history for conservation. CAB International, pp 143–162
Hutcheson J, Jones D (1999) Spatial variability of insect communities in a homogenous system: measuring biodiversity using Malaise trapped beetles in a *Pinus radiata* plantation in New Zealand. For Ecol Manage 118:93–105
Kennedy NM, Pillar WA, Pain CF (1978) Late quaternary land surfaces and geomorphic changes in the Rotorua basin, North Island, New Zealand. N Z J Sci 21:249–264
Kerr G (1999) The use of silvicultural systems to enhance the biological diversity of plantation forests in Britain. Forestry 72:191–205
Koivula M, Kukkonen J, Niemela J (2002) Boreal carabid-beetle (Coleoptera, Carabidae) assemblages along the clear-cut originated succession gradient. Biodivers Conserv 11:1269–1288
Lang A (2000) The pitfalls of pitfalls: a comparison of pitfall trap catches and absolute density estimates of epigeal invertebrate predators in arable land. J Pest Sci 73:99–106
Larochelle A, Larivière M-C (2001) Carabidae (Insecta: Coleoptera): catalogue. Fauna N Z 43:1–285
Leathwick J, Morgan F, Wilson G et al (2002) Land environments of New Zealand: a technical guide. Ministry for the Environment, Wellington
Lindenmayer DB, Hobbs RJ (2004) Fauna conservation in Australian plantation forests–a review. Biol Conserv 119:151–168
Lozon JD, MacIsaac HJ (1997) Biological invasions: are they dependent on disturbance? Environ Rev 5:131–144
Magura T, Tothmeresz B, Molnar T (2001) Forest edge and diversity: carabids along forest-grassland transects. Biodivers Conserv 10:287–300
Magura T, Tothmeresz B, Elek Z (2003) Diversity and composition of carabids during a forestry cycle. Biodivers Conserv 12:73–85
Mazurek MJ, Zielinski WJ (2004) Individual legacy trees influence vertebrate wildlife diversity in commercial forests. For Ecol Manage 193:321–334
McCune B, Mefford MJ (1999) PC-Ord: multivariate analysis of ecological data. MJM Software, Gleneden beach, Oregon
McGlone MS (1989) The Polynesian settlement of New Zealand in relation to environmental and biotic changes. N Z J Ecol 12(Suppl):115–129
Mesibov R (2005) Native species dominate the millipede fauna in a second-rotation *Pinus radiata* plantation in Tasmania, Australia. Pac Conserv Biol 11:17–22
Niemela J, Langor DW, Spence JR (1993) Effects of clear-cut harvesting on boreal ground-beetle assemblages (Coleoptera:Carabidae) in western Canada. Conserv Biol 7:551–561
Nittérus K, Gunnarson B (2006) Effect of microhabitat complexity on the local distribution of arthropods in clear-cuts. Environ Entomol 35:1324–1333
Norton DA (1998) Indigenous biodiversity conservation and plantation forestry: options for the future. N Z J Forest 43:34–39
Norton DA (2001) Is the amount and focus of ecological research in New Zealand sufficient to sustain indigenous biodiversity on private land? N Z J Ecol 25:77–82
Ogden J, Braggins J, Stretton K et al (1997) Plant species richness under *Pinus radiata* stands on the central North Island volcanic plateau, New Zealand. N Z J Ecol 21:17–29
Oxbrough AG, Gittings T, O'Halloran J et al (2005) Structural indicators of spider communities across the forest plantation cycle. For Ecol Manage 212:171–183
Parfitt RL, Percival HJ, Dahlgren RA et al (1997) Soil and solution chemistry under pasture and *radiata* pine in New Zealand. Plant Soil 191:279–290
Pawson SM (2006) Effects of landscape heterogeneity and clearfell harvest size on invertebrate biodiversity in plantation forests. Unpublished PhD Thesis. University of Canterbury, Christchurch, New Zealand
R-Team (2006) R: a language and environment for statistical computing. R-Foundation for Statistical Computing, Vienna
Rand TA, Tylianakis JM, Tscharntke T (2006) Spillover edge effects: the dispersal of agriculturally subsidised insect natural enemies into adjacent natural habitats. Ecol Lett 9:603–614
Reed DH (2004) Extinction risk in fragmented habitats. Anim Conserv 7:181–191
Roche M (1990) History of New Zealand forestry. GP Print Ltd, Wellington
Ryder HR (1948) Birds of Kaingaroa forest. N Z Bird Notes 3:20–22
Sala OE, Chapin FS, Armesto JJ et al (2000) Global biodiversity scenarios for the year 2100. Science 287:1770–1774

Southwood TRE (1994) Ecological methods: with particular reference to the study of insect populations. Chapman & Hall, London

Ter Braak CJ (1995) Ordination. In: Jongman RHG, Ter Braak CJ, Van Tongeren OFR (eds) Data analysis in community and landscape ecology. Cambridge University Press, Cambridge, pp 91–169

ter Braak CJ, Smilauer P (1999) Canoco for Windows Version 4.02. Wageningen

Terralink (2004) New Zealand land cover database (LCDB2). Terralink International Limited, Wellington, New Zealand

Tilman D, May RM, Lehman CL et al (1994) Habitat destruction and the extinction debt. Nature 371:65

Walker S, Price R, Rutledge D et al (2006) Recent loss of indigenous cover in New Zealand. N Z J Ecol 30:169–177

Ward DF, New TR, Yen AL (2001) Effects of pitfall trap spacing on the abundance, richness and composition of invertebrate catches. J Insect Cons 5:47–53

Wardle P (1991) Vegetation of New Zealand. Cambridge University Press, Cambridge

Weeks MF (1949) Bird population of exotic forests. N Z Bird Notes 3:83–84

Wilmshurst JM, McGlone MS (1996) Forest disturbance in the central North Island, New Zealand, following the 1850 BP Taupo eruption. Holocene 6:399–411

Woodcock BA, Watt AD, Leather SR (2003) Influence of management type on Diptera communities of coniferous plantations and deciduous woodlands. Agric Ecosyst Environ 95:443–452

Importance of semi-natural habitats for the conservation of butterfly communities in landscapes dominated by pine plantations

Inge van Halder · Luc Barbaro · Emmanuel Corcket · Hervé Jactel

Originally published in the journal Biodiversity and Conservation, Volume 17, No 5, 1149–1169.
DOI: 10.1007/s10531-007-9264-5

Abstract While the area of plantation forests continues to increase worldwide, their contribution to the conservation of biodiversity is still controversial. There is a particular concern on the central role played by natural habitat remnants embedded within the plantation matrix in conserving species-rich insect communities. We surveyed butterflies in maritime pine plantation landscapes in south-western France in 83 plots belonging to seven habitat types (five successional stages of pine stands, native deciduous woodlands and herbaceous firebreaks). The effect of plot, habitat and landscape attributes on butterfly species richness, community composition and individual species were analysed with a General Linear Model (GLM), partial Canonical Correspondence Analysis (CCA) and the IndVal method. The most important factors determining butterfly diversity and community composition were the presence of semi-natural habitats (deciduous woodlands and firebreaks) at the landscape scale and the composition of understorey vegetation at the plot scale. Pure effects of plot variables explained the largest part of community variation (12.8%), but landscape factors explained an additional, independent part (6.7%). Firebreaks were characterized by a higher species richness and both firebreaks and deciduous woodlands harboured species not or rarely found in pine stands. Despite the forest-dominated landscape, typical forest butterflies were rare and mainly found in the deciduous woodlands. Threatened species, such as *Coenonympha oedippus* and *Euphydryas aurinia*, were found in pine stands and in firebreaks, but were more abundant in the latter. In the studied plantation forest, the conservation of butterflies depends mainly on the preservation of semi-natural habitats, an adequate understorey management and the maintenance of soil moisture levels.

Keywords Butterflies · Communities · Deciduous woodlands · Firebreaks · Habitat · Landscape · *Pinus pinaster* · Plantation forests

I. van Halder (✉) · L. Barbaro · H. Jactel
INRA, UMR1202 Biodiversité, Gènes et Communautés, 69 Route d'Arcachon, F-33612 Cestas, France
e-mail: inge.vanhalder@pierroton.inra.fr

E. Corcket
UMR1202 Biodiversité, Gènes et Communautés, Ecologie des Communautés, Université Bordeaux 1, Avenue des Facultés, F-33405 Talence, France

E.G. Brockerhoff et al. (eds.), *Plantation Forests and Biodiversity: Oxymoron or Opportunity?*
DOI: 10.1007/978-90-481-2807-5_12

Introduction

Plantation forests with their intensive silvicultural management and simplified structure and composition are often considered less valuable for biodiversity conservation than natural forests (Hartley 2002). Many, but not all studies comparing plantations to more natural forests have indeed shown an impoverished flora and fauna in plantations (Moore and Allen 1999; Lindenmayer and Hobbs 2004; Carnus et al. 2006). However, forest management in plantation forests is not incompatible with biodiversity conservation and possibilities exist to enhance their biodiversity (Kerr 1999; Hartley 2002; Carey 2003; Carnus et al. 2006). Apart from providing a habitat, plantation forests can also have beneficial effects as landscape matrix elements by increasing the connectivity of natural forest remnants (Aberg et al. 1995) or by acting as a buffer to mitigate negative edge effects for forest interior species (Aune et al. 2005; Fischer et al. 2006). Biodiversity conservation in plantation landscapes will however also depend on the presence of more natural habitat elements, such as wetlands or late successional stages of remnant forest, within the plantation matrix (Lindenmayer and Hobbs 2004; Fischer et al. 2006).

Conserving biodiversity in plantation forests implies the identification of explanatory, environmental factors that determine patterns of species occurrences. Since species respond to environmental factors at different, interacting scales (from the micro-habitat and habitat to the landscape and regional scale) multi-scale approaches are required to analyse these causal mechanisms (Wiens 1989; Cushman and McGarigal 2002). In this study we analysed the effect of factors at both the local and landscape scale on butterfly diversity in pine plantation landscapes. Butterflies were chosen because they are easy to identify in surveys and include species with different habitat preferences and dispersal capacities and show therefore different responses to habitat and landscape features (Dennis 1992; Thomas 1995). Moreover, a large number of butterfly species are declining at an alarming rate through substantial parts of their European range and conservation measures are urgently needed (Van Swaay and Warren 1999).

During their life cycle most butterfly species need complementary resources (hostplants for larvae, nectar plants for adults, roosting-, resting- and overwintering-sites, favourable micro-climatological conditions) resulting in very direct relationships with habitat characteristics such as vegetation composition and management (Dennis et al. 2003). For many taxa, including butterflies, habitat characteristics alone are often insufficient to predict species presence or abundance and landscape characteristics can provide additional explanatory information (Mazerolle and Villard 1999; Jeanneret et al. 2003a; Krauss et al. 2003; Bergman et al. 2004; Stefanescu et al. 2004). Many butterfly studies conducted at the landscape scale have focussed on the effect of patch size and isolation and have used the equilibrium theory of island biogeography (Mac Arthur and Wilson 1967) or the metapopulation theory (Hanski 1999) to explain species richness or population dynamics, respectively (Thomas and Harrison 1992; Baguette et al. 2000; Steffan-Dewenter and Tscharntke 2000; Anthes et al. 2003). Both theories assume clearly delimited habitats surrounded by uniformly unsuitable habitat (the landscape matrix). However landscape matrices are not entirely hostile and the 'mosaic concept' (Wiens 1995; Duelli 1997) can offer an alternative to explain species richness. In this concept species richness increases with the number of biotope types per unit area, the number of patches, the edge length and the proportion of natural and semi-natural areas (Duelli 1997). Many butterfly species are found along edges or use resources in different vegetation types (Dennis et al. 2006) thus supporting the mosaic concept. Positive effect of patch density (PD) on butterfly diversity has been demonstrated by Schneider and Fry (2001) and Debinski et al. (2001). Dunning et al. (1992) described these effects of landscape context in terms of landscape complementation and supplementation, corresponding to the use of patches with non-substitutable or substitutable resources, respectively. Landscape effects found to be linked with species diversity will also depend on the scale of the analysed landscape; shorter distances will be more related to landscape complementation/supplementation and mosaic concepts (Weibull et al. 2000; Schneider and Fry 2001) and larger scales to metapopulation functioning and habitat thresholds (Bergman et al. 2004).

The aim of this study was therefore to identify key factors at both the habitat and landscape level that drive butterfly diversity in plantation forests, and that can be used by forest managers and landscape planners to maintain or restore butterfly diversity. We address the following questions:

- Do habitat types in a pine plantation landscape differ in butterfly species richness and composition?
- What is the contribution of semi-natural and open habitats such as oak woodland remnants and herbaceous firebreaks to butterfly diversity in pine plantation landscapes?
- What is the relative importance of understorey vegetation composition, habitat-type and landscape attributes on butterfly community composition?

Methods

Study area and plot selection

The study was carried out in South West France in the 'Landes de Gascogne' (Fig. 1), a region covering one million ha and dominated by plantations of native maritime pine (*Pinus pinaster*). Silvicultural management of the pine stands is intensive, including soil preparation and fertilization before seeding or planting, mechanical understorey removal and four thinning operations within the 40–50 years rotation cycle (Trichet et al. 1999). Deciduous woodlands are rare and found along rivers or as scattered patches of a few hectares. They are generally dominated by *Quercus robur*, on dry sites by *Q. pyrenaica* and along rivers by *Alnus glutinosa* and *Q. robur*. Open areas in the landscape are mainly represented by large maize fields, pine clearcuts, firebreaks and powerlines.

The whole region is covered by nutrient poor, acid podzol soils with a pH of 3.5–5.5 (Trichet et al. 1999). Differences in soil moisture have an important effect on the understorey vegetation composition in forest stands: in wet conditions *Molinia caerulea* is dominant with presence of *Erica tetralix*, intermediate conditions are characterized by dominance of *Pteridium aquilinum* and *Ulex europaeus* and in dry condition *Calluna vulgaris* and *Erica cinerea* dominate (Timbal and Maizeret 1998). Firebreaks and powerlines can have a heathland vegetation as described above or a grassland vegetation, dominated by for example *Holcus lanatus* or *Anthoxanthum odoratum*. Management of firebreaks and powerlines is very diverse. In the studied region the usefulness of firebreaks in preventing forest fires is considered doubtful and only firebreaks that are classified in a fire prevention scheme are mown once a year in summer. Other, private firebreaks are mown less often (every 4–8 years) or are progressively transformed to pine plantations. Management of powerlines is in general extensive (every 4–8 years) and aims at suppressing the regrowth of woody species, but in some cases they are mown annually. Firebreaks and powerlines typically have a width of about 15–100 m.

Within the 'Landes de Gascogne' two study sites were selected: Tagon (5,000 ha), situated 35 km southwest of Bordeaux and Solferino (10,500 ha), located 65 km to the south of

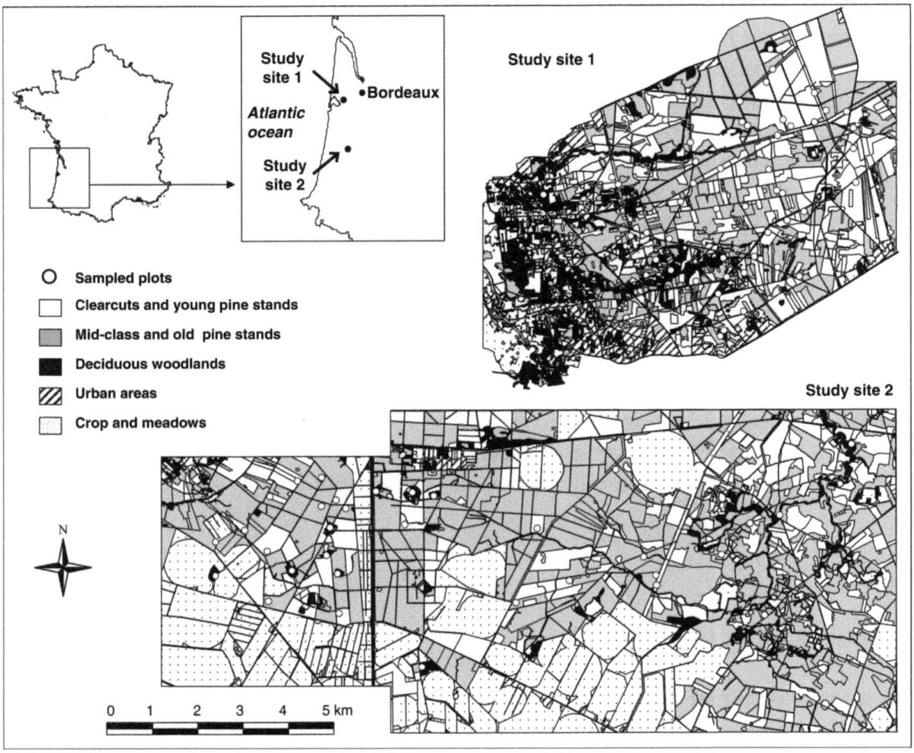

Fig. 1 Map of the two study sites in the south-west of France and location of the sampled plots within each site. The polygon boundaries represent the edges of landscape elements such as pine stand edges or edges of roads. Firebreaks are the very narrow polygons between some stands or along some roads and at the scale of the figure cannot be separately indicated since they occupy only a small percentage of the total landscape

Tagon (Fig. 1). Both sites are dominated by maritime pine plantations, present similar types of soils, and include different degrees of landscape fragmentation and heterogeneity. A total of 83 plots were selected in the two sites (Fig. 1, Table 1) belonging to seven different habitat types. These seven habitat types were defined a priori and represent the main land-use types within the forested landscape as well as being habitat types of ecological relevance to butterflies. Five of them were related to successional stages of maritime pine plantations: herbaceous clearcuts, shrubby clearcuts, young pines (canopy height <7 m), mid-class pines (canopy height 7–15 m) and older pines (canopy height >15 m). The two other habitat types were deciduous woodlands (isolated patches or riparian forests) and firebreaks or powerlines (hereafter called firebreaks). Plots with different types of understorey vegetation (humid, mesic and dry) were selected for each habitat type.

Butterfly sampling

Butterflies were recorded in the 83 plots using the line-transect method (Pollard and Yates 1993). In each plot a transect of eight sections of 50 m long was laid out and butterflies were counted within 2.5 m on each side of the transect line and 5 m ahead of the recorder. Species were identified by sight or caught and released for species difficult to identify (e.g. *Thymelicus* species). Each plot was visited four times between May 14th and September 4th 2004. Surveys were conducted between 10:00 and 18:00 h and only during appropriate

Table 1 Plot and landscape variables used in GLM and CCA analyses

Acronym	Description	FB 12 plots	PP1 10 plots	PP2 10 plots	PP3 10 plots	PP4 11 plots	PP5 11 plots	DW 19 plots
Plot variables								
Nectar	Log (number of flowers +1)	7.8	6.0	6.9	6.8	5.4	6.1	3.5
%Soil	% Bare soil	13.7	29.1	28.2	9.6	26.1	16.4	33.4
%Paqu	% *Pteridium aquilinum*	1.5	12.5	3.0	17.2	7.3	16.8	20.2
%Mcae	% *Molinia caerulea*	28.3	42.3	22.7	29.2	35.6	52.5	10.4
%Grass	% Other grass species	40.0	14.7	10.4	20.9	8.3	8.9	19.7
%Dicots	% Herbaceous dicotyledons	7.2	2.1	1.7	0.9	1.4	1.0	1.0
%Erica	% Ericaceae <0.7 m	6.9	5.1	15.7	15.4	12.7	7.4	1.0
%Umin	% *Ulex minor* <0.7 m	1.3	1.5	1.5	9.0	1.4	2.0	0.1
%Oth-her	% Other plants herbaceous layer	1.5	1.4	3.8	1.8	4.0	1.1	14.8
%Ueur	% *Ulex europaeus* >0.7 m	1.2	1.4	14.1	3.1	7.6	2.7	0.3
%Faln	% *Frangula alnus* >0.7 m	0.8	3.0	6.9	3.9	7.5	5.0	1.1
%Esco	% *Erica scoparia* >0.7 m	0.6	0.3	1.9	1.2	3.8	6.5	0.8
%Oth-shrub	% Other shrubs >0.7 m	3.2	1.7	5.8	0.5	2.5	1.5	20.3
Moisture	Soil moisture at 50 cm (classes 0–4)	1.7	2.2	1.4	1.7	1.5	1.8	1.3
Landscape variables								
%FB	% Firebreaks	7.2	1.1	1.9	2.2	3.2	2.0	2.1
%PP1	% Herbaceous clearcuts	1.7	27.0	3.1	2.1	1.7	4.4	6.4
%PP2	% Shrubby clearcuts	16.0	7.9	22.2	8.1	4.4	9.2	8.0
%PP3	% Young pine stands (<7 m)	26.8	11.1	15.6	33.2	10.0	18.2	14.6
%PP4	% Mid-class pine stands (7–15 m)	26.2	28.6	21.1	33.5	55.7	21.0	16.5
%PP5	% Older pine stands (>15 m)	10.6	17.4	27.8	14.0	13.9	32.1	17.6
%DW	% Deciduous/mixed woodland	6.9	2.6	4.7	1.5	3.0	2.4	16.5
SHDI	Shannon diversity index	1.6	1.4	1.5	1.4	1.2	1.5	1.6
SHEI	Shannon eveness index	0.75	0.71	0.74	0.68	0.60	0.75	0.73
PRD	Patch richness density	15.9	14.5	14.5	14.7	14.1	14.8	17.8
SHAPE	SHAPE index	2.1	1.9	2.0	1.9	1.9	1.9	2.0
ED	Edge density (edge length in m/ha)	229.3	180.5	203.1	193.3	184.9	193.2	224.8
PD	Patch density (patches/100 ha)	56.0	41.8	49.3	46.9	40.7	44.5	57.0

Mean values are given per habitat type. Abbreviations for habitat types: *FB* firebreaks, *PP1* herbaceous clearcuts, *PP2* shrubby clearcuts, *PP3* young pine stands (<7 m), *PP4* mid-class pine stands (7–15 m), *PP5* older pine stands (>15 m), *DW* deciduous woodlands. Landscape variables were measured in a 50-hectare circle (including the inventoried plot)

weather conditions (temperature >20°C, cloudless or just a few clouds and wind force <5 Beaufort). The order of the plots and habitat types surveyed was randomized per visiting period and visits to the two sites alternated. For data analysis the total number of individuals per species was pooled over the four visits and eight sections for each plot.

Butterfly species were classified as typical forest or non-forest species. Limits between forest and non-forest species are not strict, but we defined as forest species those species whose adults and immature stages are more often found within forests than in open habitats (Ebert and Rennwald 1991). We also attributed a European and national threat status to all native species, excluding migrants such as *Vanessa cardui* (Van Swaay and Warren 1999). For the European status we used the list of threatened species cited in the Red Data Book of European Butterflies (Van Swaay and Warren 1999). In this Data Book the IUCN criteria, which are based on population declines over a 10-year period, were adjusted to butterfly data using a roughly equivalent distribution decline over a 25-year period. Species with a decrease of at least 20% are classified as threatened, and depending on their total decrease and present distribution classed as critically endangered, endangered or vulnerable. Species with a decrease of 15–20% and a present distribution of <1% of Europe are also classified as vulnerable. For the French national status we calculated the distribution trend by dividing the number of departments where a species was not seen after 1980, but was present before 1980, by the total number of departments where the species was ever seen (Lafranchis 2000). Species with a distribution decrease of at least 30% were classified as 'nationally threatened', assuming that these species are vulnerable at a national scale. We used a less severe threshold to compensate for the lack of data in several departments (Lafranchis 2000). Butterfly species are named in the text according to Karsholt and Razowski (1996).

Plot variables

We measured a set of potential explanatory variables at both the plot and landscape scale in order to relate butterfly species richness and community composition to environment. At the plot scale three types of variables were measured: flower abundance, vegetation composition of the herbaceous and shrub layer and soil moisture (Table 1). Flower abundance was measured to estimate the availability of nectar, the most common food source for adult butterflies in temperate areas (Shreeve 1992; Ebert and Rennwald 1991). Flower abundance was estimated during each of the four butterfly surveys using the method described by Clausen et al. (2001). Only plant species known to be used by butterflies as nectar plants were noted (Ebert and Rennwald 1991; van Halder, personal observations). Flower abundance was estimated per plant family or per species for abundant and easily identifiable species. The number of flower units was estimated in every section using the following abundance classes: 1–25, 26–50, 51–100, 101–200, 201–400, 401–800 and 800–1,600 flower units (Clausen et al. 2001). For data analysis the mid-values of each class were summed over the eight sections, the four visits and the different flower species or families. Total flower abundance was log-transformed to reduce the effect of outliers and because we hypothesized a non-linear relation between butterfly and flower abundance.

At the end of the field season understorey vegetation composition was recorded in a representative section within each plot. The vegetation was divided into a herbaceous layer (<0.7 m) and a shrub layer (0.7–7 m) and for each layer the % cover of the main vegetation components was estimated as the relative foliage area projected on a horizontal plane.

Soil moisture was estimated once between May and July 2004 at two points in each plot at a depth of 50 cm using a relative scale from 0 to 4, based on tactile and visual criteria. We used this estimation method because it can be used on soil samples extracted with an auger, a very easy and quick method, whereas measurements with a probe at 50 cm depth would have needed to dig a soil profile of at least 50 cm deep. We measured volumic soil moisture using a Theta Probe type ML2x (Delta-T Devices Ltd., Cambridge, UK) with 12 replications per class to test the relationship between moisture estimates and measures. There were significant changes in measured soil moisture between our relative classes (ANOVA, $F = 78.4$, $P < 0.001$). The scale from 0 to 4 corresponds to a mean soil moisture of 4.0, 12.8, 19.6, 44.4 and 62.6%, respectively.

Landscape variables

Land-use types in the two study sites were mapped in a GIS (Arcview 3.3, ESRI) using aerial colour photos with a resolution of 50 cm as background layer. Photos dated from 2000 and 2002 for the Tagon and Solferino region, respectively, and patch attributes were verified in the field. Twelve different land-use types that could be identified on these aerial photos were mapped: the seven surveyed habitat types and five rarer land-use types: hedgerows, meadows, crops, roads and urban areas. Landscape metrics were calculated within circular buffers with a radius of 400 m (circa 50 ha), from the centre of the sampled plots, using Fragstats 3.3 in raster version and a cell size of 2.5 m (McGarigal et al. 2002). Since the surveyed patch is (partly) included in our buffer the calculated metrics represent a combination of patch and landscape characteristics (Fahrig 2003). For most butterfly species in temperate areas 400 m is considered a moderate dispersal distance (Maes and Van Dyck 1999) and could therefore reveal ecologically relevant landscape relations. Larger buffers were not analysed because overlap between buffers would increase collinearity of data. Within each buffer the percentage cover of the seven main habitat types was calculated as well as several metrics reflecting landscape heterogeneity and fragmentation (Table 1). We used the Shannon Diversity Index (SHDI), the Shannon Eveness Index (SHEI) and the Patch Richness Density (PRD) as metrics of landscape heterogeneity and the SHAPE index, the PD and Edge Density (ED) as metrics for landscape fragmentation (McGarigal et al. 2002). Studies on birds, spiders and carabids in the same area have shown an effect of landscape composition and landscape structure (patch size, ED and SHDI) on species composition and richness (Barbaro et al. 2005). Landscape effects on butterflies have been analysed in several studies showing an effect of landscape composition (Schneider and Fry 2001; Söderström et al. 2001; Jeanneret et al. 2003a; Stefanescu et al. 2004), landscape fragmentation (Schneider and Fry 2001) and landscape heterogeneity (Weibull et al. 2000; Jeanneret et al. 2003a; Krauss et al. 2003).

Data analysis

Analyses were performed at different levels of biodiversity (species richness, single species abundances and composition of species assemblages) using hierarchical sets of explanatory variables: habitat type, plot variables and landscape variables. The effect of habitat type on number of species and total abundance of butterflies was tested by a one-way ANOVA, followed by Tukey's post hoc test. Total abundance of butterflies was log-transformed to improve normality of residuals. Species richness was analysed with a general linear model (GLM), using site, habitat and their interaction as categorical variables, and plot and landscape variables as continuous variables. Quadratic terms of plot and landscape variables were added to examine the possibility of curvilinear relationships between explanatory variables and species richness. We used a forward stepwise selection procedure ($P < 0.05$ for inclusion) to build the model. The possible interaction between selected categorical and continuous variables was tested in a forward procedure with the selected variables and their interaction.

To identify species characteristic for a habitat type or a group of habitat types we used the Indicator Value (IndVal) method (Dufrêne and Legendre 1997). Indicator species can be defined as species found mostly in a certain habitat type and present in the majority of sites of that type. To incorporate these two criteria the IndVal index multiplies the relative abundance of a species in a habitat (mean abundance in a habitat divided by the sum of mean abundances in all habitats) with the frequency of occurrence in that habitat. The index is calculated for each habitat and the IndVal of a species corresponds to the largest IndVal value observed over the different habitats. The index is maximum (100%) when the individuals of a species are only observed in one habitat type and in all sites of that habitat. The IndVal of a species can be calculated for each level of a hierarchic site typology. The level where the species reaches its maximum IndVal index can be considered as the 'best' clustering level for that species (Dufrêne and Legendre 1997). This procedure distinguishes between generalist species (maximum IndVal at a higher cluster level) and stenotopic species (maximum at lower levels). In our Indval analysis we used a hierarchic site typology based on the habitat types we had distinguished a priori. The clustering of groups in the hierarchy was based on their stand structure similarity. The first level groups all sites and permits identification of species that have higher IndVals for all samples than for any sample subset (generalist species). The second level separates open habitats from forested habitats, in the next steps the open habitats are separated in herbaceous and shrubby habitats, the forested habitats in pine stands and deciduous woodlands and so on (see Fig. 3 for separations in further steps). This classification tests if species are characteristic for a specific clustering of predefined habitat types. The statistical significance of the index was estimated at each level of the hierarchy by a random reallocation procedure of plots among plot groups based on 999 permutations (Dufrêne and Legendre 1997). Species present with <5 individuals were excluded from analysis.

Canonical Correspondence Analysis (CCA) was used to relate environmental variables to species assemblages (Ter Braak 1986; Palmer 1993). CCA is an ordination technique for multi-variate direct gradient analysis in which the ordination axes of a Correspondence Analysis (CA) are constrained to be linear combinations of the environmental variables (Ter Braak 1986). The % of variance in the species data set that is 'explained' by the environmental variables can be calculated by dividing the inertia of the canonical axes by the total inertia of the CA and this % represents an overall method of CCA fit. We tested the explanatory effects of three sets of variables in separate CCA analyses: the seven habitat types, the 14 plot variables and the 13 landscape variables (see Table 1). For each set significant variables were selected in a forward stepwise procedure based on the additional variation explained by each variable ($P < 0.05$ for inclusion). Next, we combined the selected variables in one CCA model and calculated the variation explained independently and jointly by the different sets of variables by performing several partial CCA analyses (Borcard et al. 1992; Cushman and McGarigal 2002). In a partial CCA the pure effect of a variable or a group of variables is calculated after eliminating the variance due to other variables (the covariables). The different parts of the variation partitioning were calculated following the formulas given by Cushman and McGarigal (2002). Finally, we determined for each variable if it explained a significant part of variation when the variables in the two other subsets were used as covariables. Significance of the additional effect of each variable during the forward selection procedure and of the different (partial) CCA models was tested with 999 Monte Carlo permutations. In CA and (partial) CCA rare species represented by <5 individuals were omitted (Jongman et al. 1995). ANOVA and GLM were calculated with STATISTICA 7.1, CA and (partial) CCA with CANOCO 4.5 software.

Results

Species richness and abundance of individual species

A total of 2,750 individuals belonging to 44 species were recorded in the 83 plots (see Table 2). The number of species varied from 2 to 22 per plot and the number of individuals from 2 to 154. Most abundant species, with more than 200 individuals, were in decreasing

Table 2 List of butterfly species observed in the 83 plots, the total number of observed individuals per species, the mean number of individuals per habitat type, their European and national threat status and their classification as a forest species

Species	Total abundance	FB	PP1	PP2	PP3	PP4	PP5	DW	European threat status	National threat status	Forest species
Pyronia tithonus	469	11.5	1.4	2.2	9.7	6.7	5.6	3.3			
Maniola jurtina	239	7.8	0.3	0.3	0.7	2.5	0.5	5.2			
Coenonympha oedippus	236	5.3	5.5	2.2	3.6	0.6	4.5	0.2	CR	T	
Lycaena phlaeas	227	6.4	0.7	1.6	0.9	4.7	4.1	1.1			
Minois dryas	218	4.8	3.9	3.6	3.6	1.7	2.6	0.1		T	
Coenonympha pamphilus	217	5.8	3.1	4.5	3.5	0.8	1.6	0.5			
Cupido argiades	202	11.3	1.7	1.5	2.8	0.1	0.5	0.1			
Pararge aegeria	181	0.0	0.0	0.0	0.1	0.3	0.1	9.3			Forest
Euphydryas aurinia	118	7.2	0.7	0.2	0.6	0.1	0.5	0.6	VU		
Gonepteryx rhamni	95	1.3	1.5	0.8	1.2	1.2	0.6	1.3			Forest
Hipparchia statilinus	79	3.2	0.3	1.2	0.3	0.3	1.7	0.1		T	
Heteropterus morpheus	56	1.9	0.6	0.7	1.0	0.5	0.4	0.1		T	
Aricia agestis	46	0.4	0.3	0.0	0.0	0.7	0.9	1.1			
Ochlodes venata	46	0.2	0.2	0.1	1.0	0.5	1.2	0.6			
Coenonympha arcania	45	0.0	0.1	0.0	0.7	0.6	0.9	1.1			
Polyommatus icarus	43	2.7	0.0	0.3	0.1	0.1	0.2	0.2			
Thymelicus lineola	37	2.8	0.0	0.0	0.1	0.0	0.0	0.2			
Hipparchia semele	30	0.2	0.2	0.0	0.1	0.9	0.2	0.7		T	
Melitaea cinxia	17	1.4	0.0	0.0	0.0	0.0	0.0	0.0			
Colias croceus	16	1.2	0.0	0.2	0.0	0.0	0.0	0.0			
Limenitis reducta	15	0.4	0.1	0.2	0.0	0.0	0.0	0.4			Forest
Boloria selene	12	0.8	0.0	0.0	0.0	0.0	0.0	0.2		T	
Argynnis paphia	11	0.0	0.0	0.1	0.0	0.0	0.0	0.5			Forest
Lycaena alciphron	11	0.1	0.1	0.1	0.0	0.0	0.5	0.1			
Brintesia circe	10	0.3	0.1	0.1	0.0	0.0	0.2	0.1			
Thymelicus sylvestris	8	0.5	0.0	0.0	0.2	0.0	0.0	0.0			
Callophrys rubi	7	0.2	0.0	0.1	0.2	0.2	0.0	0.0			
Melitaea athalia	7	0.1	0.0	0.0	0.0	0.0	0.3	0.2			
Pyrgus malvoides	7	0.5	0.1	0.0	0.0	0.0	0.0	0.0			
Satyrium ilicis	7	0.1	0.0	0.1	0.1	0.0	0.4	0.0			Forest
Arethusana arethusa	5	0.3	0.0	0.0	0.1	0.0	0.0	0.0		T	
Neozephyrus quercus	5	0.0	0.0	0.0	0.0	0.0	0.0	0.3			Forest
Pieris rapae	5	0.1	0.0	0.1	0.0	0.0	0.0	0.2			
Lampides boeticus	4	0.1	0.2	0.0	0.1	0.0	0.0	0.0			
Erynnis tages	3	0.2	0.0	0.0	0.0	0.0	0.0	0.1			
Pieris napi	3	0.0	0.0	0.0	0.0	0.0	0.0	0.2			
Vanessa atalanta	3	0.0	0.0	0.0	0.0	0.0	0.0	0.2			
Boloria dia	2	0.2	0.0	0.0	0.0	0.0	0.0	0.0			
Celastrina argiolus	2	0.0	0.0	0.0	0.0	0.1	0.0	0.1			
Iphiclides podalirius	2	0.2	0.0	0.0	0.0	0.0	0.0	0.0			
Lasiommata megera	1	0.0	0.0	0.0	0.0	0.0	0.0	0.1			
Limenitis camilla	1	0.0	0.0	0.0	0.0	0.0	0.0	0.1			Forest
Pieris brassicae	1	0.0	0.0	0.0	0.0	0.0	0.0	0.1			
Vanessa cardui	1	0.0	0.0	0.0	0.0	0.0	0.0	0.1			

Species are ordered by their total abundance. Abbreviations for habitat types: *FB* firebreaks, *PP1* herbaceous clearcuts, *PP2* shrubby clearcuts, *PP3* young pine stands (<7 m), *PP4* mid-class pine stands (7–15 m), *PP5* older pine stands (>15 m), *DW* deciduous woodlands. European threat status: *CR* critically endangered, *VU* vulnerable, National threat status (in France): *T* threatened order *Pyronia tithonus*, *Maniola jurtina*, *Coenonympha oedippus*, *Lycaena phlaeas*, *Minois dryas*, *C. pamphilus* and *Cupido argiades*. Among the 44 recorded species seven can be characterized as typical forest-species: *Gonepteryx rhamni*, *Neozephyrus quercus*, *Satyrium ilicis*, *Limenitis camilla*, *Limenitis reducta*, *Argynnis paphia* and *Pararge aegeria* (Ebert and Rennwald 1991). Only two of them (*P. aegeria* and *G. rhamni*) were relatively abundant (181 and 95 individuals, respectively), for all others <15 individuals were observed. Two species are listed as threatened in Europe: *C. oedippus* (critically endangered) and *Euphydryas aurinia* (vulnerable) (Van Swaay and Warren 1999). With 236 and 118 individuals, respectively, these two species belong to the ten most common species observed in this study. Seven species can be considered as nationally threatened: *Heteropterus morpheus*, *Boloria selene*, *Arethusana arethusa*, *Hipparchia statilinus*, *Hipparchia semele*, *M. dryas* and *C. oedippus* (Table 2).

The mean species richness was significantly higher in the firebreaks than in all other habitat types (ANOVA, $N = 83$, $F = 5.32$, $P < 0.001$, Fig. 2). The total abundance showed the same pattern with a significantly higher mean number of individuals in the firebreaks [81.6 ± 11.7 individuals/plot (mean $\pm$ SE)] than in the other habitats (mean abundance varying from 20.2 ± 4.4 to 30.7 ± 7.9; ANOVA, $N = 83$, $F = 5.53$ and $P < 0.001$). The deciduous woodland patches had the highest mean richness of typical forest species (Fig. 2), which was significantly higher than that of the other habitat types (ANOVA, $N = 83$, $F = 4.37$ and $P < 0.001$), with the exception of the firebreaks. The number of threatened species was significantly higher in the firebreaks than in the deciduous woodland patches (ANOVA, $N = 83$, $F = 3.09$, $P = 0.009$, Fig. 2), but did not differ significantly from that in the pine stands. Forward selection of variables in GLM resulted in the selection of both habitat and landscape variables that explained 47.1% of species richness variation ($F = 7.22$, $P < 0.001$). The mean species richness per plot depended on the habitat type and was positively correlated with the availability of nectar in the understorey vegetation of the plot and the % cover of young pine stands (quadratic term) in the surrounding landscape and negatively correlated with the % herbaceous clearcuts in the landscape. In a forward selection procedure of these variables and the three interactions between habitat and the continuous variables no interaction terms were selected.

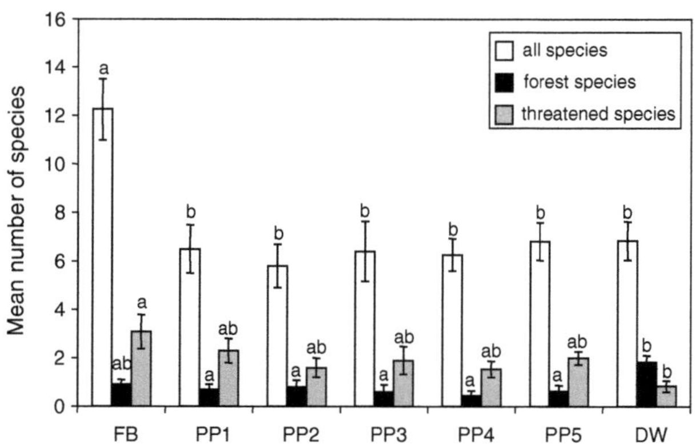

Fig. 2 Mean species richness (±standard error) per habitat type of all butterfly species, forest species and threatened species. *Bars* of the same colour sharing no *letter* are significantly different (Tukey test, $P < 0.05$). Habitat types: *FB* firebreaks, *PP1* herbaceous clearcuts, *PP2* shrubby clearcuts, *PP3* young pine stands, *PP4* mid-class pine stands, *PP5* older pine stands, *DW* deciduous woodlands

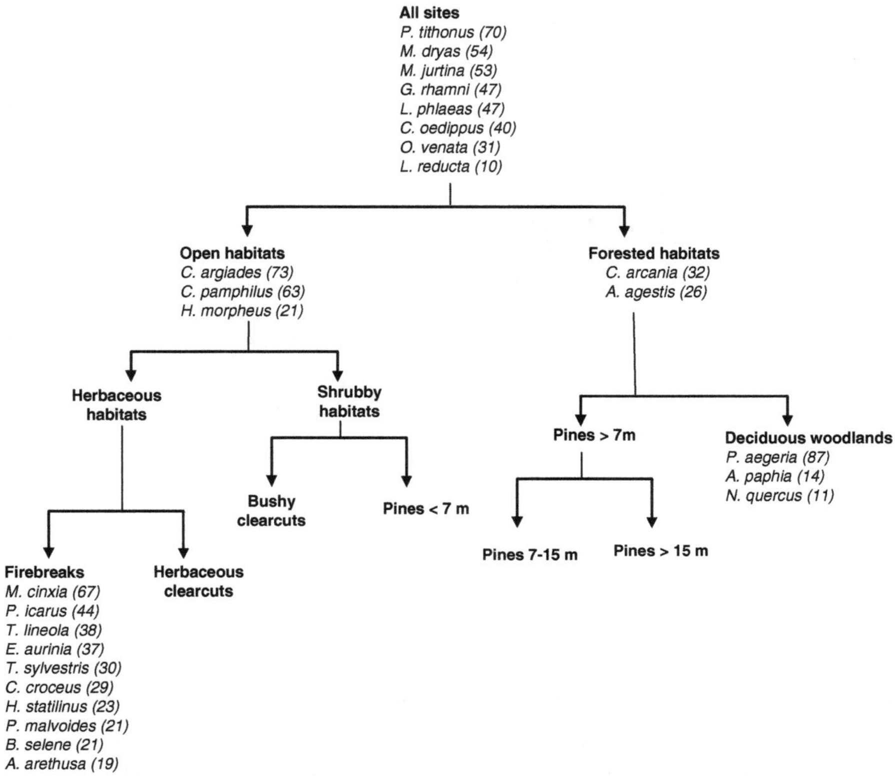

Fig. 3 Indicator species for the *different levels* of the hierarchic site typology. Species are only mentioned at the level where they have their maximum, significant indicator value (indicator value between parentheses)

Among the 33 species analysed 18 had a significant IndVal index at one or several levels of the typology and eight species had their maximum value at the first level regrouping all plots (Fig. 3). Three species had their maximum value in deciduous woodlands and ten species in firebreaks. Three species were characteristic for open sites and two for forested sites, but no species were characteristic for pine stands at lower levels of the hierarchic typology (Fig. 3).

Composition of species communities

The eigenvalues of the first two axes of a CA on a 33 species × 83 plots matrix, were, respectively, 0.62 and 0.45; further axes had an eigenvalue of 0.29 or less (total inertia of CA was 3.83). The first axis separated the deciduous woodland plots from the other plots; the second axis did not show a clear separation between the different pine stands, clearcuts and firebreaks (Fig. 4).

In the CCA with seven habitat types as environmental variables, three significant variables were retained (Table 3), which explained together 17.4% of total CA inertia ($P = 0.001$). The first axis (eigenvalue 0.44) was correlated with the deciduous woodland habitat; the second axis was mainly related to firebreaks and had an eigenvalue of only 0.16.

In the CCA with 14 plot variables as environmental variables, six significant variables (Table 3) were retained by the forward selection procedure, which explained 27.5% of total variation ($P = 0.001$). The eigenvalues of the first two axes were 0.45 and 0.33, respectively.

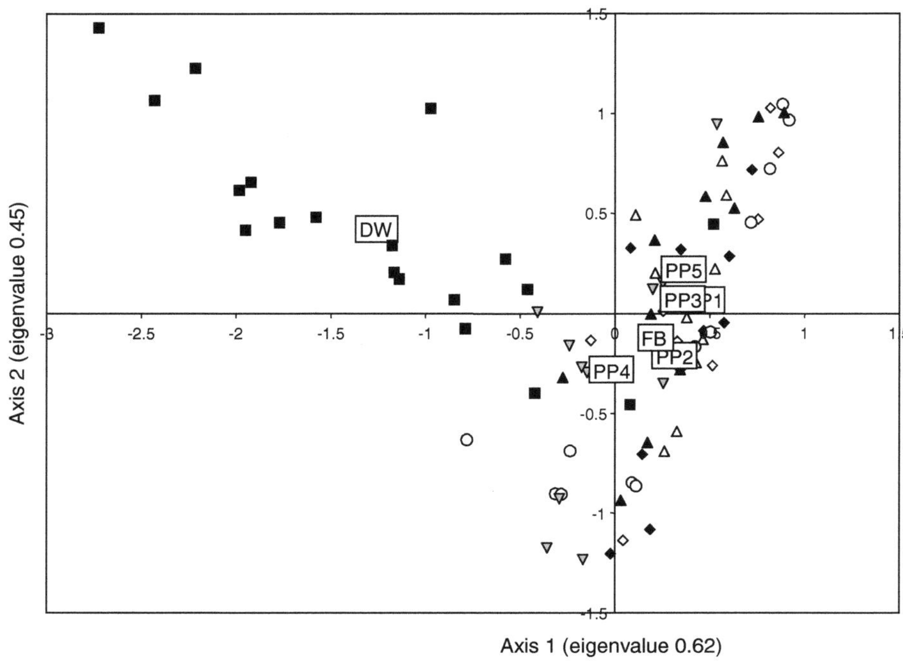

Fig. 4 Correspondence Analysis ordination of the 83 inventoried plots (axes 1 and 2). The position of the name of each habitat type indicates the mean position of plots belonging to that habitat type. *Different symbols* indicate the position of the different plot types. Plot types: *FB* firebreaks (*white circles*), *PP1* herbaceous clearcuts (*white diamonds*), *PP2* shrubby clearcuts (*black diamonds*), *PP3* young pine stands (*white triangles*), *PP4* mid-class pine stands (*grey, inversed triangles*), *PP5* older pine stands (*black triangles*), *DW* deciduous woodlands (*black squares*)

The second axis opposed sites dominated by *M. caerulea* to sites with a high cover of herbaceous dicotyledons and *U. europaeus*, and a higher nectar abundance.

Forward selection procedures of landscape variables in CCA resulted in the selection of four significant landscape variables (Table 3), which explained 18.0% of total variation ($P = 0.001$). The first two CCA axes had eigenvalues of 0.28 and 0.19. The first axis was correlated with the amount of deciduous woodlands in the landscape, the second axis opposed landscapes with a high cover of firebreaks to landscapes with a high cover of shrubby clearcuts.

Canonical Correspondence Analysis with these 13 selected variables combined explained 41.7% of species variation ($P = 0.001$). Examination of the CCA plot (Fig. 5) shows that the first axis opposed forest species such as *P. aegeria, A. paphia, N. quercus* and *L. reducta* which were associated with deciduous woodlands (DW and %DW) to species found in pine stands and firebreaks (e.g. *C. oedippus*, *M. dryas* and *C. argiades*). Best correlated with the second axis were the percentage cover of *M. caerulea* (%Mcae) in the plot and the percentage shrubby clearcuts (%PP2) in the landscape on the positive side of this axis and the percentage cover of herbaceous dicotyledons (%Dicots), of *U. europaeus* (%Ueur), the flower abundance (Nectar) and the percentage of firebreaks in the landscape (%FB) on the negative side. The second axis is therefore mainly correlated with the vegetation composition in pine stands and firebreaks. Species such as *C. oedippus, H. morpheus* and *M. dryas* were found in open pine stands and firebreaks with a high cover of *M. caerulea* and located in landscapes with a high cover of clearcuts. Species at the opposite end of the

Table 3 Selected variables per variable subset in order of selection during the stepwise selection procedure in CCA, the additional variance explained by each variable at the time of inclusion, the % variance explained by each variable subset and the % variation explained by each variable using the variables in the two other subsets as covariables (partial CCA) and the associated probability (*P*)

Variable	Additional % explained in forward selection	% Variation explained per variable set	% Explained when two other subsets used as covariables	*P*
Habitat type		17.4		
DW (deciduous woodlands)	11.2		1.9	0.002
FB (firebreaks)	4.2		2.2	0.001
PP4 (mid-class pine stands)	1.9		1.0	ns
Plot variables		27.5		
%Mcae (% *Molinia caerulea)*	9.3		5.1	0.001
%Oth-shrub (% other shrubs)	8.0		1.8	0.018
Nectar	4.0		1.9	0.007
%Ueur (% *Ulex europaeus)*	2.4		2.7	0.001
%Paqu (% *Pteridium aquilinum)*	2.1		2.1	0.006
%Dicots (% herbaceous dicotyledons)	1.7		2.5	0.002
Landscape variables		18.0		
%DW (% deciduous/mixed woodlands)	6.8		2.4	0.003
%PP2 (% shrubby clearcuts)	5.1		1.0	ns
%FB (% firebreaks)	3.9		1.6	0.025
ED (edge density)	2.2		2.3	0.002

ns Not significant ($P > 0.05$)

second axis were more abundant in stands with *U. europaeus* or were found in plots (mostly firebreaks) with higher % dicotyledons. The third axis (eigenvalue 0.21) opposed the firebreaks (axis positively correlated with FB, %FB and ED) to the pine stands (correlated with PP4, %Ueur). Positively associated with this axis were for example *E. aurinia, T. lineola* and *Melitaea cinxia*.

The decomposition of the variation in independent and confounded effects of the three variable subsets is shown in Fig. 6. The pure effects of plot variables, habitat-types and landscape features accounted for 12.8, 5.1 and 6.7% of variation, respectively (all significant $P = 0.001$). All variable subsets provided an independent, additional contribution to the explained variation, but the independent effect of plot variables was the most important. Analysis of the independent effect of each variable, after controlling for the variation explained by the other two subsets, revealed a significant effect of most variables (Table 3). Only the habitat type mid-class pines (PP4) and the landscape variable % shrubby clearcuts (%PP2) were no longer significant.

Discussion

Conserving biodiversity in plantation forests is becoming increasingly necessary because the area of planted forests continues to increase worldwide. In Europe, for example, the area of plantation forests augmented from 8.6 to 10.5 million hectares in the period of 1990–2005 (FAO 2007). One conservation option is to improve biodiversity within stands by adapting stand management (Kerr 1999; Lindenmayer and Hobbs 2004). However,

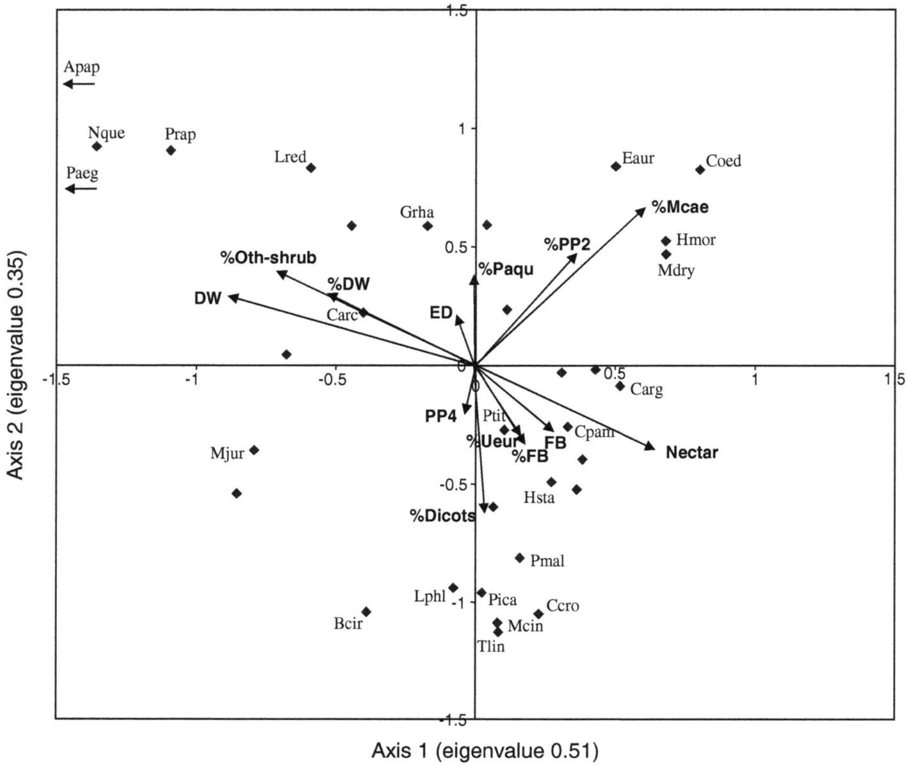

Fig. 5 Canonical Correspondence Analysis ordination biplot (axes 1 and 2) with plot, habitat and landscape variables represented by *arrows* and butterfly species by *diamonds*. Names of butterfly species are indicated only for species that are explained for more than 25% by the CCA. For legend of environmental variables see Table 1. Butterfly species: Apap *Argynnis paphia*, Bcir *Brintesia circe*, Carc *Coenonympha arcania*, Carg *Cupido argiades*, Ccro *Colias croceus*, Coed *Coenonympha oedippus*, Cpam *Coenonympha pamphilus*, Eaur *Euphydryas aurinia*, Grha *Gonepteryx rhamni*, Hmor *Heteropterus morpheus*, Hsta *Hipparchia statilinus*, Lphl *Lycaena phlaeas*, Lred *Limenitis reducta*, Mcin *Melitaea cinxia*, Mdry *Minois dryas*, Mjur *Maniola jurtina*, Nque *Neozephyrus quercus*, Paeg *Pararge aegeria*, Pica *Polyommatus icarus*, Pmal *Pyrgus malvoides*, Prap *Pieris rapae*, Ptit *Pyronia tithonus*, Tlin *Thymelicus lineola*

large-scale intensive stand management may impede the presence of many species, and the role of semi-natural habitat remnants within plantation landscapes may be essential (Lindenmayer and Hobbs 2004; Fischer et al. 2006).

Importance of semi-natural habitats in plantation landscapes

This study confirms the importance of semi-natural habitats for butterflies in pine plantation landscapes. Both herbaceous firebreaks and deciduous woodlands were characterized by the presence of butterfly species not or rarely found in pine stands. Firebreaks were more species-rich than the other habitat types and several butterfly species were almost exclusively found in firebreaks (e.g. *M. cinxia, P. icarus, T. lineola* and *E. aurinia*). The higher species richness of firebreaks might be largely attributable to their more diverse herbaceous vegetation, providing a greater and more diverse flower abundance (nectar) and hostplants

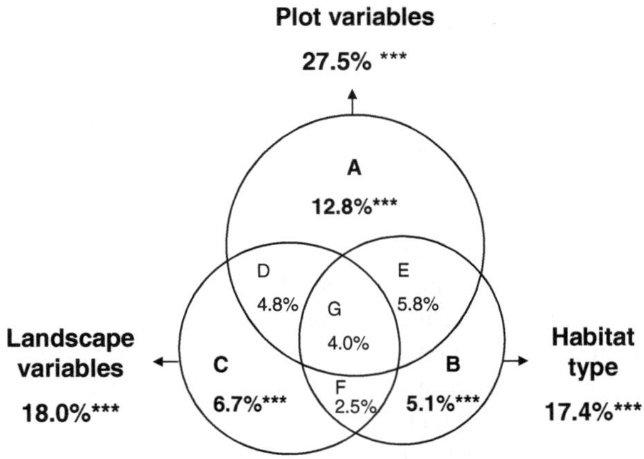

Fig. 6 Decomposition of the variance in butterfly community structure explained by plot variables, habitat type and landscape variables in independent and confounded effects using several partial CCA's. Parts A, B and C represent the independent effects of plot, habitat and landscape variables, respectively, parts D, E, F and G indicate the joint effects. Significance levels are based on 999 Monte Carlo permutations: $***P = 0.001$. The area of each *circle* is proportional to the variance explained by that group of variables. The total variance explained by the three sets of variables is 41.7%

not or rarely found in forest stands, such as *Plantago lanceolata* and herbaceous Fabaceae. Micro-climate and especially high insolation alone does not seem to explain differences in butterfly richness since open areas such as clearcuts had a lower species richness than firebreaks.

Deciduous woodlands were also characterized by the presence of several characteristic species. *P. aegeria* was very typical for deciduous plots and together with *A. paphia* and *N. quercus* formed a group of species associated with deciduous woodlands. Although our study was performed in a well-forested region the number and abundance of typical forest species was low and these species were mainly present in deciduous woodland patches. This study confirms thereby the fact that coniferous forests do not represent a suitable habitat type for most forest butterflies (Ebert and Rennwald 1991). Deciduous woodlands, on the contrary, provide hostplants for butterfly species feeding on broadleaved trees (e.g. *Quercus* sp. for *N. quercus*), have a more diverse herbaceous vegetation (with *Viola* sp. for *A. paphia*), offer a more varied structure for mate finding behaviour and probably provide more spatial variation in micro-climate than pine plantations. These differences between deciduous woodlands and pine stands are due to their different tree composition but also to their different management. Plantation stands are typically characterized by a uniform and intensive management, whereas management of deciduous woodlands is more variable in time and space allowing a greater structural diversity.

Although the butterfly communities of pine stands seem to resemble to those of firebreaks (Fig. 4), they harbour only half the number of species compared to firebreaks and no characteristic species. Apparently butterfly communities of pine stands represent an impoverished version of firebreak communities. The different successional stages of pine stands show no clear differences in butterfly community composition but several species (e.g. *C. oedippus*, *M. dryas* and *O. venata*) were less abundant in mid-class pines than in young and older pine stands because of higher canopy cover.

Relative effects of plot and landscape variables

Comparing the independent effect of plot, habitat and landscape variables revealed clearly the important effect of understorey vegetation on butterfly communities. Butterfly species show preferences for certain vegetation types and specific growing conditions of their hostplants and it is therefore logical that local factors are the most important for this taxonomic group (Thomas et al. 2001). The composition of understorey vegetation explained differences in butterfly community structure that were not explained by habitat type. The most important plot variable was the *M. caerulea* cover. This grass species dominates in sites with a high soil moisture and the second axis of the CCA is explained by a gradient in vegetation composition related to soil moisture. A group of species (*C. oedippus*, *H. morpheus* and *M. dryas*) was positively associated with sites dominated by *M. caerulea*, their main hostplant in the studied region. Cover by *U. europaeus* and by dicotyledons were best correlated with the opposite side of the second axis. The *U. europaeus* cover is however not directly related to the butterfly species (as hostplant or nectar plant) but moderate cover by this shrub characterizes drier pine stands, with butterfly species such as *L. phlaeas* and *H. statilinus*. Cover by dicotyledons, that may be nectar- or hostplants for several species, was higher in herbaceous firebreaks than in other habitat types and was associated with the presence of *L. phlaeas, P. icarus, T. lineola, M. cinxia* and *C. croceus*.

The proportion of explained variance in CCA is low, but this is a common feature in multi-variate analysis of ecological communities (e.g. Jeanneret et al. 2003b; Titeux et al. 2004; Aviron et al. 2005; Schweiger et al. 2005). The aim of CCA is to identify important environmental variables and even low percentages might be informative (Ter Braak 1986). By introducing more environmental variables, the proportion of explained variance will necessarily increase, but for a meaningful analysis the number of environmental variables should not be more than c.10% of the number of plots (Lebreton et al. 1988). A part of the unexplained variance in our study may be due to variables that were not measured, such as intra-plot variation in vegetation composition and canopy cover, or differences in management regimes.

Landscape attributes explained an independent part both in partial CCA analysis and in GLM modelling, thereby confirming that different organization levels should be considered when explaining species abundance patterns (Wiens 1989). In partial CCA the only significant landscape composition variables were the percentage cover of deciduous woodlands and firebreaks. Since the surveyed plot was included in the calculation of landscape metrics, this effect of habitat amount can either be an effect of habitat patch size or an effect of landscape supplementation, i.e. the use of several, similar patches within a landscape (Dunning et al. 1992). Larger patches or more patches tend to supply a greater diversity of environmental conditions and support more species. A positive effect of woodland area within 1 km on butterfly diversity was also demonstrated by Shreeve and Mason (1980) and by Baz and Garcia-Boyero (1995). Edge density explained also a significant, independent part of community variation. A high ED may be positive for species using herbaceous strips along stand edges or for multi-habitat species (habitat complementation) (Duelli 1997). A positive or negative effect of increased ED and the associated fragmentation will however also depend on the observed species and the studied landscape type. Butterfly species characteristic for large woodlands are probably sensitive to fragmentation, but these species are very rare or absent in our study area.

The relatively low percentage of variation explained by landscape variables (18.0%, independent effect 6.7%) can be due to several factors. Possible landscape effects may be masked by the important variation in plot types (different stand types and understorey vegetation) compared to the variation in landscapes. To study landscape effects more accurately, we suggest the selection of the same plot type in a range of landscapes differing in composition and structure (Bergman et al. 2004). This may also permit to separate effects of landscape composition and fragmentation. Analysing the landscape at larger scales might also reveal additional effects. However, butterfly studies that examined the effect of different buffer sizes show contradictory results (Weibull et al. 2000; Krauss et al. 2003; Bergman et al. 2004). Differences in landscape types and their associated key factors may be responsible for these contradictions. Finally, landscape analysis also depends on the accuracy and choice of the patch typology. A patch typology based on a combination of stand type and understorey vegetation might have better described difference in habitat quality for the studied butterfly species. It would also have allowed the establishment of species (or guild) specific habitat maps (Li and Wu 2004) and to reveal more or less isolated habitat patches that do not appear in the current typology. Such a typology can however not be based exclusively on aerial photos.

The presence of threatened species

Threatened butterfly species, such as *C. oedippus*, *M. dryas*, *H. morpheus* and *H. statilinus*, were observed both in pine stands and firebreaks, but they were more abundant in the latter. Firebreaks can therefore function as an essential reservoir/source in the landscape. Wahlberg et al. (2002) demonstrated in Finland that the continued presence of meadows was necessary for the survival of *E. aurinia*, a species occurring both in meadows and in clearcuts. Firebreaks may play the same role in our dynamic landscape. It seems however likely that the large areas of pine stands play a role as alternative habitat and refugium for species occurring both in firebreaks and pine stands, that they improve landscape connectivity and that they buffer the semi-natural habitats (Aberg et al. 1995; Lindenmayer and Franklin 2002; Lindenmayer and Hobbs 2004; Aune et al. 2005). The presence of threatened butterfly species in a landscape dominated by pine plantations argues for their potential conservation value. The 'Landes de Gascogne' forest is characterized by oligotrophic habitat conditions occurring over large areas; conditions that tend to disappear under agricultural and urbanization pressure elsewhere. Typical butterflies of nutrient poor habitats are therefore threatened in several European countries, but are still occurring regularly in the studied region.

Nevertheless, this study does not show the possible negative effects of pine plantations on butterfly species present before the massive afforestation carried out in the 1850s, when the landscape was dominated by large, mainly wet heathlands. It seems likely that several butterfly species might have seriously declined as a consequence of the huge habitat transformation. Some of these species such as *Maculinea alcon*, *Plebejus argus* and *P. idas* are very rare in the Landes de Gascogne forest and survive nowadays in isolated areas of heathland vegetation (military zones and some firebreaks). The nowadays relatively rich firebreaks may thus represent an impoverished version of the original species pool of large heathlands.

How to improve butterfly diversity in plantation forests

The enhancement of biodiversity within plantation forests should include measures to promote woodland habitats for forest species, but should also include measures to retain rare or specialist species of pre-planting habitats (Oxbrough et al. 2006). In the study area, forest butterflies were mainly found in deciduous woodlands and their presence was correlated with the extent of these woodlands, suggesting that habitat thresholds may exist. Bergman et al. (2004), for example, showed a sharp increase in occupancy probability for several butterfly species when the cover of deciduous forest/semi-natural grassland was higher than values between 2 to 12%. Our analyses do not provide an estimation of how much deciduous woodland is needed to conserve characteristic species. Species such as *A. paphia* and *L. reducta* were more abundant in larger deciduous woodlands, but these were also the more varied and humid riparian forests so that the effect of habitat quality and quantity are difficult to separate.

In our study area rare or threatened species were most abundant in firebreaks and the conservation and management of open spaces can be considered as an important technique of nature conservation within plantations (Gittings et al. 2006). In the Republic of Ireland, for example, all grant-aided afforestation should contain 5–10% open space (Gittings et al. 2006). Recommendations for an optimal width of firebreaks for butterflies are difficult to give, based on the results of our study. For the 12 surveyed firebreaks, with a width varying from 15–90 m, no significant relation between species richness and firebreak width could be demonstrated. Oxbrough et al. (2006) showed for ground dwelling spiders that open spaces of <15 m wide did not support an open spider fauna due to the influence of the tree canopy. This suggests that firebreaks of 15 m may already be large enough to harbour a fauna of open spaces. The minimal width depends also on the neighbourhood of the firebreak (e.g. bordered by a high-forest stand or by a road) and its orientation, which will affect the light conditions (Ferris and Carter 2000). The actual management of firebreaks in the studied area is very variable (varying from annual mowing to about once every 8 years) and this variation is partly responsible for their diverse butterfly composition. The vegetation diversity within firebreaks can be increased and the temporarily negative impact of management reduced by a more varied management regime within firebreaks. For wide forest rides Ferris and Carter (2000) recommend a system with three different intervention frequencies. Comparable systems could be used in firebreaks, creating a more natural forest edge and by maintaining the largest part of the firebreak as herbaceous vegetation with a varied structure and composition.

The fact that composition of understorey vegetation explained the largest part of butterfly community composition implies that management within pine stands (e.g. removal of shrub layer, thinning, soil preparation before planting) and other habitats will directly affect butterfly diversity. In the Landes de Gascogne forest butterfly composition was most strongly influenced by differences in vegetation composition related to soil humidity. Maintaining existing humidity gradients and conserving the wet areas in the landscape are therefore decisive measures in conserving butterfly diversity in all habitat types, especially because silvicultural and agricultural practices tend to decrease soil moisture. Within-stand variation in canopy cover or understorey vegetation was not measured in this study, but field observations showed that butterflies were more abundant in gaps or in parts of stands with a lower canopy cover. Maintaining this variation within pine stands will therefore be profitable for butterflies. The positive effect of more open pine stands on butterfly species richness and composition has also been demonstrated in *Pinus ponderosa* (Waltz and Covington 2004) and *Pinus edulis/Juniperus monosperma* forests (Kleintjes et al. 2004). This effect was attributed to a higher light intensity (Waltz and Covington 2004) or to an increase in understorey cover (Kleintjes et al. 2004). Field observations also suggested the importance of variation in understorey vegetation composition. Most of the observed butterflies depend on herbaceous plants as hostplants, but some use shrubs or trees. Stand management trying to create a varied understorey vegetation dominated by herbs but with presence of some shrubs or deciduous trees seems most beneficial for butterflies. A more varied management of understorey vegetation (managing only one row out of two as observed in some stands) will favour vegetation diversity.

Conclusion

Three factors appear to have an important effect on butterfly richness and community composition in the studied pine plantation landscape: the presence of deciduous woodlands, the presence of firebreaks and the variation in understorey vegetation, related to both soil moisture and management practices. Explanatory factors measured at the local scale (plot vegetation and habitat type) explained the largest part of community variation, but landscape factors explained an additional, independent part. This confirms the importance of multi-scale analyses to explain patterns of biodiversity. Our study demonstrates the importance of interstitial habitats at the landscape level and shows that stand management can influence butterfly diversity, mainly by maintaining a diverse herbaceous layer.

Acknowledgements We wish to thank Audrey Lugot, Manon Dupuich, Karine Payet and Zoé Delépine for their help during the field work. Annie Ouin and Marc Dufrêne gave valuable comments on an earlier version of the manuscript, Carlos Lopez-Vaamonde has kindly checked the English language and two anonymous reviewers provided useful comments and corrections. This study was financed by the European Union, ERDF-Interreg Atlantic Area, FORSEE project.

References

Aberg J, Jansson G, Swenson JE, Angelstam P (1995) The effect of matrix on the occurrence of hazel grouse (*Bonasa bonasia*) in isolated habitat fragments. Oecologia 103:265–269

Anthes N, Fartmann T, Hermann G, Kaule G (2003) Combining larval habitat quality and metapopulation structure—the key for successful management of pre-alpine *Euphydryas aurinia* colonies. J Insect Conserv 7:175–185

Aune K, Jonsson BG, Moen J (2005) Isolation and edge effects among woodland key habitats in Sweden: is forest policy promoting fragmentation? Biol Conserv 124:89–95

Aviron S, Burel F, Baudry J, Schermann N (2005) Carabid assemblages in agricultural landscapes: impacts of habitat features, landscape context at different spatial scales and farming intensity. Agric Ecosyst Environ 108:205–217

Baguette M, Petit S, Queva F (2000) Population spatial structure and migration of three butterfly species within the same habitat network: consequences for conservation. J Appl Ecol 37:100–108

Barbaro L, Pontcharraud L, Vetillard F, Guyon D, Jactel H (2005) Comparative responses of bird, carabid, and spider assemblages to stand and landscape diversity in maritime pine plantation forests. Ecoscience 12:110–121

Baz A, Garcia-Boyero A (1995) The effects of forest fragmentation on butterfly communities in central Spain. J Biogeogr 22:129–140

Bergman KO, Askling J, Ekberg O, Ignell H, Wahlman H, Milberg P (2004) Landscape effects on butterfly assemblages in an agricultural region. Ecography 27:619–628

Borcard D, Legendre P, Drapeau P (1992) Partialling out the spatial component of ecological variation. Ecology 73:1045–1055

Carey AB (2003) Biocomplexity and restoration of biodiversity in temperate coniferous forest: inducing spatial heterogeneity with variable-density thinning. Forestry 76:127–136

Carnus JM, Parrotta J, Brockerhoff E, Arbez M, Jactel H, Kremer A, Lamb D, O'Hara K, Walters B (2006) Planted forests and biodiversity. J For 104:65–77

Clausen HD, Holbeck HB, Reddersen J (2001) Factors influencing abundance of butterflies and burnet moths in the uncultivated habitats of an organic farm in Denmark. Biol Conserv 98:167–178

Cushman SA, McGarigal K (2002) Hierarchical, multi-scale decomposition of species-environment relationships. Landsc Ecol 17:637–646

Debinski DM, Ray C, Saveraid EH (2001) Species diversity and the scale of the landscape mosaic: do scales of movement and patch size affect diversity? Biol Conserv 98:179–190

Dennis RLH (1992) The ecology of butterflies in Britain. Oxford University Press, Oxford, UK
Dennis RLH, Shreeve TG, Van Dyck H (2003) Towards a functional resource-based concept for habitat: a butterfly biology viewpoint. Oikos 102:417–426
Dennis RLH, Shreeve TG, Van Dyck H (2006) Habitats and resources: the need for a resource-based definition to conserve butterflies. Biodivers Conserv 15:1943–1966
Duelli P (1997) Biodiversity evaluation in agricultural landscapes: an approach at two different scales. Agric Ecosyst Environ 62:81–91
Dufrêne M, Legendre P (1997) Species assemblages and indicator species: the need for a flexible asymmetrical approach. Ecol Monogr 67:345–366
Dunning JB, Danielson BJ, Pulliam HR (1992) Ecological processes that affect populations in complex landscapes. Oikos 65:169–175
Ebert G, Rennwald E (1991) Die schmetterlinge Baden-Württembergs. Band 1 Tagfalter I. Verlag Eugen Ulmer, Stuttgart
Fahrig L (2003) Effects of habitat fragmenation on biodiversity. Annu Rev Ecol Evol Syst 34:487–515
FAO (2007) State of the Word's Forests 2007. Food and Agriculture Organisation of the United Nations, Rome
Ferris R, Carter C (2000) Managing rides, roadsides and edge habitats in lowland forests. For Comm Bull 123:1–78
Fischer J, Lindenmayer DB, Manning AD (2006) Biodiversity, ecosystem function, and resilience: ten guiding principles for commodity production landscapes. Front Ecol Environ 4:80–86
Gittings T, O'Halloran J, Kelly T, Giller PS (2006) The contribution of open spaces to the maintenance of hoverfly (Diptera, Syrphidae) biodiversity in Irish plantation forests. For Ecol Manage 237:290–300
Hanski I (1999) Metapopulation ecology. Oxford University Press, New York
Hartley MJ (2002) Rationale and methods for conserving biodiversity in plantation forests. For Ecol Manage 155:81–95
Jeanneret P, Schupbach B, Luka H (2003a) Quantifying the impact of landscape and habitat features on biodiversity in cultivated landscapes. Agric Ecosyst Environ 98:311–320
Jeanneret P, Schupbach B, Pfiffner L, Herzog F, Walter T (2003b) The Swiss agri-environmental programme and its effects on selected biodiversity indicators. J Nat Conserv 11:213–220
Jongman RGH, Ter Braak CJF, Van Tongeren OFR (1995) Data analysis in community and landscape ecology. Cambridge University Press, Cambridge
Karsholt O, Razowski J (1996) The Lepidoptera of Europe. A distributional checklist. Apollo Books, Stenstrup
Kerr G (1999) The use of silvicultural systems to enhance the biological diversity of plantation forests in Britain. Forestry 72:191–205
Kleintjes PK, Jacobs BF, Fettig SM (2004) Initial response of butterflies to an overstory reduction and slash mulching treatment of a degraded pinon-juniper woodland. Restor Ecol 12:231–238
Krauss J, Steffan-Dewenter I, Tscharntke T (2003) How does landscape context contribute to effects of habitat fragmentation on diversity and population density of butterflies? J Biogeogr 30:889–900
Lafranchis T (2000) Les papillons de jour en France, Belgique et Luxembourg et leurs chenilles. Biotope, Mèze (France)
Lebreton JD, Chessel D, Prodon R, Yoccoz N (1988) L'analyse des relations espèce-milieu par l'analyse canonique des correspondances. I.-Variables de milieu qualitatives. Acta Oecol 9:53–67
Li HB, Wu JG (2004) Use and misuse of landscape indices. Landsc Ecol 19:389–399
Lindenmayer DB, Franklin JF (2002) Conserving forest biodiversity: a comprehensive multiscaled approach. Island Press, Washington, Covelo, London
Lindenmayer DB, Hobbs RJ (2004) Fauna conservation in Australian plantation forests—a review. Biol Conserv 119:151–168
Mac Arthur RH, Wilson EO (1967) The theory of island biogeography. Princeton University Press, Princeton, NJ
Maes D, Van Dyck H (1999) Dagvlinders in vlaanderen—ecologie, verspreiding en behoud. Stichting Leefmilieu/Antwerpen i.s.m. Instituut voor Natuurbehoud en Vlaamse Vlinderwerkgroep, Brussel
Mazerolle MJ, Villard MA (1999) Patch characteristics and landscape context as predictors of species presence and abundance: A review. Ecoscience 6:117–124
McGarigal K, Cushman SA, Neel MC, Ene E (2002) FRAGSTATS: spatial pattern analysis program for categorical maps. Computer software program produced by the authors at the University of Massachusetts, Amherst. Available at the following web site: www.umass.edu/landeco/research/fragstats/fragstats.html
Moore SE, Allen HL (1999) Plantation forestry. In: Hunter MLJ (ed) Maintaining biodiversity in forest ecosystems. Cambridge Univesity Press, New York
Oxbrough AG, Gittings T, O'Halloran J, Giller PS, Kelly TC (2006) The influence of open space on ground-dwelling spider assemblages within plantation forests. For Ecol Manage 237:404–417

Palmer MW (1993) Putting things in even better order—the advantages of canonical correspondence-analysis. Ecology 74:2215–2230

Pollard E, Yates TJ (1993) Monitoring butterflies for ecology and conservation. Chapman & Hall, London

Schneider C, Fry GLA (2001) The influence of landscape grain size on butterfly diversity in grasslands. J Insect Conserv 5:163–171

Schweiger O, Maelfait JP, Van Wingerden W, Hendrickx F, Billeter R, Speelmans M, Augenstein I, Aukema B, Aviron S, Bailey D, Bukacek R, Burel F, Diekotter T, Dirksen J, Frenzel M, Herzog F, Liira J, Roubalova M, Bugter R (2005) Quantifying the impact of environmental factors on arthropod communities in agricultural landscapes across organizational levels and spatial scales. J Appl Ecol 42: 1129–1139

Shreeve TG (1992) Adult behaviour. In: Dennis RLH (ed) The ecology of butterflies in Britain. Oxford University Press, Oxford, UK

Shreeve TG, Mason CF (1980) The number of butterfly species in woodlands. Oecologia 45:414–418

Söderström B, Svensson B, Vessby K, Glimskär A (2001) Plants, insects and birds in semi-natural pastures in relation to habitat and landscape factors. Biodivers Conserv 10:1839–1863

Stefanescu C, Herrando S, Paramo F (2004) Butterfly species richness in the north-west Mediterranean basin: the role of natural and human-induced factors. J Biogeogr 31:905–915

Steffan-Dewenter I, Tscharntke T (2000) Butterfly community structure in fragmented habitats. Ecol Lett 3:449–456

Ter Braak CJF (1986) Canonical correspondence analysis: a new eigenvector technique for multivariate direct gradient analysis. Ecology 67:1167–1179

Thomas CD (1995) Ecology and conservation of butterfly metapopulations in the fragmented British landscape. In: Pullin AS (ed) Ecology and conservation of butterflies. Chapman & Hall, London

Thomas CD, Harrison S (1992) Spatial dynamics of a patchily distributed butterfly species. J Anim Ecol 61:437–446

Thomas JA, Bourn NAD, Clarke RT, Stewart KE, Simcox DJ, Pearman GS, Curtis R, Goodger B (2001) The quality and isolation of habitat patches both determine where butterflies persist in fragmented landscapes. Proc R Soc Lond B Bio 268:1791–1796

Timbal T, Maizeret C (1998) Biodiversité végétale et gestion durable de la forêt landaise de pin maritime: bilan et évolution. Rev For Fr 5:403–424

Titeux N, Dufrene M, Jacob JP, Paquay M, Defourny P (2004) Multivariate analysis of a fine-scale breeding bird atlas using a geographical information system and partial canonical correspondence analysis: environmental and spatial effects. J Biogeogr 31:1841–1856

Trichet P, Jolivet C, Arrouays D, Loustau D, Bert D, Ranger J (1999) Le maintien de la fertilité des sols forestiers landais dans le cadre de la sylviculture intensive de pin maritime. Revue bibliographique et identification des pistes de recherche. Etude et Gestion des Sols 6:197–214

Van Swaay CAM, Warren MS (1999) Red data book of European butterflies (Rhopalocera). Council of Europe Publishing, Strasbourg

Wahlberg N, Klemetti T, Hanski I (2002) Dynamic populations in a dynamic landscape: the metapopulation structure of the marsh fritillary butterfly. Ecography 25:224–232

Waltz AEM, Covington WW (2004) Ecological restoration treatments increase butterfly richness and abundance: mechanisms of response. Restor Ecol 12:85–96

Weibull AC, Bengtsson J, Nohlgren E (2000) Diversity of butterflies in the agricultural landscape: the role of farming system and landscape heterogeneity. Ecography 23:743–750

Wiens JA (1989) Spatial scaling in ecology. Funct Ecol 3:385–397

Wiens JA (1995) Habitat fragmentation: island v landscape perspectives on bird conservation. Ibis 137:97–104

Relevance of exotic pine plantations as a surrogate habitat for ground beetles (Carabidae) where native forest is rare

Lisa A. Berndt · Eckehard G. Brockerhoff · Hervé Jactel

Originally published in the journal Biodiversity and Conservation, Volume 17, No 5, 1171–1185.
DOI: 10.1007/s10531-008-9379-3

Abstract Plantation forests are of increasing importance worldwide for wood and fibre production, and in some areas they are the only forest cover. Here we investigate the potential role of exotic plantations in supporting native forest-dwelling carabid beetles in regions that have experienced extensive deforestation. On the Canterbury Plains of New Zealand, more than 99% of the previous native forest cover has been lost, and today exotic pine (*Pinus radiata*) plantations are the only forest habitat of substantial area. Carabids were caught with pitfall traps in native kanuka (*Kunzea ericoides*) forest remnants and in a neighbouring pine plantation, grassland and gorse (*Ulex europaeus*) shrubland. A total of 2,700 individuals were caught, with significantly greater abundance in traps in young pine, grassland and gorse habitats than in kanuka and older pine. Rarefied species richness was greatest in kanuka, a habitat that supported two forest specialist species not present in other habitat types. A critically endangered species was found only in the exotic plantation forest, which also acts as a surrogate habitat for most carabids associated with kanuka forest. The few remaining native forest patches are of critical importance to conservation on the Canterbury Plains, but in the absence of larger native forest areas plantation forests are more valuable for carabid conservation than the exotic grassland that dominates the region.

Keywords Biodiversity · Carabidae · Exotic species · Habitat fragmentation · Habitat loss · Threatened species

L. A. Berndt (✉)
Scion, Private Bag 3020, Rotorua 3046, New Zealand
e-mail: lisa.berndt@scionresearch.com

E. G. Brockerhoff
Scion, P.O. Box 29237, Fendalton, Christchurch 8540, New Zealand
e-mail: eckehard.brockerhoff@scionresearch.com

H. Jactel
INRA, 69 Route d'Arcachon, 33612 Cestas Cedex, France
e-mail: herve.jactel@pierroton.inra.fr

E.G. Brockerhoff et al. (eds.), *Plantation Forests and Biodiversity: Oxymoron or Opportunity?*
DOI: 10.1007/978-90-481-2807-5_13

Introduction

The area of plantation forests is increasing worldwide, against the trend of a declining natural forest cover (FAO 2001). Although the potential role of plantation forests in the conservation of biological diversity is controversial (Rosoman 1994; Norton 1998; Hartley 2002; Brockerhoff et al. 2008), examples of such forests from many countries have shown they can provide habitat for indigenous species from various taxa (Geldenhuys 1997; Humphrey et al. 2000; Humphrey et al. 2002; Brockerhoff et al. 2003; Maunder et al. 2005). Thus, plantation forests may be important for the conservation of biodiversity, particularly in areas where natural forests are lost or fragmented.

Habitat loss and fragmentation pose major threats to biodiversity in forest ecosystems (Wilson 1988; Saunders et al. 1991; Murcia 1995). This is the case in New Zealand which was mostly forested in pre-human times, but human colonisation since the thirteenth century AD (Wilmshurst and Higham 2004) has led to large-scale deforestation, and today natural forests occupy only about 20% of the land area (McGlone 1989). New Zealand was identified as one of the global biodiversity hotspots in an assessment that considered richness, endemicity, and loss and fragmentation of habitat (Myers et al. 2000). New Zealand has very high levels of endemicity; approximately 81% of the vascular plants, 63% of terrestrial vertebrates (Myers et al. 2000), and 90–97% of the insects are endemic (Atkinson and Cameron 1993; Taylor 1997; McGuiness 2007).

The Canterbury Plains in the South Island is one of the regions of New Zealand that suffered particularly severe forest loss (McEwen 1987). Around 700 years of human-induced disturbance has reduced the natural vegetation on the formerly forested Plains to <0.5% of the total area (Brockerhoff et al. 2001; Leathwick 2001). Only a few natural forest remnants remain on the Plains, and most are kanuka (*Kunzea ericiodes*) forest. All surviving remnants are small fragments measuring <20 ha (Meurk et al. 1995). Only two Canterbury Plains kanuka remnants, Bankside Scientific Reserve and Eyrewell Scientific Reserve, each measuring about 2 ha, are protected as part of the conservation estate (Molloy 1970; Molloy and Ives 1972; Meurk et al. 1995). These kanuka remnants are of particular conservation value because they support a scarce, drought-adapted flora and some rare and threatened species (Ecroyd and Brockerhoff 2005).

At the start of European settlement in the 1850s kanuka forest covered 8,000–12,000 ha in the Eyrewell region of the Canterbury Plains (Molloy and Ives 1972). Today, exotic vegetation dominates this landscape, mainly as pastoral grassland, crop land, and some planted forest. The only forest of significant area in the Eyrewell region is the 7,000 ha Eyrewell Forest, a plantation of Monterey pine, *Pinus radiata*. Because of the poor soils, this area was considered unsuitable for agriculture and therefore a plantation was established from 1928, mostly on land previously cleared of kanuka forest (Molloy and Ives 1972). Kanuka persists in much of the plantation as an understorey (Brockerhoff et al. 2003; Ecroyd and Brockerhoff 2005).

Here we investigate the role that exotic plantation forests play in providing surrogate forest habitat for indigenous invertebrate communities in areas where the natural forest cover has been lost or fragmented. Apart from our main objective to investigate the comparative importance of plantations as habitat for carabids, we also wanted to conduct a survey of carabids in the Eyrewell region where at least one endangered species had been reported. Carabid beetles were selected for study as they are relatively well known in New Zealand and world-wide, are easily sampled using pitfall traps, are threatened by habitat loss (McGuiness 2007), and are also widely used as biodiversity indicators (Butterfield et al. 1995; Allegro and Sciaky 2003). We compare pitfall-trapped carabid beetles in native kanuka forest remnants with those caught in exotic pine plantation in the Eyrewell region on the Canterbury Plains, New Zealand. Comparisions were also made with carabid assemblages in exotic grassland, the dominant vegetation type in the area, and exotic gorse (*Ulex europeaus*) shrubland.

Methods

Study area

The alluvial Canterbury Plains cover an area of about 759,000 ha bound by the Southern Alps in the west and by the Pacific Ocean and the volcanic hills of Banks Peninsula in the east (Fig. 1). The Eyrewell region is a gently sloping plain to the north of the Waimakariri River, ranging from 65 m altitude at the eastern end, to 220 m at Eyrewell Reserve in the west. Soils are moderate- to well-draining, droughty, light and stony (Molloy and Ives 1972; Meurk et al. 1995). Average annual rainfall is about 800 mm (Ecroyd and Brockerhoff 2005).'Kanuka' habitat consisted of three remnants of low forest with a dense to partially open canopy composed solely of kanuka: the 2 ha Eyrewell Scientific Reserve (for habitat description see Meurk et al. 1995; Ecroyd and Brockerhoff 2005) near the western end of Eyrewell Forest, another kanuka remnant of about 1 ha located on private land nearby, and a remnant of about 10 ha near the eastern end of Eyrewell Forest. Kanuka forest in this area was up to 7.4 m in height with an understorey of shrubs and herbs (Ecroyd and Brockerhoff 2005).

Pine plantations in New Zealand are managed with rotations of about 27 years during which, under favourable conditions, trees can grow to a height of 40 m and a diameter at breast height of 70 cm or more. With these growth rates such plantations quickly develop forest-like habitat conditions with regard to microclimate, light conditions, and tree stature. Eyrewell Forest was established from the late 1920s. Young stands labeled 'young pine' were in third rotation *P. radiata* that had been harvested and replanted 2–6 years prior to sampling. The stocking rate was ca. 1,000 trees per ha and trees were ca. 2–5 m in height, with an open canopy and an understorey dominated by naturally regenerated exotic grasses and forbs. The 'old pine' stands we studied were planted 21 to 27 years prior to sampling, trees were ca. 22–26 m tall, pruned and thinned to a stocking rate of approximately 250 stems per ha and had a DBH of 35–53 cm. The understorey vegetation of older stands was dominated by either kanuka and/or gorse, with varying cover and abundance, and numerous native and exotic plant species were shared with the kanuka forest remnants (Brockerhoff et al. 2003, Ecroyd and Brockerhoff 2005).

'Grassland' sites were located along fencelines or in ungrazed paddocks of sheep and dairy farms between the eastern and central, and the central and western sections of Eyrewell Forest. Ungrazed locations were chosen to avoid interference by livestock and irrigation equipment. Additional sites were located in ungrazed grassland adjacent to the kanuka forest inside Eyrewell Reserve.

'Gorse' sites were located in exotic gorse and broom (*Cytisus scoparius*) shrubland present in a narrow strip along the southern boundary of the central block of Eyrewell Forest.

Carabid sampling

Pitfall trapping was conducted over two summer seasons (2000/2001 and 2002/2003) in the five habitat types (kanuka, old pine, young pine, grassland and gorse) in the Eyrewell region. Trap locations are shown on Fig. 1 and an overview of the sampling effort in different

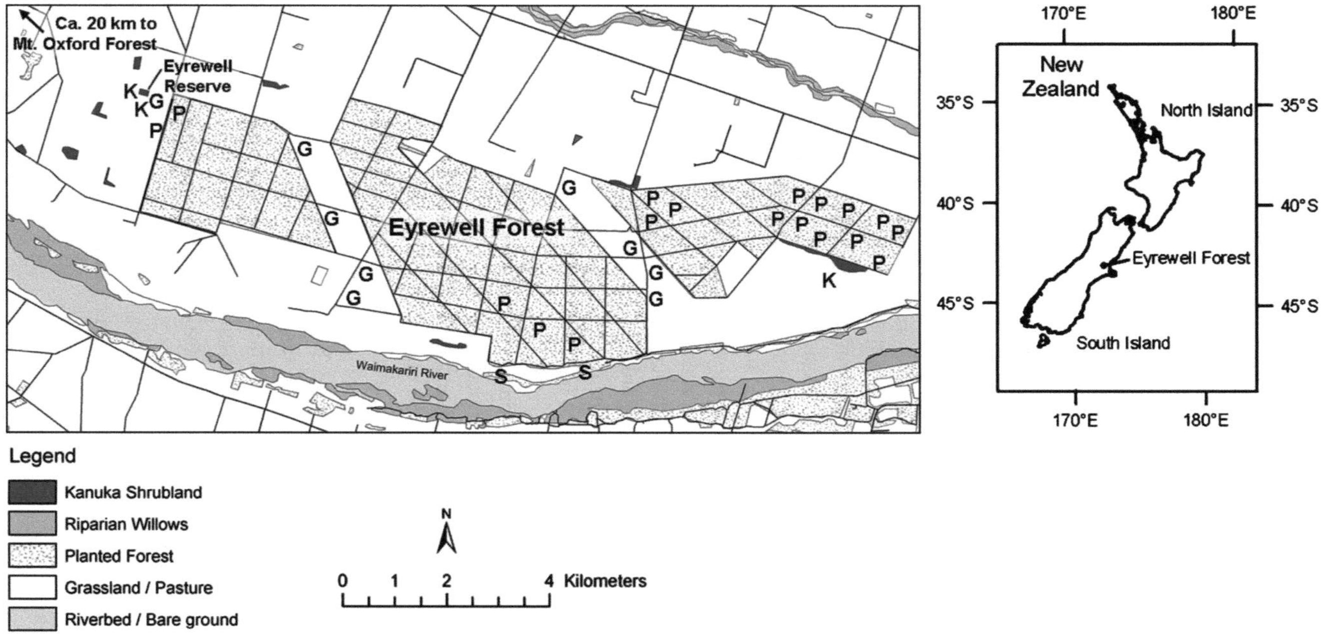

Fig. 1 Map of the study area with pitfall trap locations in kanuka remnants ('K'), plantation forest ('P', both young and old pine habitats), grassland and pasture areas ('G'), and exotic shrubland ('S') (note, most symbols (K, P, G, S) represent several traps, where individual trap symbols would overlap). Total trap numbers are indicated in Table 1

habitats is given in Table 1. Traps were usually positioned at least 50 m apart except in the kanuka and gorse areas where, due to their smaller area, traps were as close as 30 m. Note that sampling effort per area of available habitat was greater in kanuka and gorse due to the considerably smaller total area of both these habitat types relative to the area of pine plantation. Sampling focussed on forest habitats because we were interested in forest species that would be affected by deforestation.

Pitfall traps made from 750 ml polypropylene cups were installed such that the opening (diameter 110 mm) was level with the surrounding surface. To increase trap efficiency, two white intersecting guide panels (1.2 × 0.1 m) were installed over the pitfall traps. A white plastic rain cover (150 × 150 mm), secured with large pebbles, was placed on top of the guide panels above the trap opening. Traps were filled with about 200 ml of preservative (70% water, 30% monoethylene glycol, with salt and soap added) and were changed approximately every 2 weeks in 2000/2001, and monthly in 2002/2003. All carabid specimens were transferred to 70% ethanol and sorted to morphospecies for subsequent identification using keys (Larochelle and Lariviere 2001), named museum specimens, and specialist advice (see acknowledgements).

Analysis

Data were pooled across sampling dates and converted to carabids per 100 trap days (relative abundance) prior to analysis. Each trap was considered a replicate and the number of traps are shown in Table 1. ANOVAs with post hoc Fisher's LSD tests (SYSTAT 9, SPSS Inc. 1998) were used to compare log(x + 1)-transformed mean relative abundance per trap, and mean percent of individuals that were native. Sample-based rarefaction curves (EstimateS 7.0.0, Colwell 2004), made without replacement and re-scaled to show individuals on the x-axis, were used to compare total species richness and native species richness between habitats (Gotelli and Colwell 2001). This comparison was made by bisecting the rarefaction curves at the smallest total number of individuals caught in any habitat (i.e., at the end point of the kanuka curve).

Species associated with each habitat were identified using the indicator species analysis of Dufrene and Legendre (1997), carried out in PC-ORD 4.01, (McCune and Mefford 1999). Unconstrained (Principle Co-ordinate Analysis, PCoA) and constrained (CAP Analysis) ordinations, were performed using the CAP program (Anderson 2003) to explore relationships between the five habitats in the pitfall trapped carabids. Chi-squared distances were used in the ordination to emphasise differences in composition (Anderson and Willis 2003). PCoA was chosen as an indirect gradient analysis, allowing overall patterns in the data cloud, as well as within-group variability, to be assessed. CAP is a discriminant analysis, based on the a priori habitat groupings, which more clearly shows location differences in the data cloud than the PCoA on which the CAP calculations are based (Anderson and

Table 1 Trapping effort in each habitat type during the southern hemisphere summers of 2000/2001 and 2002/2003

Habitat type	Trap days (# traps) (14 Nov 2000–19 Mar 2001)	Trap days (# traps) (31 Oct 2002–6 Mar 2003)	Total trap days
Kanuka	1928 (16)	1397 (13)	3325
Old pine	1446 (12)	9764 (80)	11210
Young pine	1446 (12)	1358 (12)	2804
Grassland	362 (3)	1623 (15)	1985
Gorse	na	900 (8)	900
Total	5182 (43)	15042 (128)	20224

Willis 2003). Product-moment correlation coefficients (CAP programme) were used to identify species involved in determining group differences (i.e., species with high positive or negative correlations with the CAP axes) (Anderson and Willis 2003).

Results

Thirteen native and five exotic carabid species were caught during this study (Table 2). Of the total of 2700 individuals, 45% belonged to native species. Both native and exotic species occurred in all habitats but with greatly differing proportions (below). All of the native species caught were endemic to New Zealand, including three local endemics (*Megadromus antarcticus* (Chaudoir), *Oregus crypticus* Pawson and *Holcaspis brevicula* Butcher) restricted to the Canterbury region (Larochelle and Lariviere 2001; Pawson et al. 2003).

Species composition

Generalist species were a feature of the carabid fauna of the Eyrewell region, with 9 of the 18 species present in three or more of the habitats (Table 2). In particular the medium to large, flightless native species *Meg. antarcticus*, *Metaglymma moniliferum* Bates and

Table 2 Species composition, relative abundance (carabids per 100 trap days), indicator values (IndVal)

Species and origin	Characteristics	Carabids per 100 trap days (IndVal)				
		Kanuka	Old pine	Young pine	Grassland	Gorse
Native						
Megadromus antarcticus	L, NF, G	1.63 (8)	0.80 (3)	8.74 (52*)	1.26 (4)	4.46 (26)
Metaglymma moniliferum	M, NF, G	1.15 (5)	1.73 (7)	0.93 (4)	3.41 (18)	8.34 (54*)
Scopodes fossulatus	S, ?, G	0.11 (2)	0.01 (0)	0.10 (2)	0.33 (13)	
Oregus crypticus	M, NF, G	0.09 (0)	0.58 (7)	1.07 (22)	0.14 (0)	1.34 (26)
Demetrida dieffenbachia	S, NF, G	0.06 (0)	0.90 (32*)	0.22 (3)		
Cicindela dunedensis	M, OF, O	0.03 (0)		0.55 (36*)		
Holcaspis intermittens	M, NF, F	0.03 (3)				
Holcaspis elongella	M, NF, F	0.03 (3)				
Mecyclothorax rotundicollis	S, FF, G	0.03 (0)		0.11 (0)	3.48 (53*)	
Holcaspis brevicula	S, NF, F?			0.04 (4)		
Hypharpax antarcticus	S, FF, O			0.36 (33*)	0.05 (1)	
Platynus macropterus	M, OF, F				0.05 (1)	0.33 (22*)
Notagonum feredayi	S, NF, G				0.05 (6)	
Exotic						
Hypharpax australis	S, FF, O	0.06 (0)		38.62 (81*)	8.84 (8)	
Haplanister crypticus	S, FF, G	0.03 (1)	0.03 (1)	0.03 (1)	0.05 (2)	
Anisodactylus binotatus	S, FF, O	0.03 (0)		0.18 (14)		
Laemostaenus complanatus	M, OF, G		0.05 (0)		6.65 (11)	
Gnathaphanus melbournensis	S, ?, ?			2.65 (21)	0.91 (11)	0.11 (0)
Total individuals trapped		109	459	1493	508	131
Native species		9	5	9	8	4
Exotic species		3	2	4	4	1
All species		12	7	13	12	5

*IndVal significant at $P < 0.01$, and total number of individuals and species caught in each habitat. Behavioural and biological characteristics are indicated by letters: approximate size (S = small <10 mm, M = medium 10–20 mm, L = large >20 mm); flight ability (NF = not flighted, OF = occasional flier, FF = frequent or regular flier); habitat preferences (G = generalist, O = open habitat, F = forest) ('?' indicates uncertainty or lack of knowledge). Behavioural information from Larochelle and Lariviere 2001

Table 3 Mean relative abundance (carabids per 100 trap days, ±S.E.) and percent of carabids caught that were native in each habitat type. Means sharing a letter do not differ significantly between habitats (α = 0.05)

Habitat type	Mean relative abundance per trap (±SE)	Mean % native carabids (±SE)
Kanuka	3.3 (±0.48) c	96.1 (±2.1) a
Old pine	4.1 (±0.29) c	98.0 (±0.9) a
Young pine	53.6 (±7.18) a	27.3 (±2.6) c
Grassland	25.2 (±12.01) b	71.0 (±7.0) b
Gorse	14.6 (±3.35) b	99.2 (±0.8) a

O. crypticus were ubiquitous (Table 2). The six exotic species were mostly of European and Australian origin: *Hypharpax australis* (Dejean) (Australian), *Haplanister crypticus* Moore (unknown origin), *Anisodactylus binotatus* (Fabricius) (Holarctic, Oriental, North African), *Laemostaenus complanatus* (Dejean) (European), *Clivina vagans* Putzeys (Australian) (Larochelle and Lariviere 2001), and *Gnathaphanus melbournensis* (Laporte de Castelnau) (Australian) (Larochelle and Lariviere 2005).

Of the 12 species caught in the kanuka habitat, two generalist species, *Meg. antarcticus* and *Met. moniliferum* made up 85% of individuals (Table 2). Nine of the 12 species were rare within this habitat, each represented by less than 0.1 individuals per 100 trap days (Table 2). The exotic species *Hy. australis, Ha. crypticus* and *A. binotatus* accounted for only 4% of the carabid individuals caught in kanuka. No species found was a significant indicator for this habitat, as determined using the IndVal procedure (Table 2), despite the presence of two species, *Holcaspis intermittens* (Chaudoir) and *Holcaspis elongella* (White), that were unique to the kanuka habitat.

The pine plantation habitats supported 14 species in total, with seven species caught in old pine, and 13 in young pine (Table 2). The native species *Demetrida dieffenbachii* (White) was a significant indicator species of old pine (Table 2). In the young pine habitat, the native species *Meg. antarcticus, Cicindela dunedensis* Laporte de Castelnau and *Hypharpax antarcticus*(Laporte de Castelnau) were significant indicators, along with the exotic species *Hy. australis* . The pine plantation had 10 species in common with kanuka, six of them native (Table 2). An additional two native species (*Ho. brevicula* and *Hy. antarcticus*) were caught in the pine habitat but not in kanuka. Exotic species were a dominant feature of the young pine carabid fauna. *Hypharpax australis* was caught in particularly high numbers (Table 2), making the main contribution to the dominance of exotics in young pine (see Table 3).

Of the 12 species caught in grassland, the exotic species *H. australis* and *L. complanatus* were the most common (Table 2). The native species *Mecyclothorax rotundicollis* (White) and *Met. moniliferum* were also common in the traps (Table 2), with the former being a significant indicator of the grassland habitat (Table 2). Only one species, *Notagonum feredayi* (Bates), was caught solely in the grassland habitat (Table 2). Two native species (*Platynus macropterus* (Chaudoir) and *N. feredayi*) were caught in the grassland habitat but not in the kanuka or pine habitats.

Only five species were caught in the gorse habitat, although the trap catch suggests that the abundance of carabids was relatively high (Table 3). The native species *Meg. moniliferum* and *P. macropterus* dominated the trap catch, with the latter a significant indicator species of the gorse habitat (Table 2).

Relative abundance

Total relative abundance of carabids in the pitfall traps was 13 times greater in the young pine habitat than in kanuka or old pine (ANOVA, $P < 0.001$, Table 3). Relative abundance in grassland and gorse was also significantly greater than that in kanuka or old pine (ANOVA, $P < 0.001$, Table 3). Of the total number of carabids caught in kanuka, old pine and gorse habitats, over 96% were from native species (Table 3). The proportion of native carabids in these habitats was significantly greater than that in grassland, with the smallest proportion of natives found in young pine (ANOVA, $P < 0.001$, Table 3).

Species richness

When total rarefied species richness was compared between habitats (at the end point of the kanuka curves; i.e., at ca. 90 individuals; Fig. 2a), kanuka had the greatest richness, followed by grassland, young pine, old pine and gorse. The rarefaction of native species only (Fig. 2b) showed the same trend, although kanuka, young pine and grassland were more similar in this analysis than when exotics were included. Error bars (95% confidence intervals) overlapped on both graphs, however, indicating that there were no significant differences in species richness among habitats.

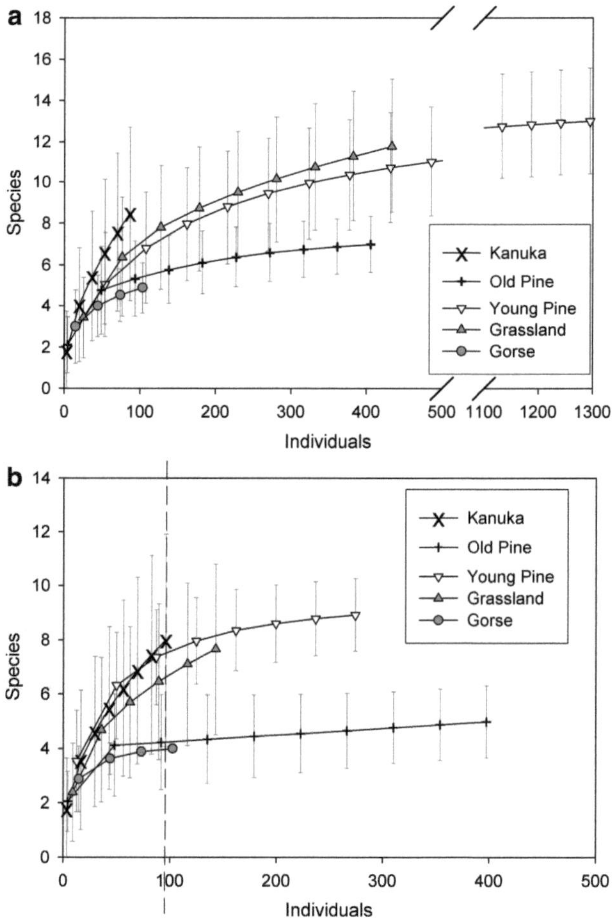

Fig. 2 Sample-based rarefaction curves (±95% CI) for all species (**a**) and for native species (**b**) in each habitat, re-scaled to show individuals on the x-axis. Species richness is compared at the end point for the kanuka samples, indicated by the dashed line

Ordination

The unconstrained ordination (PCoA, Fig. 3a) showed no separation between habitat types, with all ellipses overlapping. This reflects the dominance of generalist species in the fauna. Constraining the ordination using the CAP technique (Fig. 3b) separated young pine from kanuka, old pine and gorse along Axis 1. This separation was driven by the abundance of *Hy. australis* in young pine (Table 2), which was negatively correlated with Axis 1 (Fig. 4). Grassland traps were not tightly clustered, with the 95% confidence ellipse overlapping all other habitat types. However, the majority of grassland traps were located further along Axis 2 than traps from the other four habitat types, which appears to be driven by the positive correlation of *Mec. rotundicollis* with this axis (Fig. 4).

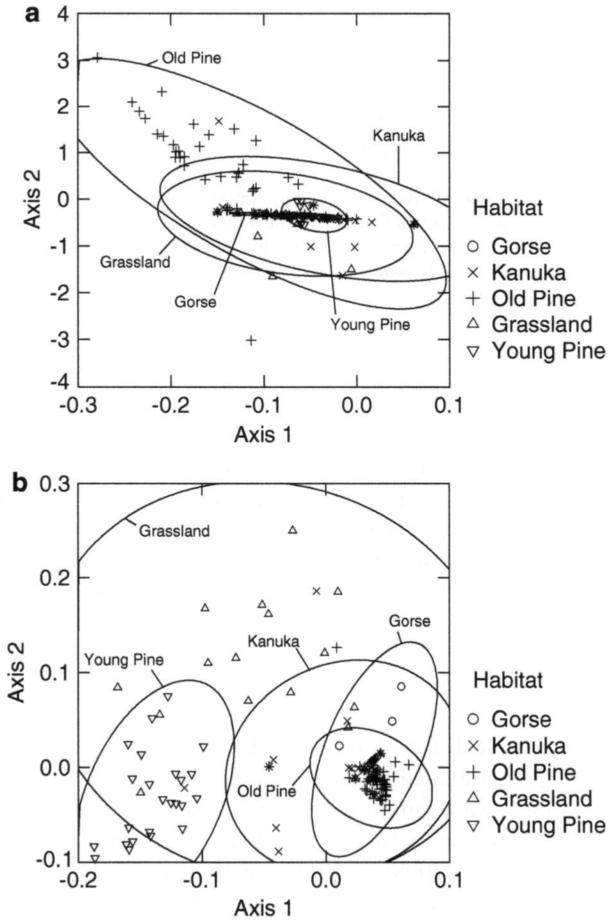

Fig. 3 (**a**) Unconstrained principle co-ordinate analysis of the five habitat types. One kanuka trap was an outlier with an Axis 1 score of 14.7 and was included in the ordination, but is not shown on the graph. (**b**) Constrained principle co-ordinate analysis (CAP analysis) of the five habitat types. Both graphs were calculated using chi-squared distances and ellipses are Gaussian bivariate, calculated in SYSTAT 9 (SPSS Inc. 1998), with a probability set at 0.95

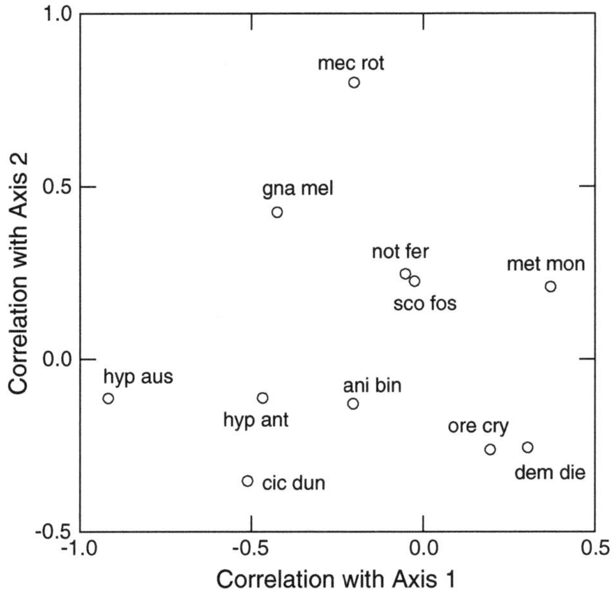

Fig. 4 Correlation of species with CAP axes 1 and 2. Only species with correlations >0.2 are shown (ani bin: *Anisodactylus binotatus*; cic dun: *Cicindela dunedensis*; dem die: *Demetrida dieffenbachii*; gna mel: *Gnathaphanus melbournensis*; hyp ant: *Hypharpax antarcticus*; hyp aus: *Hypharpax australis*; mec rot: *Mecyclothorax rotundicollis*; met mon: *Metaglymma moniliferum*; not fer: *Notagonum feredayi*; ore cry: *Oregus crypticus*; sco fos: *Scopodes fossulatus*)

Discussion

The carabid assemblages of the five habitat types showed distinct similarities, with three medium to large, flightless, generalist species common in all sites (*Meg. antarcticus, Met. moniliferum, O. crypticus*). This reflects the recent history of disturbance in the area, as discussed below. On closer inspection, however, many differences between the carabid assemblages are apparent, with the carabids of the plantation habitats most similar overall to those found in kanuka.

Habitat comparisons

The kanuka and old pine habitats showed similarities in the species composition, as shown by the ordination. Relative abundance was also similar between these habitat types and exotic carabids were rare in both habitats. Similar results were obtained in a similar study in another region of New Zealand where a greater proportion of natural forest remains (Pawson et al. 2008). Native New Zealand forests harbour few exotic beetle species, either due to resistance of the native beetle community to invasion, or due to the lack of habitat suitable for exotic species (Harris and Burns 2000). Exotic invertebrate species in New Zealand tend to be abundant in human modified environments such as towns, agricultural crops and exotic grassland (Watt 1975) whereas, so far, there are few invasive species that are adapted to mature forest, whether exotic or native.

Young pine had the greatest number of carabid species in common with kanuka, although the dominance of exotic species made the carabid assemblages distinct, as shown by the ordination. The high abundance of the small, flighted exotic species *H. australis* is similar to the situation found in young plantation forests in Europe, where this guild of carabids are often common (Magura et al. 2002; Butterfield 1997). Overall, the young pine habitat was richest in species and number of individuals, reflecting a pattern observed by other workers (Hubner and Baumgarten 2005; Magura et al. 2003; Pawson et al. 2008).

Grassland shared the dominant generalist carabid species with other habitats, but like the young pine habitat, exotic species were numerically dominant and overall species richness was relatively high. Pitfall traps in grassland were in diverse locations within the grassland habitat, and this habitat variability may explain the lack of clustering of these traps in the ordination (Fig. 4). Trap samples closer to the edge habitats, or ecotones, may have greater diversity than expected for the habitat type, due to the overlap in species from neighbouring habitats (Downie et al. 1996; Harris and Burns 2000; Molnar et al. 2001).

The gorse shrubland habitat was comparatively species poor but it had a high abundance of the three main generalist species (*Meg. antarcticus, Met. moniliferum* and *O. crypticus*), the only species shared with the kanuka habitat. Although this suggests that gorse shrubland may be of limited value for carabid conservation, the gorse shrubland we sampled was relatively young and of limited extent, which may have influenced its species composition. By contrast, Harris et al. (2004) found that native insect diversity was similar between gorse shrubland and kanuka/manuka (*Leptospermum scoparium*) shrubland of the same age and they considered gorse to be valuable habitat for many native species.

Vegetation history and the carabid fauna

To establish the role of Eyrewell Forest in carabid conservation it is useful to first consider a benchmark for comparison. The present carabid fauna of the Eyrewell region is likely to have been influenced by the history of the vegetation. The arrival of Polynesians, beginning in the thirteenth century AD (Wilmshurst and Higham 2004), is thought to have increased the frequency of fires on the Plains (Molloy and Ives 1972; McGlone 1983). This human-induced disturbance regime combined with the poor soils of the Eyrewell region reduced the extent of the original kanuka forests (Ecroyd and Brockerhoff 2005) leaving a mosaic of kanuka low forest and shrubland, interspersed with native grassland (Molloy and Ives 1972). The level of disturbance of these habitats increased further after the arrival of European settlers around 1850 who transformed the landscape into one dominated by agriculture. The dynamic nature of this environment since human arrival in New Zealand would have suited generalist carabid species, such as *Meg. antarcticus*, *Met. moniliferum* and *O. crypticus*, which were common in all habitats in this study.

The carabid fauna of the original forests of the Eyrewell region, prior to human-induced disturbance, may have included a greater abundance of forest specialist species than were found in this study. Forest specialists often disappear with increasing disturbance and fragmentation of forests (Barbaro et al. 2005) due to habitat loss and edge effects (e.g., Molnar et al. 2001). Two species (*Ho. intermittens* and *Ho. elongella*) caught in very low numbers in kanuka remnants in this study may be relicts of the original forest fauna. These species have been caught in high numbers in pitfall traps in native southern beech (*Nothofagus* spp.) forest in the nearby foothills of the Southern Alps (Berndt and Brockerhoff unpublished data). Based on these observations, the kanuka forest carabid assemblage we found probably does not fully reflect the original fauna. However, the observed carabid assemblage of the current kanuka remnants are all that remains for comparison on the Canterbury Plains.

Eyrewell Forest and carabid conservation

Although the canopy layer of the plantation forest is a true 'mono-culture', there is a relatively rich understorey vegetation that shares many species with the kanuka remnants in the area (Brockerhoff et al. 2003; Ecroyd and Brockerhoff 2005). Thus, similarities in the shady forest environment and understorey vegetation between the kanuka and the plantation forests probably explain some of the similarities between the respective carabid assemblages we observed. In addition, although there were earlier, failed attempts to use the area now occupied by Eyrewell Forest for agriculture, much of the forest was initially planted into land that had recently been cleared of kanuka forest (Molloy and Ives 1972), at a time when the conservation values of this forest type were not appreciated. This would certainly have allowed greater survival of forest carabid species than if the land went through a pastoral phase before planting.

Within the plantation forest environment there are a range of habitat types. These, to some degree, mimic the original heterogeneous kanuka landscape which, in the 1850s, was described as islands of tall, dense kanuka surrounded by shorter and sparser growth (Molloy and Ives 1972). Eyrewell Forest is managed in even-aged strips 120 m wide (Sommerville 1980) creating a mosaic of age classes which corresponds somewhat to the patchy nature of the original kanuka landscape. Although the main agents of disturbance differ (human-induced fire disturbance in the kanuka landscape (Molloy and Ives 1972) versus harvesting in the plantation) both resulted in a similar patchy forest landscape. This similarity between the historical and modern landscapes may make the plantation forest suitable habitat for conservation of much of the current native carabid fauna of the Eyrewell region. Butterfield et al. (1995) also found similarities between the carabid fauna of plantation forests and surrounding native moorland habitat when all stages of the forestry cycle were considered. The landscape heterogeneity provided by these cycles in plantation forests provides habitat for a greater diversity of carabid species than even-aged forests (Jukes et al. 2001).

In addition to providing habitat for the native carabid fauna as a whole, Eyrewell Forest is the sole known habitat for the critically endangered carabid *Ho. brevicula.* Only 10 specimens of this local endemic have ever been found (Brockerhoff et al. 2005). *Holcaspis brevicula* has been caught in pine stands of all age classes, from recently planted to close to harvest (Brockerhoff et al. 2005). This suggests that this species is able to persist in a habitat with relatively frequent disturbance. No *Ho. brevicula* have ever been found in any of the small kanuka remnants in the area despite the greater trapping intensity in kanuka remnants than in the plantation forest (this study, Brockerhoff et al. 2005), so Eyrewell Forest is playing a critical role in the conservation of this species, in the absence of kanuka habitat of substantial area. It is possible that similar situations apply to other plantation forests. Such plantations are not necessarily recognisable as such because plantation establishment may have occurred a long time ago. For example, Norway spruce forests that are widespread in many parts of central Europe are rarely representative of the natural vegetation, but their value as forest habitat is generally acknowledged (e.g., Hubner and Baumgarten 2005).

Conclusions

The few remaining kanuka remnants on the Canterbury Plains represent important examples of a formerly widespread flora and fauna. Because they have limited legal protection and are under threat from fire, land development and grazing (Meurk et al. 1995; Ecroyd and Brockerhoff 2005), it is clear, therefore, that this habitat type and its associated fauna are seriously endangered in this region. Protection and, wherever possible, restoration of natural forest and scrub communities should be priorities in this region. However, given prevailing land use pressures, Eyrewell Forest and other plantation forest represent significant substitute forest habitat for native species. Such situations are likely to occur elsewhere in areas where losses of native forest were substantial. Although natural forests are clearly preferable for the conservation of most forest species, exotic plantation forests may provide valuable habitat that could be important for such species which cannot survive in agricultural or other modified, open habitats (see also Brockerhoff et al. 2008—this issue). Interestingly, this plantation forest itself could become a threatened habitat. This is because of a recent trend towards conversion of plantation forests to exotic grassland for dairy production in parts of New Zealand where this is economically viable. If Eyrewell Forest was converted to exotic grassland, there would be a considerable loss of native carabid species from the area, probably including the extinction of *Ho. brevicula.*

Acknowledgements We thank Rowan Emberson, Alison Evans, Peter Johns, André Larochelle, Alan Leckie, Sylvia McLaren, Steve Pawson and Tanja Weis for assistance with trapping or identification of beetles. Helpful comments on an earlier draft were provided by Steve Pawson and two anonymous reviewers. Funding was provided by the New Zealand Foundation for Research Science and Technology under C04806 and C04X0214 and by the Department of Conservation. Thanks also to Carter Holt Harvey Forests, the Department of Conservation, and several farmers for access to field sites.

References

Allegro G, Sciaky R (2003) Assessing the potential role of ground beetles (Coleoptera, Carabidae) as bioindicators in poplar stands, with a newly proposed ecological index (FAI). For Ecol Manage 175:275–284. doi:10.1016/S0378-1127(02)00135-4

Anderson MJ (2003) CAP: a FORTRAN computer program for canonical analysis of principle coordinates. Department of Statistics, University of Auckland, New Zealand

Anderson MJ, Willis TJ (2003) Canonical analysis of principal coordinates: a useful method of constrained ordination for ecology. Ecology 84:511–525. doi:10.1890/0012-9658(2003)084[0511:CAOPCA]2.0.CO;2

Atkinson IAE, Cameron EK (1993) Human influence on the terrestrial biota and biotic communities of New Zealand. Trends Ecol Evol 8:447–451. doi:10.1016/0169-5347(93)90008-D

Barbaro L, Pontcharraud L, Vetillard F, Guyon D, Jactel H (2005) Comparative responses of bird, carabid and spider assemblages to stand and landscape diversity in maritime pine plantation forests. Ecoscience 12:110–121. doi:10.2980/i1195-6860-12-1-110.1

Brockerhoff EG, Ecroyd CE, Langer ER (2001) Biodiversity in New Zealand plantation forests: policy trends, incentives and the state of our knowledge. N Z J For 46:31–37

Brockerhoff EG, Ecroyd CE, Leckie AC, Kimberley MO (2003) Diversity and succession of adventive and indigenous vascular understorey plants in *Pinus radiata* plantation forests in New Zealand. For Ecol Manage 185:307–326. doi:10.1016/S0378-1127(03)00227-5

Brockerhoff EG, Berndt LA, Jactel H (2005) Role of exotic pine forests in the conservation of the critically endangered New Zealand ground beetle *Holcaspis brevicula* (Coleoptera: Carabidae). N Z J Ecol 29:37–43

Brockerhoff EG, Jactel H, Parrotta JA, Quine CP, Sayer J (2008) Plantation forests and biodiversity: oxymoron or opportunity? Biodivers Conserv. doi:10.1007/s10531-008-9380-x

Butterfield J (1997) Carabid community succession during the forestry cycle in conifer plantations. Ecography 20:614–625. doi:10.1111/j.1600-0587.1997.tb00430.x

Butterfield J, Luff ML, Baines M, Eyre MD (1995) Carabid beetle communities as indicators of conservation potential in upland forests. For Ecol Manage 79:63–77. doi:10.1016/0378-1127(95)03620-2

Colwell RK (2004) EstimateS: statistical estimation of species richness and shared species from samples. Version 7. http://viceroy.eeb.uconn.edu/estimates

Downie IS, Coulson JC, Butterfield JEL (1996) Distribution and dynamics of surface-dwelling spiders across pasture-plantation ecotone. Ecography 19:29–40. doi:10.1111/j.1600-0587.1996.tb00152.x

Dufrene M, Legendre P (1997) Species assemblages and indicator species: the need for a flexible asymmetrical approach. Ecol Monogr 67:345–366

Ecroyd CE, Brockerhoff EG (2005) Floristic changes over 30 years in a Canterbury Plains kanuka forest remnant, and comparison with adjacent vegetation types. N Z J Ecol 29:279–290

FAO (2001) State of the World's forests. Food and Agriculture Organization of the United Nations, Rome. 2001. Forest Plantation Thematic Papers

Geldenhuys CJ (1997) Native forest regeneration in pine and eucalypt plantations in Northern Province, South Africa. For Ecol Manage 99:101–115. doi:10.1016/S0378-1127(97)00197-7

Gotelli NJ, Colwell RK (2001) Quantifying biodiversity: procedures and pitfalls in the measurement and comparison of species richness. Ecol Lett 4:379–391. doi:10.1046/j.1461-0248.2001.00230.x

Harris RJ, Burns BR (2000) Beetle assemblages of kahikatea forest fragments in a pasture-dominated landscape. N Z J Ecol 24:57–67

Harris RJ, Toft RJ, Dugdale JS, Williams PA, Rees JS (2004) Insect assemblages in a native (kanuka—*Kunzea ericoides*) and an invasive (gorse—*Ulex europaeus*) shrubland. N Z J Ecol 28:35–48

Hartley MJ (2002) Rationale and methods for conserving biodiversity in plantation forests. For Ecol Manage 155:81–95. doi:10.1016/S0378-1127(01)00549-7

Hubner C, Baumgarten M (2005) Early effects of forest regeneration with selective and small scale clear-cutting on ground beetles (Coleoptera, Carabidae) in a Norway spruce stand in Southern Bavaria (Höglwald). Biodivers Conserv 14:1989–2007. doi:10.1007/s10531-004-2528-4

Humphrey JW, Newton AC, Pearce AJ, Holden E (2000) The importance of conifer plantations in northern Britain as a habitat for native fungi. Biol Conserv 96:241–252. doi:10.1016/S0006-3207(00)00077-X

Humphrey JW, Ferris R, Jukes MR, Peace AJ (2002) The potential contribution of conifer plantations to the UK Biodiversity Action Plan. Bot J Scotl 54:49–62

Jukes MR, Peace AJ, Ferris R (2001) Carabid communities associated with coniferous plantations in Britain: the influence of site, ground vegetation and stand structure. For Ecol Manage 148:271–286. doi:10.1016/S0378-1127(00)00530-2

Larochelle A, Lariviere M-C (2001) Carabidae (Insecta: Coleoptera): catalogue. Fauna N Z 43:285

Larochelle A, Lariviere M-C (2005) Harpalini (Insecta: Coleoptera: Carabidae: Harpalinae). Fauna N Z 53:160

Leathwick JR (2001) New Zealand's potential forest pattern as predicted from current species-environment relationships. N Z J Bot 39:447–464

Magura T, Elek Z, Tóthmérész B (2002) Impacts of non-native spruce reforestation on ground beetles. Eur J Soil Biol 38:291–295. doi:10.1016/S1164-5563(02)01162-7

Magura T, Tóthmérész B, Elek Z (2003) Diversity and composition of carabids during a forestry cycle. Biodivers Conserv 12:73–85 doi:10.1023/A:1021289509500

Maunder C, Shaw W, Pierce R (2005) Indigenous biodiversity and land use—what do exotic plantation forests contribute? N Z J For 50:20–26

McCune B, Mefford MJ (1999) PC-ORD. Multivariate analysis of ecological data. Version 4.01. MjM Software, Gleneden Beach

McEwen WM (1987) Ecological regions and districts of New Zealand. Third Revised Edition in Four 1:500 000 Maps, Department of Conservation, Wellington, N.Z., Publication No. 5

McGlone MS (1983) Polynesian deforestation of New Zealand: a preliminary synthesis. Archaeol Oceania 18:1–10

McGlone MS (1989) The Polynesian settlement of New Zealand in relation to environmental and biotic changes. N Z J Ecol 12:115–129

McGuiness CA (2007) Carabid beetle conservation in New Zealand. J Insect Conserv 11:31–41. doi:10.1007/s10841-006-9016-y

Meurk CD, Bellingham P, Macmillan B (1995) The last kanuka landscape on the Canterbury Plains? Canterbury Bot Soc J 29:11–24

Molloy BPJ (1970) Bankside—a new scientific reserve on the Canterbury Plains. Proceedings of New Zealand Ecological Society, vol 17. pp 47–51

Molloy BPJ, Ives DW (1972) Biological reserves of New Zealand 1. Eyrewell scientific reserve, Canterbury. N Z J Bot 10:673–700

Molnar T, Magura T, Tothmeresz B, Elek Z (2001) Ground beetles (Carabidae) and edge effect in oak-hornbeam forest and grassland transects. Eur J Soil Biol 37:297–300. doi:10.1016/S1164-5563(01)01103-7

Murcia C (1995) Edge effects in fragmented forests: implications for conservation. Trends Ecol Evol 10:58–62. doi:10.1016/S0169-5347(00)88977-6

Myers N, Mittermeier RA, Mittermeier CG, da Fonseca GAB, Kent J (2000) Biodiversity hot spots for conservation priorities. Nature 403:853–858 Medline. doi:10.1038/35002501

Norton DA (1998) Indigenous biodiversity conservation and plantation forestry: options for the future. N Z For 43:34–39

Pawson SM, Emberson RM, Armstrong KF, Paterson AM (2003) Phylogenetic revision of the endemic New Zealand carabid genus *Oregus* Putzeys 1868 (Coleoptera: Carabidae: Broscini). Invertebr Syst 17:625–640. doi:10.1071/IS02042

Pawson SM, Brockerhoff EG, Meenken ED, Didham RK (2008) Non-native plantation forests as alternative habitat for native forest beetles in a heavily modified landscape. Biodivers Conserv. doi:10.1007/s10531-008-9363-y
Rosoman G (1994) The plantation effect. Greenpeace, Auckland
Saunders DA, Hobbs RJ, Margules CR (1991) Biological consequences of ecosystem fragmentation. Conserv Biol 5:18–32. doi:10.1111/j.1523-1739.1991.tb00384.x
Sommerville A (1980) Wind stability: forest layout and siliviculture. N Z J For Sci 10(3):476–501
SPSS Inc. (1998) SYSTAT 9. SPSS Incorporated, Chicago, IL, USA
Taylor R (1997) State of New Zealand's environment 1997. The Ministry for the Environment, Wellington
Watt JC (1975) The terrestrial insects. In: Kuschel G (eds) Biogeography and ecology in New Zealand. Dr. W. Junk, The Hague, pp 507–535
Wilmshurst JM, Higham TFG (2004) Using rat-gnawed seed to independently date the arrival of Pacific rats and humans in New Zealand. Holocene 14(6):801–806. doi:10.1191/0959683604hl760ft
Wilson EO (1988) Biodiversity. National Academy Press, Washington D.C., USA

Stand-level management of plantations to improve biodiversity values

Jason Cummings · Nick Reid

Originally published in the journal Biodiversity and Conservation, Volume 17, No 5, 1187–1211.
DOI: 10.1007/s10531-008-9362-z

Abstract As conservation reserves expand, the likelihood that they will capture areas degraded by previous land use increases. Ecological restoration of such areas will therefore play an increasing role in biodiversity conservation. On the New South Wales North Coast, recent expansion in the conservation estate has captured over 300 softwood and hardwood plantations, many with understoreys dominated by exotic weeds. Here we present an overview of the practices we have adopted in managing flooded gum (*Eucalyptus grandis*) plantations infested with lantana (*Lantana camara*) to enhance their biodiversity value. Experiments designed to overcome barriers limiting regeneration of native forest in conjunction with measurement of soil and plant responses yielded insights into the management of former timber plantations for biodiversity. Canonical Correspondence Analysis indicated that the level of canopy retention (or logging intensity) within sites consistently explained the greatest amount of variation in plant community composition (32–38% post-treatment). Thinning and burning stimulated regeneration of native species. Retained canopy cover was proportional to the richness or abundance of native woody shrubs, understorey trees and native perennial herbs, indicating that management intensity can be varied to promote a range of conservation values. A state-and-transition model summarising purported management actions and likely outcomes for these plantations is presented. This is the first time plantations have been managed solely for biodiversity. Logging income means that plantation restoration can be cost-neutral, and the positive influence of a cover crop of trees means that plantation management may generally be manipulated to promote biodiversity conservation.

J. Cummings · N. Reid
Ecosystem Management, University of New England, Armidale, NSW 2351, Australia

N. Reid
e-mail: nrei3@une.edu.au

Present Address:
J. Cummings (✉)
Minerals Council of Australia, P.O. Box 4497, Kingston, ACT, Australia
e-mail: jason.cummings@minerals.org.au

E.G. Brockerhoff et al. (eds.), *Plantation Forests and Biodiversity: Oxymoron or Opportunity?*
DOI: 10.1007/978-90-481-2807-5_14

Keywords State-and-transition model · Plantations · Biodiversity conservation · Restoration · *Eucalyptus grandis* · *Lantana camara* · National park · Australia

Introduction

Internationally, plantation establishment for oil and fibre production is increasing, while the clearing of native forests continues (Dudley et al. 2005). Accordingly, the role and importance of plantations in landscape management and restoration have received greater attention of late (e.g. Hartley 2002; Lamb 2005). It is likely that a variety of plantation types will become more important for biodiversity conservation, through either (1) plantations being captured in the reserve estate, (2) plantation establishment being used as a tool for restoring parts of protected areas, or (3) privately established plantations demonstrating conservation values through corridor or buffer values. Improved stand-level management of plantations can increase their value from a conservation perspective (Keenan et al. 1997; Lindenmayer et al. 2003), but how to achieve this in a cost-effective manner is often unclear.

On the North Coast of New South Wales (NSW), Australia, recent expansion of the conservation estate has captured over 300 softwood and hardwood plantations, many dominated by introduced weeds in the understorey. Bongil Bongil National Park (BBNP) is a case in point. It was gazetted in 1995 to conserve coastal wetland, littoral rainforest and heath communities, but its land parcelling included degraded areas. At the time of gazettal, plantations of native hardwoods (principally flooded gum *Eucalyptus grandis*) and introduced softwoods (slash pine *Pinus elliottii*) occupied 16% of the park. The BBNP Plan of Management (POM) prioritises the restoration of eucalypt plantations to enhance biodiversity and habitat values: 'management of [eucalypt plantations] will aim to return the areas to a natural condition' using experiments trialling 'thinning, clearing, planting and different fire regimes' (NPWS 1999).

While thinning plantations may enhance biodiversity values (Keenan et al. 1997), no study has examined the impact on biodiversity of a range of thinning regimes in plantations. Gap creation is used in forests to stimulate regeneration of timber species (e.g. SFNSW 1995), and can have positive effects on understorey biodiversity. In North American Douglas fir (*Pseudotsuga menziesii*) plantations, silvicultural treatment has a significant effect on understorey plant diversity, with ferns and graminoids responding positively to thinning (Thomas et al. 1999). In Canadian white pine (*Pinus strobus*) stands, thinning increases shrub diversity and decreases the range of regenerating trees (Wetzel and Burgess 2001). Thinning of silvertop ash (*E. sieberi*) forest in Victoria, Australia, increases the cover of herbaceous species five times (Bauhus et al. 2001). Therefore, gap creation in mature dense plantations may well enhance understorey biodiversity, but the specific impacts are likely to vary with the type, scale and intensity of intervention and be difficult to predict.

The homogeneous nature of plantations and the disturbance associated with their establishment often provides a uniform habitat for one or a few weedy understorey species to dominate. Many plantations in BBNP and elsewhere in coastal northern NSW have an understorey dominated by lantana (*Lantana camara*), a Neotropical scrambling shrub introduced to Australia as an ornamental. It spreads by frugivorous birds distributing the seeds, as well as by local vegetative expansion (Totland et al. 2005). Lantana can arrest plant succession (Webb et al. 1972) and reduce plant species richness in rainforest (Fensham et al. 1994), and it appears to do this in the flooded gum-dominated plantations in BBNP by forming a dense, near-monoculture in the understorey. This phenomenon is not unusual. Understorey dominance by introduced species has resulted in arrested succession and poor regeneration of native tree and understorey species in plantations elsewhere (Harrington and Ewell 1997; de la Cretaz and Kelty 2002). The introduced pasture grass, broad-leaf paspalum (*Paspalum wettsteinii*), blocks regeneration of native species in coastal blackbutt (*E. pilularis*) and Sydney blue gum (*E. saligna*) plantations elsewhere in BBNP (Villa et al. 2005). Thus, removal of the lantana understorey in combination with thinning of dense plantation trees should increase structural and compositional heterogeneity, and provide an opportunity for native species to establish, increasing the ability of the plantations to resist future weed invasion (Fox and Fox 1986; Attiwill 1994).

State-and-transition models are useful for describing and interpreting ecosystem dynamics (Whisenant 1999; Hobbs and Harris 2001). They describe the combinations of ecological conditions and management required for transitions between different ecological states (Westoby et al. 1989). A stable state is a more or less intransient configuration of ecosystem condition with only minor variation through time (Westoby et al. 1989). Transitions, triggered by natural events or management, represent relatively rapid changes between different stable states. Thresholds in environmental conditions and management identify when such transitions can occur. For plants, such thresholds exist at the species (Cropper 1993) and community levels (Holling 1973; Friedel 1991). Transitions can be linear, non-linear or discontinuous changes between states (Scheffer and Carpenter 2003). Transient or intermediate states can also be recognised between two stable ecosystem configurations. Once a threshold has been crossed and a rapid transition towards a more degraded state has occurred, restoration back towards a less degraded state requires management intervention (Hobbs and Norton 1996). Thresholds that trigger transitions may comprise abiotic characteristics (e.g. soil chemistry, hydrology) or biotic interactions (e.g. plant competition, herbivory; Whisenant 1999). After transition across a threshold, management actions must remove the abiotic and biotic barriers maintaining the system in the degraded state if restoration is to be successful (Hobbs and Norton 1996; Whisenant 1999). State-and-transition models are useful in ecological restoration as they summarise the management interventions and natural events that alter systems (Hobbs and Norton 1996). Abiotic and biotic barriers to restoration need to be identified and removed or overcome to promote transitions between states (Whisenant 1999). Small-scale experiments can be used to identify restoration barriers (Lake 2001), so that management can confidently target the actions required at landscape scale.

In partnership with the NSW National Parks and Wildlife Service, we adopted an adaptive experimental approach to managing the flooded gum-dominated plantations in BBNP to improve their conservation value. We used experiments to distinguish potential abiotic and biotic barriers limiting transitions towards more favourable goal communities. Goal communities were interpreted broadly to be any vegetation with a native plant-dominated understorey and preferably greater diversity in the overstorey than just flooded gum, although even-aged forests of this species occur naturally in the area (Baur 1962; Curtin et al. 1991; NPWS 1999).

As far as we are aware, this is the first time plantations have been managed solely for biodiversity, and the work provides interesting insights in this regard. For instance, logging means that the ecological restoration of plantations may be cost-neutral or revenue-positive. Our objectives in this study were to: (1) demonstrate that stand-level management of plantations influences the regeneration of weeds and native plants; (2) summarise patterns of regeneration in the post-disturbance, early-establishment phase; and (3) synthesise our understanding of the BBNP plantation experiments in the form of a state-and-transition model for the management of these plantations for biodiversity conservation.

Methods

Site description

BBNP is located on the mid-North Coast of NSW and was approximately 1,300 ha at the time of this study. Neighbouring land uses include forestry, grazing, rural-residential and low-density urbanisation. Subtropical Coffs Harbour has a warm and humid climate with an average daily minimum and maximum temperature of 14.0°C and 23.2°C, respectively (1943–2001; BOM 2002). From August 1999 to August 2002, Coffs Harbour received less than the long-term average rainfall in 26 of 37 months.

Most areas within BBNP were selectively logged before 1960, and some higher elevation areas were cleared for cattle pasture. Mining of mineral sands was conducted from the 1960s until the mid-1970s: clear-felling of vegetation on coastal sands occurred within the mining path. In 1972, flooded gum plantations were established on sand-mine tailings, former cattle pasture and in cleared native forest. The plantations were unmanaged until gazettal of the land in 1995 as BBNP. Failure of some plantations on the sand mass occurred (Cummings et al. 2005), and elsewhere self-thinning resulted in high variability in tree size and timber quality. Three 1.5-ha sites with an understorey dominated by lantana and representative of a large proportion of the BBNP flooded gum plantations, were arbitrarily selected for this ecological restoration experiment. Sites 1 and 2 were located on slate-derived, clay soils and Site 3 was on the sand mass. These sites were originally either wet sclerophyll forest or littoral rainforest on the sand mass, or wet sclerophyll (moist coastal hardwood) forest on the clay soils. Sites 1 and 2 had a north–west aspect and were established on cattle pasture and in cleared native forest, respectively. Site 3 was flat and established after sand mining (see Cummings et al. 2007 for further details regarding the plantations).

Experimental design, monitoring and statistical procedures

At each site, a factorial split-plot experiment was established, with six 100 × 25 m plots split into four 25 × 25 m sub-plots. Each of the six plots was allocated one of the following three canopy treatments: (1) control (no plantation trees removed); (2) thinned (removal of 7 out of every 8 trees); and (3) clear felled (all plantation trees removed). Tree removal occurred in February 2000 and approximated a 'thinning from below' regime (SFNSW n.d.). Each site was burnt in April 2000 to reduce slash, kill or retard regenerating lantana, and promote germination of the seed bank (e.g. *Acacia* and *Dodonaea* spp; Floyd 1966). Each sub-plot was allocated to one of six understorey treatments, with even representation of sub-plot treatments within main plots: (a) no further management; (b) annual weed control; rainforest seedlings planted (c) with and (d) without annual weed control; and wet sclerophyll seedlings planted (e) with and (f) without annual weed control. Since the seedlings failed to establish, the design was reduced to weed control versus no weed control amidst the natural regeneration (Cummings et al. 2007). Annual post-thinning weed control was implemented on a weed-specific basis, with all common weed species sprayed with glyphosate.

Near each site in undisturbed lantana-infested flooded gum plantation, a permanent control sub-plot (25 × 25 m) was established at the beginning of the experiment to monitor variation in response variables in the absence of overstorey and understorey manipulations. Vegetation sampling was conducted at the sub-plot level in treated sites and control sub-plots, and occurred before and after treatment implementation. Canopy cover was measured using a crown cover densiometer (Forest Densiometers, Oklahoma). Because the canopy treatments were not uniform (due to the post-burning death of trees), canopy thinning was not analysed as a discrete factor. Instead, canopy cover was used as a covariate in analyses at the sub-plot level. Understorey vegetation was sampled using six 2 × 2 m quadrats in each sub-plot in August–September 1999 (pre-treatment), August–September 2000 (3 months post-treatment), February 2001 (9 months post-treatment) and August–September 2001 (15 months post-treatment) (see Cummings et al. 2007 for further details). In 2002, two 5 × 5 m quadrats were sampled in each sub-plot to compare native woody regeneration in control and treated sub-plots with planted seedling survival. Only densities of woody plants >1 m in height were counted. Soil samples were collected pre-treatment (August 1999) and 3 months (August 2000) and 15 months (August 2001) post-treatment and analysed for organic matter content, pH_{H_2O}, electrical conductivity, total phosphorus (P), total nitrogen (N) and concentrations of ammonium (NH_4^+), nitrate (NO_3^-), magnesium (Mg^{2+}), potassium (K^+), sodium (Na^+) and calcium (Ca^{2+}) (using methods in Rayment and Higginson 1992).

Canonical Correspondence Analysis (CCA; ter Braak 1986) was used to examine relationships between soil chemical variables, canopy cover, weed control and plant community composition in 2001 (15 months post-treatment). The vegetation input matrix consisted of the cover of plant species recorded more than four times in all 66 treated sub-plots. Environmental data included the 11 soil chemistry variables, canopy cover and weed control. The amount of variation in the plant communities explained by the environmental data was determined by summing the eigenvalues for the first two axes (ter Braak and Smilauer 1998). The marginal and conditional relative importance of environmental variables in determining species composition was examined using automatic forward selection and Monte Carlo permutation tests (ter Braak and Verdonschot 1995). In forward selection, each environmental variable is ranked based on the proportion of community variation described by it when added as the first term to the model (marginal effects). The environmental variable that describes the most variation is then used as the first variable and each other variable is ranked according to the additional variation they explain (conditional effects, ter Braak and Verdonschot 1995). Only significant conditional effects are reported.

Plant species were allocated to functional groups in order to summarise plant response to plantation treatments and construct a state-and-transition model for plantation management (see Appendix 1 for species allocations and Cummings et al. 2007 for details). Main effects (e.g. canopy cover, weed control) were tested against the site by plot interaction, using a split-plot analysis. A mixed effects model was used, utilising the restricted maximum likelihood algorithm to estimate variance parameters (Patterson and Thompson 1971; Gilmour et al. 2002). To ensure that the assumptions of normality and even dispersion were not violated, diagnostic plots were constructed: residuals versus fitted values were plotted for dispersion, and residuals versus standard normal quantiles for normality (Quinn and Keough 2002). For analyses with a normal error distribution, natural log or square root transformations were often required if assumptions of analysis of variance were violated with the original scale. Predicted means (see Johnson and Omland 2004) are presented with 95% confidence intervals, and were back-transformed as required. The MASS (Venables and Ripley 2002) and nlme (Pinheiro et al. 2003) libraries were used with the statistical program *R* for all analyses (Ihaka and Gentleman 1996).

Results

Some 186 species in 74 families of vascular plant were recorded in treated plantation sites and control sub-plots from 1999 to 2002 (Appendix 1). Each functional group consisted of at least three families. Four groups (herbaceous perennials, canopy trees, woody shrubs and climbers, and grass-like monocots) accounted for 61% of species (26, 13, 11 and 11%, respectively). None of the species were 'threatened' in the region (i.e. listed under state or commonwealth legislation), but many are constituents of locally and regionally threatened ecosystems, such as Littoral Rainforest and Sub-tropical Coastal Floodplain Forest (both listed as 'Endangered Ecological Communities' in NSW: NPWS 1999; NSW Scientific Committee 2004a, b).

Species richness averaged 5.3 per 4-m^2 quadrat and mean total cover was 93% in plantation sites in 1999. Before treatment, the understorey vegetation was dominated by introduced woody shrubs, monocots and grass-like monocots, averaging 48, 12 and 12%, respectively, across sites (Fig. 1). Typically, these species included lantana (*Lantana camara*, Verbenaceae), *Chyrsanthemoides monolifera* (Asteraceae), *Lomandra* sp. (Xanthorrhoeaceae), one of the climbers (*Ripogonum album*, Smilaceae or *Smilax australis*, Smilaceae) and a grass (either *Entolasia marginata* or *Imperata cylindrica*, Poaceae). After treatment, the understorey floristics changed markedly. Understorey species richness more than doubled, while total cover declined from 93% to 60%. Species richness and abundance of the majority of functional groups also changed markedly. By 15 months, grass-like monocots dominated the understorey, with 42% cover (Fig. 1). Perennial herbs accounted for 28% cover whilst other monocots accounted for 7% (Fig. 1). This largely native herbaceous community comprised *Oplismenus imbecillis* and *Entolasia stricta* (Poaceae) or, on the sandy site, *Cyperus trinervis* (Cyperaceae) and *Imperata cylindrica*. Both soil types supported *Viola hederaceae* (Violaceae), *Centella asiatica* (Apiaceae), *Lepidosperma laterale* (Cyperaceae), *Lomandra* sp. (Smilaceae) and *Ripogonum album* (Cyperaceae). These species are typical of early succession regeneration in locally occurring forest

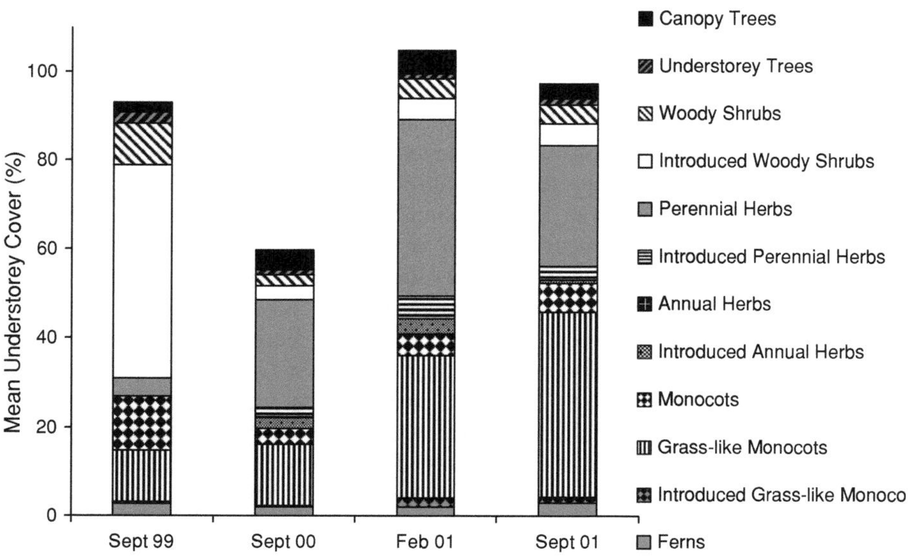

Fig. 1 Mean cover of functional groups of vascular plants in treated sub-plots over time

communities (including Sub-tropical Coastal Floodplain Forest). The treatments were intended to promote these taxa as well as potential canopy species and understorey trees and shrubs. In terms of total cover, the woody taxa were less abundant after only 15 months (Fig. 1). However, 6 years after treatment, the native woody species had grown up and dominated the treated sites (Plates 1, 2).

Absolute values of the abundance of plant functional groups (i.e. richness, density and cover) varied across sites, but the patterns in plant regeneration with respect to canopy cover were always similar (Cummings 2004; Cummings et al.2007). Although understorey cover was dominated by grasses and perennial herbs 15 months after treatment (Fig. 1, 2), grass cover was maximised in sub-plots with reduced canopy cover, while native understorey herbs were favoured by retained canopy cover (Fig. 2). Where grass cover increased over the 15-month post-treatment period, native canopy tree density, richness and density of native understorey trees and perennial herb richness declined (Fig. 2, 3). Similarly, the richness, cover (Fig. 3) and density (not shown) of native woody shrubs and climbers, as well as understorey trees, was greatest at high levels of canopy retention by the end of the

Plate 1 A view of Site 1 prior to treatment imposition, showing the dense mature flooded gum overstorey and dense lantana understorey

Plate 2 The same view in 2005, 6 years after treatment imposition. Note the reduction in overstorey canopy and plantation tree trunks, the acacia-dominated understorey, and grass groundcover

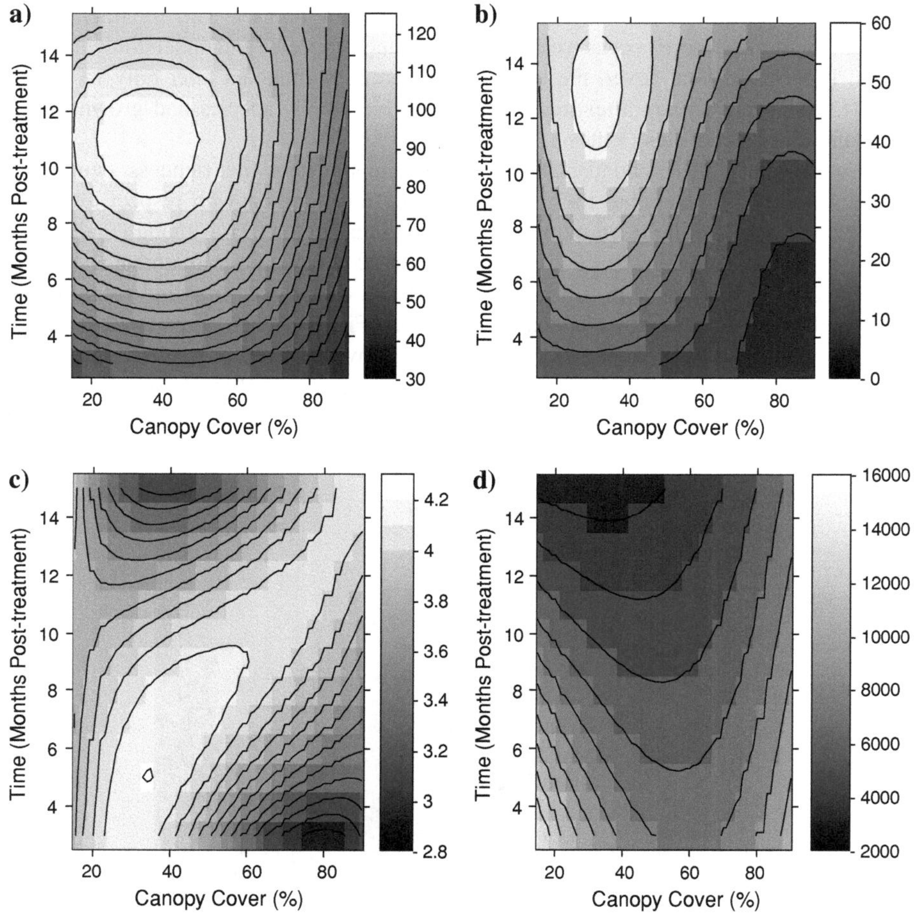

Fig. 2 Response of (**a**) total understorey vegetation cover (%); (**b**) grass-like monocot cover (% cover); (**c**) perennial forb species richness (species/quadrat); and (**d**) potential canopy tree species density (stems/ha), to retained canopy cover in the post-logging regeneration phase in 2000–2001

post-treatment sampling period. Woody weed cover increased in a non-linear manner with declining canopy cover in the absence of weed control, but control suppressed woody weed regeneration evenly across the canopy retention scale (Fig. 4). These understorey floristic patterns are summarised in Fig. 5.

By 2001, regeneration of native species in treated sites was far more abundant than in control sub-plots which remained lantana-dominated (Fig. 6), and the species richness of the native regeneration in treated sub-plots was five times greater. The natural regeneration of native woody stems stimulated by logging and burning was also much more dense than could have been achieved by tubestock planting of wet sclerophyll and rainforest species (Fig. 6), even if most of the planted seedlings had prospered rather than succumbing to drought and wallaby browsing.

Fifteen months after treatment, 38% of the variance in vegetation community composition across sites was explained by differences in soil chemistry and canopy cover (Table 1; Fig. 7). Alone, soil organic matter explained the largest amount of variation between sites, due to differences between the sand and clay substrates. When all

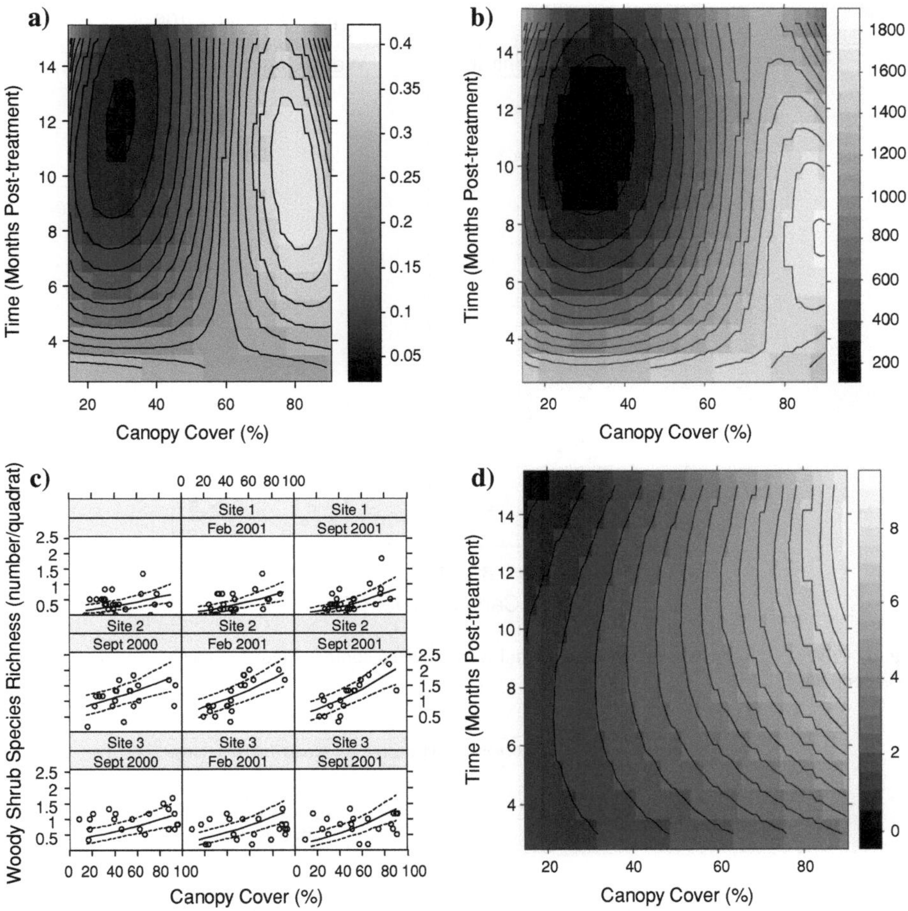

Fig. 3 Response of (**a**) understorey tree species richness (species/quadrat); (**b**) understorey trees species density (stems/ha); (**c**) species richness of shrubs and woody climbers (species/quadrat); and (**d**) cover (%) of shrubs and woody climbers cover, to retained canopy cover in treated plantations in 2000–2001. Predicted means are presented with 95% confidence limits on panel plots

independent variables were considered together, however, soil potassium and canopy cover emerged as the only significant conditional effects. Environmental variables explained as much or almost as much floristic variation within as between sites. Alone, canopy cover explained most variation in each site and was the only significant explanatory variable at Sites 1 and 2 (clay sites). On the sand, however, ammonium and total phosphorus were the only significant conditional effects at Site 3 when all independent variables were considered together. Weed control had no significant effect on plant community composition (species present and their cover) within or between sites, at this early stage of regeneration.

State-and-transition model

A state-and-transition model was developed for the ecological restoration of flooded gum plantations in BBNP (Fig. 8), based on our experimental results (Cummings 2004;

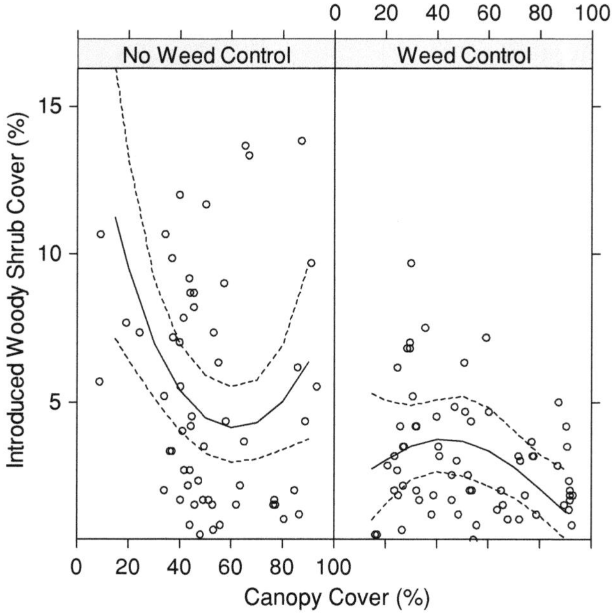

Fig. 4 Cover (%) of introduced shrubs with and without weed control. Predicted means are presented with 95% confidence limits

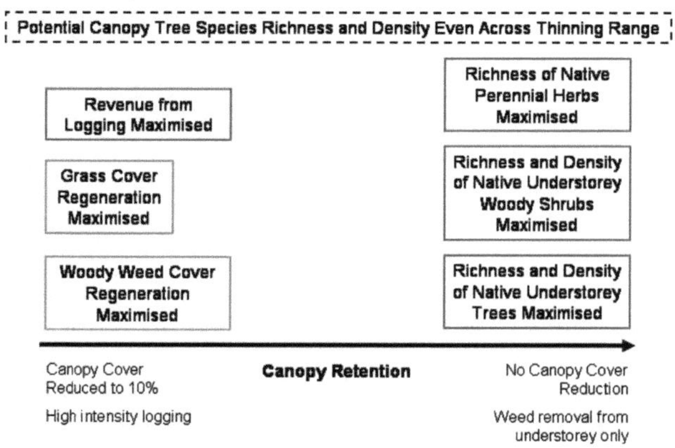

Fig. 5 Summary of responses of the cover, richness and density of the main plant functional groups to thinning and burning of flooded gum plantations infested with lantana (adapted from Cummings et al. 2007)

Cummings et al. 2005, 2007) and the literature. Our formulation of states and transitions is based on the following:

- Lantana-infested flooded gum plantation and failed plantation (blady grass-dominated grassland on sand-mined sites) are stable states
- A transition threshold consisting of abiotic (soil) limitations is evident where plantation establishment failed and woody canopy cover averages 15% or less
- There is no evidence of an abiotic transition threshold being crossed when mature plantation is thinned down to a canopy cover of 15%

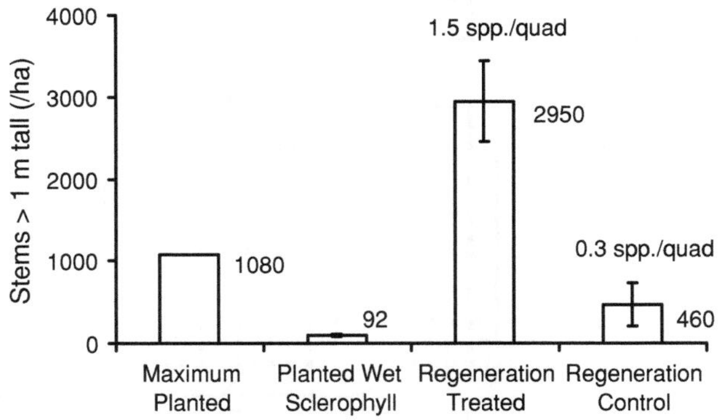

Fig. 6 Actual and potential native woody plant establishment in 2002. The histogram bars represent stem density in four situations: (i) if all planted seedlings had established ('Maximum Planted'); (ii) actual best-case establishment of planted seedlings ('Planted Wet Sclerophyll'); (iii) the natural regeneration measured in treated sub-plots ('Regeneration Treated') and (iv) untreated control sub-plots ('Regeneration Control'). Data are observed means ± one standard error. The number of native woody species/quadrat is indicated for treated and undisturbed (control) sub-plots. Actual density figures are given adjacent to each bar

Table 1 Environmental explanation of vegetation community composition between and within sites, 15 months post-treatment (September 2001)

Analysis	Total variation explained in all eigenvalues (%)	Marginal effect and variation explained (%)	Conditional effects and additional variation explained (%)
All sites	37.7	Organic matter (47)**	Potassium (8)** Canopy cover (6)**
Site 1	31.7	Canopy cover (10)**	NA
Site 2	32.2	Canopy cover (16)**	NA
Site 3	37.7	Canopy cover (26)**	Ammonium (12)* Total phosphorus (11)*

*** $P < 0.001$, ** $P < 0.01$, * $P < 0.05$, n.s. = non-significant, NA = not applicable due to non-significant conditional effects

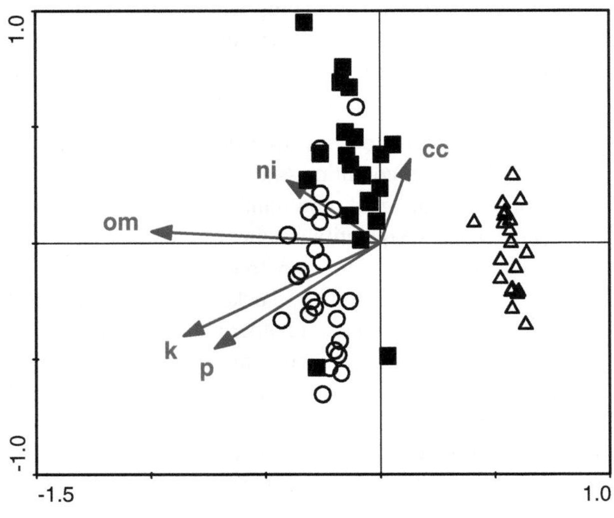

Fig. 7 Bi-plot of first two CCA axes relating differences in sub-plot understorey vegetation composition with soil chemistry, weed control and canopy cover variables in September 2001 (15 months post-treatment; only the 5 most influential variables are displayed). Environmental variables: soil organic matter (om), soil potassium (k), soil total phosphorus (totp), soil nitrate (ni) and canopy cover (cc). Sites 1 (○), 2 (■) and 3 (Δ)

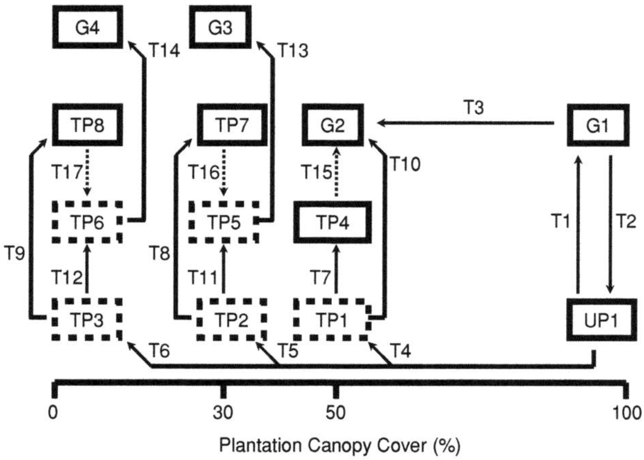

Fig. 8 State-and-transition model summarising management actions to enhance the biodiversity of lantana-infested flooded gum plantations on the NSW mid-North Coast. Stable states (——); intermediate states (- - -). See Table 2 for the catalogue of individual states and transitions

- Thinning plantations in conjunction with understorey weed control influences understorey regeneration patterns
- The level of thinning intensity (or its converse, canopy retention) influences the responses of different plant functional groups
- The species richness, density and cover of regenerating native woody shrubs and trees in the understorey decline as the flooded gum canopy is reduced
- There is a concomitant increase in grass cover with reduced flooded gum canopy
- As grass cover increases, understorey tree establishment declines
- In the absence of weed control after logging and burning, the cover of woody weeds (e.g. lantana) increases in accelerating fashion with declining canopy cover
- The unchecked regeneration of lantana results in its return as the dominant understorey species in a stable state, inhibiting native species regeneration (Dutton et al. 1989; Fensham et al. 1994)
- The rate of regeneration of all woody species increases in a linear manner with declining canopy cover
- The resilience of the former forest communities at these sites is such that some of the constituent species can regenerate upon removal of the biotic barriers to restoration (competition from understorey woody weeds and plantation trees)
- Irrespective of whether these sites follow a successional pathway characterised by initial floristic composition or rainforest dynamics, canopy trees recruit and establish in the created gaps (e.g. Kooyman 1996)

Each state and transition has been described in Table 2. State UP1 represents the stable degraded plantation condition (an overstorey monoculture with a woody weed-dominated understorey). Four goal communities, G1–G4, are recognised. These are primarily differentiated by the quantity amount of plantation trees left in the canopy. G1 represents a stable state in which plantation trees dominate the canopy and native trees and shrubs dominate the understorey (e.g. Tucker and Murphy 1997). Such a state has considerable conservation value compared to other plantation states, given that forest dominated by pure, even-aged flooded gum occurs naturally in the area (Curtin et al. 1991), and does not require further manipulation for biodiversity conservation. States G2–G4 are differentiated by the amount of initial thinning of plantation trees, and the resources required subsequently to manage the sites towards these states. Goal communities for flooded gum plantations in BBNP are dictated by soil type, elevation, nearby vegetation and fire hazard, and include

Table 2 Description of states and transitions (Fig. 8) and the management required to affect transitions

Catalogue of states	Catalogue of transitions (T)
UP1—Unmanaged Plantation: stable state of *L. camara* infested eucalypt plantation, no canopy species diversity, little or no mid-storey structure or diversity. Canopy cover of 100% plantation trees	T1: Removal of *L. camara* from understorey via crushing and burning and subsequent herbicide of regenerating weeds
G1—Unthinned Plantation Goal Community: stable state of eucalypt plantation with native tree and shrub understorey. Canopy cover of 100% plantation trees, low grass cover regeneration	T2: Gradual dominance by *L. camara* where control is insufficient and native species were unable to compete
TP1—Thinned Plantation: transient state of 50% canopy cover by plantation trees. Understorey comprised medium numbers of native and introduced shrub and tree seedlings, with medium grass cover regeneration	T3: Stem injection or natural death of plantation trees
TP4—Thinned Plantation with Lantana Understorey: stable state of *L. camara* dominated understorey, with 50% plantation tree canopy cover	T4: Removal of 50% of the canopy, crushing and burning of *L. camara* and slash in understorey
G2—Thinned Plantation Goal Community: stable state of a goal community. Sclerophyll or rainforest trees in canopy with native spp. dominating understorey. Plantation tree canopy cover of 50%	T5: Removal of 70% of the canopy, crushing and burning of *L. camara* and slash in understorey
TP2—Thinned Plantation: transient state of 30% canopy cover by plantation trees. Understorey comprised low numbers of native and introduced shrub and tree seedlings, with high grass cover regeneration	T6: Removal of 100% of the canopy, crushing and burning of *L. camara* and slash in understorey
TP5—Thinned Plantation with Native Understorey: transient state of 30% canopy cover by plantation trees. Understorey native shrubland, grass cover reduced, canopy tree seedlings present and developing	T7: Lack of weed control results in *L. camara* regrowth
TP7—Thinned Plantation with *Lantana* Understorey: stable state of *L. camara* dominated understorey, with 30% plantation tree canopy cover	T8: Lack of weed control results in *L. camara* regrowth
G3—Thinned Plantation Goal Community: stable state of a goal community. Sclerophyll or rainforest trees in canopy with native spp. dominating understorey. Plantation tree canopy cover of 30%	T9: Lack of weed control results in *L. camara* (on clay soils) or *I. cylindrica* (on sand soils) regrowth
TP3—Thinned Plantation: transient state of 0% canopy cover of plantation trees. Understorey comprised low numbers of native and introduced shrub and tree seedlings, with high grass cover regeneration	T10: Limited post-thinning woody weed control and growth of regenerating canopy and understorey tree seedlings into space created from plantation tree removal
TP6—Thinned Plantation with Native Understorey: transient state of 0% canopy cover by plantation trees. Understorey native shrubland, grass cover reduced, canopy tree seedlings present and developing	T11: Post-thinning control of woody weeds and grass, with exclusion of wallabies, allows establishment of limited regenerating shrubs and understorey trees

Table 2 continued

Catalogue of states	Catalogue of transitions (T)
TP8—Thinned Plantation with *Lantana* or *Imperata* Understorey: stable state of *L. camara* (on clay soils) or *I. cylindrica* (on sand soils) dominance, with 0% plantation tree canopy cover	T12: Greatest post-thinning control of woody weeds and grass, with exclusion of wallabies, allows establishment of limited regenerating shrubs and understorey trees. Supplementary plantings may be required
G4—Thinned plantation Goal Community: stable state of a goal community. Sclerophyll or rainforest trees in canopy with native spp. dominating understorey. Plantation tree canopy cover of 0%	T13: Woody weed control allows recruitment and establishment of rainforest or sclerophyll canopy trees
	T14: Woody weed control allows recruitment and establishment of rainforest or sclerophyll canopy trees
	T15–17: Represent possible transitions should *L. camara* or *I. cylindrica* be allowed to re-dominate; after re-doing initial weed control, planting seedlings and fencing to exclude wallabies

threatened ecological communities such as Littoral Rainforest on sand and Sub-tropical Coastal Floodplain Forest on low-elevation clay soils (NPWS 1999; NSW Scientific Committee 2004a, b). However, for the purposes of this investigation and NPWS's plantation management program in the Park, vegetation that is not weed infested, is more structurally diverse, and has a greater representation of native species in the understorey than the pre-treatment state, is considered an appropriate goal. Through adaptive management, the conservation goals for treated plantation sites can be successively reviewed, refined and updated, as better information becomes available and societal expectations evolve.

G1 sites naturally occur to a limited extent in BBNP without management intervention, and may be an appropriate goal for weed-infested sites where thinning is not possible or viable. Understorey weed control alone, particularly with the use of fire to stimulate regeneration, should lead to a change in dominance in the understorey from introduced to native species (Transition 1). It is likely that continued weed control will be required until the native understorey is dense and stable enough to resist introduced species establishment and dominance (Transition 2). Alternatively, where a native understorey already exists (G1), stem injection of herbicide might be used to reduce plantation tree dominance and allow other native canopy species to emerge (Transition 3), promoting forest structural and compositional heterogeneity.

States G2, G3 and G4 represent goal communities with a native understorey and a canopy of greater floristic and structural diversity than G1. The first stage in accelerating the progression of the degraded plantations (UP1) towards these is the removal of plantation trees by thinning to appropriate densities and burning the slash (Transitions 4, 5 and 6). For each of these pathways, a transient state is identified, immediately after thinning and burning, with a flush of native tree and shrub seedlings and grass regeneration. Density, cover and richness of native understorey trees and shrubs decrease with canopy removal, while grass and woody weed regeneration increase with canopy removal (e.g. an intermediate level of thinning; Fig. 5). Therefore, desirable native woody seedlings are subjected to the greatest competition in the transitional plantation state, TP3, as indicated by the poor establishment of understorey trees at maximum grass cover. Native woody regeneration is subjected to successively less understorey competition in TP2 and TP1, and to least inhibition by herbaceous and woody weed growth where the canopy is undisturbed.

The abundance of canopy and understorey trees with moderate retention of canopy cover may allow quicker establishment of goal community, G2 (Transition 10), without the intermediate establishment of native shrubland that requires further recruitment of canopy and understorey trees (TP5 and TP6, Transitions 11 and 12). Although the rate of regeneration is maximised by reducing canopy (e.g. total cover and understorey structure), managing the composition of the regeneration and controlling lantana requires increased management inputs. With lower rates of regeneration of native shrubs and trees at reduced canopy levels, more effort may be required for planted seedling establishment and protection from browsing (Transitions 11–14) to progress towards goal communities. For example, clear felling plantation trees on the sand mass (Transition 6 on sandy soils) risks the loss of a large amount of nutrient capital through leaching and crossing an abiotic threshold, leading to blady grass grassland. Significantly greater management inputs are then required to re-establish woody cover than had the threshold not been crossed (Transition 5 on sand). Woody weed regeneration also increases with canopy reduction, adding to the need for more intensive management of regeneration with increased logging.

In each transient plantation state (TP1–3), failure to control lantana regeneration is likely to result in its re-establishment as the dominant understorey species. This would result in stable treated plantation states, TP4, TP7 or TP8, the only difference from the pre-treatment condition (UP1) being a reduced cover of plantation trees. Establishment of these states fails to capture the benefits of native seedling regeneration after logging and burning. Moreover, if regenerating trees and shrubs do not set seed before lantana returns, the likelihood of obtaining adequate native regeneration (Transitions 15–17) a second time is reduced owing to seed bank exhaustion, requiring considerable investment in planted seedlings and their protection from wallabies.

Barriers to regeneration, or restoration thresholds (Whisenant 1999), can be identified for several transitions. Transitions 1 and 2 cross the biotic barrier of lantana dominance in the understorey. Transition 3 overcomes the biotic barrier to regeneration of canopy tree diversity by thinning the plantation trees. Transitions 4 and 5 simultaneously overcome lantana and plantation tree dominance, initiating ecological restoration for both the understorey and canopy levels. Transition 6 is similar to Transitions 4 and 5 but risks crossing an abiotic threshold, through erosion on clay slopes or reduction of soil organic matter and leaching of nutrients in sandy soils. Transitions 15 and 16 are hypothesised and are similar to Transitions 1 and 2 in that competition from understorey weeds is reduced. Transition 17 rectifies both the abiotic and biotic barriers to redirect succession ultimately towards a goal community (Cummings et al. 2005).

Discussion and management implications

Irrespective of soil type and historical management differences, level of plantation thinning was the main influence on plant regeneration patterns in our experiments. Approximately 38% of the variation in plant species composition between sites was explained by environmental variables. This amount of explained variation is comparable to other direct gradient analyses of vegetation disturbed in space and time, for instance French pastures (28%; Lavorel et al. 1998), Australian *Sphagnum* bogs (32%; Clarke and Martin 1999) and NSW managed eucalypt and corymbia forests (up to 34%; Pharo and Beattie 2001). In Spanish plantations, silvicultural management was more important than soil correlates in determining understorey vegetation composition (Rubio et al. 1999). Within sites in the present experiment, canopy cover consistently explained the greatest amount of variation in community composition (32–38% post-treatment). King (1985) also noted the importance of crown cover in the regeneration of wet sclerophyll hardwood species (*Eucalyptus* spp. and *Lophostemon confertus*) on the NSW mid-North Coast.

Whereas grass regeneration was favoured by canopy reduction in these plantations, the richness and density of native understorey trees, woody shrubs and perennial herbs increased with canopy retention. Similar responses have been reported previously. In Sydney blue gum (*Eucalyptus saligna*) plantations in Hawaii, recruitment and survival of native seedlings was inversely related to suppression by a dense cover of introduced grass (Harrington and Ewell 1997). Grass and fern cover was negatively correlated with abundance and species richness of woody regeneration across a range of Costa Rican plantation species (Powers et al. 1997). Grass density in regenerating Amazonian rainforest affected richness of woody perennial regeneration, resulting in a limited number of early pioneer species although overall density was unchanged (Parrotta et al. 1997). Canopy cover has been related to the occurrence and growth of shade-tolerant and intolerant species in NSW rainforest (Floyd 1991), northern Queensland (Turton and Duff 1992; Tucker and Murphy 1997) and Amazonian rainforest (Parrotta et al. 1997) and in NSW coastal wet sclerophyll systems (opening the understorey canopy; King 1985). Understorey trees and woody shrubs and climbers in this study were able to regenerate better in sub-plots with greater canopy retention, indicating that our forest plantations are responding in a similar manner to rainforest and wet sclerophyll forest in the region and elsewhere. If plantation response to management interventions generally mimics natural forest recovery, plantations may generally be managed for improved stand biodiversity outcomes based on the response of regionally analogous forest ecosystems.

The state-and-transition model highlights the importance of native regeneration and post-thinning weed control. While ecological restoration may be accelerated by removing most of the plantation canopy, greater management inputs are required to avoid weed suppression of native regeneration and to direct the system towards goal communities. Should an abiotic transition barrier be crossed, as hypothesised for the sites where plantation establishment failed, broad-scale restoration will be particularly costly due to the expense of topsoil reconditioning and seedling establishment and protection from browsing wallabies. This begs the question why thin the canopy of these plantations, in the first place? Disregarding the issue of the logging revenue, there are several ecological and pragmatic reasons why canopy reduction should continue to be used in the restoration of lantana-infested hardwood plantations in BBNP and elsewhere. (1) Logging usefully crushes the lantana understorey, enabling burning to stimulate native plant germination and recruitment, and subsequent weed control. It is important to capture this opportunity quickly for the restoration of these plantations for biodiversity conservation, since some of the seed bank is likely to have survived for the past three to four decades, from the time when native forest occupied these sites. If the existing plantation is allowed to grow and self-thin slowly, much or all of the original seed bank will be extinguished by the passage of time and the diversity of regeneration will be limited to species capable of medium to long-distance dispersal. (2) Logging creates canopy gaps in otherwise even-aged plantations, promoting structural diversity in the canopy and understorey, an uneven-aged forest, and eventually floristic diversity in the mid-storey and canopy. (3) At a landscape scale, canopy reduction at a range of intensities promotes a mosaic of habitats. (4) Logging also accelerates understorey regeneration, further enhancing the habitat mosaic at landscape scale.

Whilst an understanding of the ecology of the natural forests surrounding plantation sites is essential to manage for biodiversity conservation, socio-economic considerations will invariably also be important. Although ecological restoration is often costly, the cost per hectare for initiating succession towards positive biodiversity outcomes in these plantations was low, due to the sale of the timber removed by logging and the presence of a viable native seed bank. The success of the latter was probably due to the recency of native forest occupation at some sites, proximity to seed sources and native forest in the surrounding landscape, the efficiencies gained from simultaneously removing canopy cover (and capturing income) and crushing the weed understorey, and the stimulatory effect of burning the logging slash and crushed understorey on native seedling recruitment. In some ways, the flooded gum plantation trees act as a cover crop for the regeneration of more natural forest systems, and promote restoration of more complex forest communities in the absence of further disturbance. The conservation values of plantations have been noted previously (e.g. Parrotta 1995; Haggar et al. 1997; Keenan et al. 1997; Parrotta 1999; Yirdaw and Luukkanen 2003), and are due in part to the fact that plantations are intermediate between pasture and forest systems. The cover crop of trees in plantations means they may generally be less difficult and costly to restore to diverse forest than herbaceous-dominated systems, since the sub-canopy microclimate can aid in facilitation of forest understorey regeneration.

Our research highlights the utility of adopting a state-and-transition approach by identifying management actions that represent opportunities and potential hazards. Continued monitoring is required to improve on the model, test the longer-term transitions and confirm whether the predicted stable states develop. Future disturbances such as unplanned fire need to be incorporated. The model also needs to be tested at a broader scale by implementing treatments in a range of sites with varying degrees of understorey degradation. Irrespective of the accuracy of the model, our results show that the stand-level management of plantations can improve their value for biodiversity conservation (as suggested by Lindenmayer et al. 2003). We suspect that many plantations may be managed relatively cheaply (at least in part for conservation), for instance through the sale of logs, and anticipate that plantation management will play an increasing role in biodiversity conservation at stand and landscape scales.

Acknowledgements Dr Carl Grant initiated this research, which was jointly funded by the NSW NPWS and the Australian Research Council (project number C19906697). The authors thank Dr Ian Davies (Statistics, UNE), and Martin Smith, Glenn Storrie, Alan Jeffery, John O'Gorman and Dr Tony Fleming of the NSW National Parks and Wildlife Service, Department of Environment and Climate Change, for their support of the research. Three anonymous reviewers and the guest editors are thanked for their comments and improvements of the manuscript. Wiley-Blackwell are acknowledged for providing permission to re-publish parts or all of Figures 2, 3 and 4.

Appendix

Species identified in flooded gum plantation sub-plots between 1999 and 2002, and the functional groups to which they were assigned

Family	Species	Common	Functional group
ACANTHACEAE	*Adiantum hispidulum*	Rough maidenhair fern	Ferns
ACANTHACEAE	*Brunoniella australis*	Blue trumpet	Herbaceous perennials
ACANTHACEAE	*Cheilanthes distans*	Bristly cloak fern	Ferns
ACANTHACEAE	*Cheilanthes sieberi* subsp. sieberi	Rock fern	Ferns
ACANTHACEAE	*Pseuderanthemum variabile*	Pastel flower	Herbaceous perennials

Appendix continued

Family	Species	Common	Functional group
AGAVACEAE	*Cordyline stricta*	Narrow palm lily	Woody shrubs and climbers
ANACARDIACEAE	*Euroshinus falcata* var. falcata	Ribbonwood	Trees—potential canopy
APIACEAE	*Centella asiatica*		Herbaceous perennials
APIACEAE	*Hydrocotyle laxiflora*	Stinking pennywort	Herbaceous perennials
APIACEAE	*Oreomyrrhis eriopoda*		Herbaceous perennials
APOCYNACEAE	*Ervatamia angustisepala*	Banana bush	Woody shrubs and climbers
APOCYNACEAE	*Parsonsia straminea*	Common silkpod	Woody shrubs and climbers
ARACEAE	*Alocasia brisbanensis*	Cunjevoi	Herbaceous perennials
ARACEAE	*Gymnostachys anceps*	Settler's flax	Non-grass monocots
ARALIACEAE	*Astrotricha latifolia*	Broad leaf star hair	Woody shrubs and climbers
ARALIACEAE	*Polyscias sambucifolia* subsp. A	Elderberry panax	Understorey trees
ARECACEAE	*Archontophoenix cunninghamiana*	Bangalow palm	Trees—potential canopy
ASCLEPACEAE	**Asclepias currasiva*	Blood flower	Introduced herbaceous perennials
ASCLEPIADACEAE	**Gomphocarpus fruticosus*	Narrow-leaved cotton bush	Introduced herbaceous perennials
ASCLEPIADACEAE	*Marsdenia* sp.		Herbaceous perennials
ASTERACEAE	**Ageratina adenophora*	Crofton weed	Introduced herbaceous perennials
ASTERACEAE	**Ambrosia artemisiifolia*	Annual ragweed	Introduced herbaceous annuals
ASTERACEAE	**Baccharis halimifolia*	Groundsel bush	Introduced herbaceous perennials
ASTERACEAE	**Bidens pilosa*	Cobblers pegs	Introduced herbaceous perennials
ASTERACEAE	**Chrysanthemoides monilifera*	Bitou bush	Introduced woody shrubs
ASTERACEAE	**Cirsium vulgare*	Spear thistle	Introduced herbaceous annuals
ASTERACEAE	**Conzya bonariensis*	Flaxleaf fleabane	Introduced herbaceous annuals
ASTERACEAE	**Conzya parva*		Introduced herbaceous annuals
ASTERACEAE	**Crassocephalum crepidioides*	Thickhead	Introduced herbaceous annuals
ASTERACEAE	**Gamochaeta spicata*	Spiked cudweed	Introduced herbaceous perennials
ASTERACEAE	**Hypochaeris radicata*	Flatweed, Catsear	Introduced herbaceous perennials
ASTERACEAE	**Senecio madagascariensis*	Fireweed	Introduced herbaceous annuals
ASTERACEAE	**Sonchus oleraceae*	Common sowthistle	Introduced herbaceous annuals

Appendix continued

Family	Species	Common	Functional group
ASTERACEAE	*Chrysocephalum apiculatum*	Common everlasting, yellow buttons	Herbaceous perennials
ASTERACEAE	*Euchiton* sp.	Cudweed	Herbaceous annuals
ASTERACEAE	*Ozothamnus diosmifolius*	White dogwood	Woody shrubs and climbers
ASTERACEAE	*Ozothamnus rufescens*		Woody shrubs and climbers
ASTERACEAE	*Senecio* sp.		Herbaceous perennials
ASTERACEAE	*Sigesbeckia orientalis* ssp. orientalis	Indian weed	Herbaceous perennials
ASTERACEAE	*Vernonia cinerea var. cinerea*		Herbaceous perennials
BLECHNACEAE	*Blechnum* sp.		Ferns
BLECHNACEAE	*Doodia aspera*	Rasp fern	Ferns
BRASSICACEAE	*Cardamine paucijuga*	Bitter-cress	Herbaceous annuals
CALLITRICHACEAE	*Callitriche* sp.		Herbaceous perennials
CAMPANULACEAE	*Wahlenbergia communis*	Tufted bluebell	Herbaceous perennials
CAMPANULACEAE	*Wahlenbergia stricta*	Tall bluebell	Herbaceous perennials
CARYOPHYLLACEAE	**Paronychia brasiliana*	Chilean whitlow wort	Introduced herbaceous perennials
CARYOPHYLLACEAE	*Drymaria cordata*	Tropical chickweed	Herbaceous annuals
CASUARINACEAE	*Allocasuarina littoralis*	Black she-oak	Trees—potential canopy
CELASTRACEAE	*Celastrus subspicata*		Woody shrubs and climbers
COMMELINACEAE	*Commelina cyaneae*	Native wandering jew	Herbaceous perennials
CONVOLVOLACEAE	*Calystegia marginata*	Forest bindweed	Herbaceous perennials
CONVOLVULACEAE	*Dichondra repens*	Kidney weed	Herbaceous perennials
CONVOLVULACEAE	*Polymeria calycina*		Herbaceous perennials
CYATHEACEAE	*Cyathea australis*	Rough treefern	Ferns
CYPERACEAE	**Cyperus sesquiflorus*		Introduced grass-like monocots
CYPERACEAE	*Carex inversa*	Knob sedge	Grass-like monocots
CYPERACEAE	*Cyperus congestus*		Grass-like monocots
CYPERACEAE	*Cyperus polystachyus*		Grass-like monocots
CYPERACEAE	*Cyperus trinervis*		Grass-like monocots
CYPERACEAE	*Isolepis nodosa*	Nobby club-rush	Grass-like monocots
CYPERACEAE	*Lepidosperma laterale*		Non-grass monocots
DENNSTAEDTIACEAE	*Histiopteris incisa*	Bat's wing fern	Ferns
DENNSTAEDTIACEAE	*Pteridium esculentum*	Bracken	Ferns
DILLENIACEAE	*Hibbertia dentata*	Twining guinea flower	Herbaceous perennials
DILLENIACEAE	*Hibbertia scandens*	Climbing guinea flower	Herbaceous perennials
DIOSCOREACEAE	*Dioscorea transversa*	Native yam	Herbaceous Perennials
ELAEOCARPACEAE	*Elaeocarpus reticulatus*	Blueberry Ash	Understorey trees
EUPHORBIACEAE	*Breynia cernua*	Coffee bush	Woody shrubs and climbers

Appendix continued

Family	Species	Common	Functional group
EUPHORBIACEAE	*Glochidion ferdinandii*	Cheese tree	Trees—potential canopy
FABACEAE	**Senna X floribunda*	Smooth senna	Introduced woody shrubs
FABACEAE	*Derris involuta*	Native derris	Woody shrubs and climbers
FABACEAE	*Glycine clandestina*		Herbaceous perennials
FABACEAE	*Glycine microphylla*		Herbaceous perennials
FABACEAE	*Glycine tabacina*		Herbaceous perennials
FABOIDEAE	**Trifolium* sp.	Clover	Introduced herbaceous perennials
FABOIDEAE	*Desmodium rhytidophyllum*	Rusty tick-trefoil	Herbaceous perennials
FABOIDEAE	*Desmodium varians*	Slender tick-trefoil	Herbaceous perennials
FABOIDEAE	*Hovea acutifolia*		Woody shrubs and climbers
GERANIACEAE	*Geranium homeanum*		Herbaceous perennials
GERANIACEAE	*Geranium solanderi*	Native geranium	Herbaceous perennials
IRIDACEAE	**Sisyrinchium* sp. *A*	Scourweed	Introduced herbaceous annuals
JUNCACEAE	*Juncus* sp.		Grass-like monocots
LAURACEAE	*Cryptocarya rigida*	Rose Maple, forest maple	Trees—potential canopy
LAURACEAE	*Cryptocarya triplinervis*	Three-veined cryptocarya	Trees—potential canopy
LILIACEAE	**Lilium formosanum*		Introduced herbaceous perennials
LOBELIACEAE	*Pratia puspurascens*	Whiteroot	Herbaceous perennials
LUZURIAGACEAE	*Eustrephus latifolius*	Wombat Berry	Herbaceous perennials
LUZURIAGACEAE	*Geitnoplesium cymosum*	Scrambling lily	Herbaceous perennials
MALVACEAE	**Modiola caroliniana*	Red-flowered mallow	Introduced herbaceous annuals
MALVACEAE	**Setaria pumila*	Yellow foxtail, Pale pigeon grass	Introduced grass-like monocots
MALVACEAE	**Sida rhombifolia*	Paddy's lucerne	Introduced herbaceous perennials
MELIACEAE	*Synoum glandulosum*	Scentless rosewood	Understorey trees
MENISPERMACEAE	*Sarcopetalum harveyanum*	Pearl vine	Woody shrubs and climbers
MENISPERMACEAE	*Stephania japonica* var. discolor	Snake vine	herbaceous perennials
MIMOSACEAE	*Acacia maidenii*	Maiden's wattle	Trees—potential canopy
MIMOSACEAE	*Acacia melanoxylon*	Blackwood	Trees—potential canopy
MIMOSOIDEAE	*Acacia irrorata*	Green wattle	Trees—Potential Canopy
MIMOSOIDEAE	*Acacia sophoraea*	Coastal wattle	Woody shrubs and climbers
MONIMIACEAE	*Wilkia huegeliana*	Veiny wilkiea	Understorey Trees
MORACEAE	*Ficus coronata*	Creek sandpaper fig	Understorey Trees
MORACEAE	*Maclura cochinchinensis*	Cockspur thorn	Woody shrubs and climbers

Appendix continued

Family	Species	Common	Functional group
MYRTACEAE	*Acmena smithii*	Lilly pilly	Trees—potential canopy
MYRTACEAE	*Callistemon salignus*	White bottlebrush	Understorey trees
MYRTACEAE	*Eucalyptus grandis*	Flooded gum	Trees—potential canopy
MYRTACEAE	*Eucalyptus pilularis*	Blackbutt	Trees—potential canopy
MYRTACEAE	*Eucalyptus saligna*	Sydney blue gum	Trees—potential canopy
MYRTACEAE	*Lophostemon confertus*		Trees—potential canopy
MYRTACEAE	*Melaleuca quinquenervia*	Paperbark	Trees—potential canopy
MYRTACEAE	*Pilidiostigma glabrum*		Understorey trees
MYRTACEAE	*Rhodamnia rubescens*	Scrub turpentine	Trees—potential canopy
MYRTACEAE	*Syncarpia glomulifera*	Turpentine	Trees—potential canopy
MYRTACEAE	*Syzygium australe*	Brush cherry	Understorey trees
OLEACEAE	*Notelaea* sp.		Woody shrubs and climbers
OLEACEAE	*Notelaea longifolia*	Large mock-olive	Trees—potential canopy
OLEACEAE	*Notelaea venosa*	Veined mock-olive	Understorey trees
ONAGRACEAE	*Epilobium* sp.	Willowherb	Herbaceous perennials
OXALIDACEAE	*Oxalis chnoodes*		Herbaceous perennials
OXALIDACEAE	*Oxalis exilis*		Herbaceous perennials
PASSIFLORACEAE	**Passiflora edulis*	Edible passionfruit	Introduced herbaceous perennials
PASSIFLORACEAE	**Passiflora subpeltata*	White passionflower	Introduced herbaceous perennials
PHORMIACEAE	*Dianella caerulea var. producta*	Flax lily	Herbaceous perennials
PHYTOLACCACEAE	**Phytolacca octandra*	Inkweed	Introduced herbaceous perennials
PITTOSPORACEA	*Billardiera scandens*	Appleberry	Herbaceous perennials
POACAEA	**Sporobolus indicus*	Parramatta grass	Introduced grass-like monocots
POACAEA	*Cymbopogon refractus*	Barb-wire grass	
POACAEA	*Entolasia marginata*	Bordered panic	Grass-like monocots
POACAEA	*Entolasia stricta*	Wiry panic	Grass-like monocots
POACAEA	*Microlaena stipoides*		Grass-like monocots
POACAEA	*Microstegium nudum*		Grass-like monocots
POACAEA	*Oplismenus imbecillis*		Grass-like monocots
POACAEA	*Paspalidium distans*		Grass-like monocots
POACEAE	**Andropogon virginicus*	Whiskey grass	Introduced grass-like monocots
POACEAE	**Axonopus compressus*	Broad-leaved carpet grass	Introduced grass-like monocots
POACEAE	**Melinis repens*	Red natal grass	Introduced grass-like monocots
POACEAE	*Austrostipa* sp.		Grass-like monocots
POACEAE	*Digitaria parviflora*	Small-flowered finger grass	Grass-like monocots
POACEAE	*Eragrostis brownii*	Brown's lovegrass	Grass-like monocots

Appendix continued

Family	Species	Common	Functional group
POACEAE	*Eragrostis elongata*	Clustered lovegrass	Grass-like monocots
POACEAE	*Imperata cylindrica*	Blady grass	Grass-like monocots
POACEAE	*Oplismenus aemulus*		Grass-like monocots
POACEAE	*Paspalum wettsteinii*	Broad-leaved paspalum	Grass-like monocots
POACEAE	*Themeda australis*	Kangaroo grass	Grass-like monocots
POLYGONACEAE	*Muehlenbeckia gracillima*		Herbaceous perennials
PRIMULACEAE	**Anagallis arvensis*	Scarlet pimpernell	Introduced herbaceous perennials
PROTEACEAE	*Banksia integrifolia* var. integrifolia	Coastal banksia	Trees—potential canopy
PROTEACEAE	*Persoonia stradbrokensis*	Geebung	Understorey trees
PTERIDACEAE	*Pteris tremula*	Tender Brake	Ferns
RHAMNACEAE	*Alphitonia excelsa*	Red ash	Trees—potential canopy
ROSACEAE	*Rubus hillii*	Molucca bramble	Herbaceous perennials
ROSACEAE	*Rubus parvifolius*	Native raspberry	Herbaceous perennials
ROSACEAE	*Rubus rosifolius*	Rose-leaf bramble	Herbaceous perennials
RUBIACEAE	*Morinda jasminoides*	Morinda	Woody shrubs and climbers
RUBIACEAE	*Psychotria loniceroides*	Hairy psychotria	Understorey trees
RUTACEAE	*Acronychia* sp.		Trees—potential canopy
SAPINDACAEA	*Guioa semiglauca*	Wild quince	Understorey trees
SAPINDACEAE	*Alectryon coriaceus*	Beach alectryon	Trees—potential canopy
SAPINDACEAE	*Cupaniopsis anarcioides*	Tuckeroo	Trees—potential canopy
SAPINDACEAE	*Dodonaea triquetra*	Large-leaf hop-bush	Woody shrubs and climbers
SAPINDACEAE	*Jagera pseudorhus*	Foambark Tree	Trees—potential canopy
SCROPHULARIACEAE	*Veronica plebeia*	Trailing speedwell	Herbaceous perennials
SMILACEAE	*Ripogonum album*	White supplejack	Non-grass monocots
SMILACEAE	*Smilax australis*	Sarsaparilla	Non-grass monocots
SMILACEAE	*Smilax glyciphylla*	Sweet sarsaparilla	Non-grass monocots
SOLANACEAE	**Physalis peruviana*	Cape gooseberry	Introduced herbaceous perennials
SOLANACEAE	**Solanum mauritianum*	Wild tobacco bush	Introduced woody shrubs
SOLANACEAE	**Solanum nigrum*	Black-berry nightshade	Introduced woody shrubs
SOLANACEAE	*Duboisia myoporoides*	Corkwood	Trees—potential canopy
SOLANACEAE	*Solanum americanum*	Glossy nightshade	Herbaceous perennials
SOLANACEAE	*Solanum aviculare*	Kangaroo apple	Herbaceous perennials
SOLANACEAE	*Solanum densevestitum*		Woody shrubs and climbers
SOLANACEAE	*Solanum prinophyllum*	Forest nightshade	Herbaceous perennials
STERCULIACEAE	*Commersonia fraseri*	Brush Kurrajong	Understorey trees
STERCULIACEAE	*Rulingia dasyphylla*	Kerrawang	Understorey trees
STERCULIACEAE	*Seringia arborescens*		Woody shrubs and climbers
ULMACEAE	*Trema aspera*	Native peach	Understorey trees

Appendix continued

Family	Species	Common	Functional group
VERBENACEAE	**Lantana camara*		Introduced woody shrubs
VERBENACEAE	**Verbena bonariensis*	Purpletop	Introduced herbaceous perennials
VERBENACEAE	*Clerodendrum floribundum*	Lolly bush	Woody shrubs and climbers
VERBENACEAE	**Verbena gaudidiacdii*		Introduced herbaceous perennials
VIOLACEAE	*Hybanthus enneaspermus* subsp. stellarioides	Spade flower	Herbaceous perennials
VIOLACEAE	*Viola hederaceae*	Ivy-leaved violet	Herbaceous perennials
VITACEAE	*Cayratia clematidea*	Slender grape	Herbaceous perennials
VITACEAE	*Cissus hypoglauca*	Giant water vine	Woody shrubs and climbers
VITACEAE	*Tetrastigma nitens*	Three-leaf water vine	Woody shrubs and climbers
XANTHORRHOEACEAE	*Lomandra* sp.		Non-grass monocots
ZINGIBERACEAE	*Alpinia caerulea*	Native ginger	Herbaceous perennials

* Introduced species

References

Attiwill PM (1994) The disturbance of forest ecosystems: the ecological basis for conservative management. For Ecol Manage 63:247–300

Bauhus J, Aubin I, Messier C, Connell M (2001) Composition, structure, light attenuation and nutrient content of the understorey vegetation in a *Eucalyptus sieberi* regrowth stand 6 years after thinning and fertilisation. For Ecol Manage 144:275–286

Baur GN (1962) Forest vegetation in northeastern New South Wales. Forestry Commission NSW, Research Note No. 8

Bureau of Meteorology (2002) Climate averages for Station: 059040 COFFS HARBOUR MO. Commonwealth Bureau of Meteorology, http://www.bom.gov.au

Clarke PJ, Martin ARH (1999) *Sphagnum* peatlands of Kosciuszko National Park in relation to altitude, time and disturbance. Aust J Bot 47:519–536

Cropper SC (1993) Management of endangered plants. Australian Nature Conservation Agency. CSIRO Publishing, Canberra

Cummings J (2004) Ecological restoration of *Imperata cylindrica* grasslands and *Lantana camara* infested *Eucalyptus grandis* plantations in Bongil Bongil National Park, north coast of NSW. Unpubl. PhD Thesis, University of New England, Armidale

Cummings J, Reid N, Davies I, Grant C (2005) Adaptive restoration of sand-mined areas for biological conservation. J Appl Ecol 42:160–170

Cummings J, Reid N, Davies I, Grant C (2007) Experimental manipulation of restoration barriers in abandoned eucalypt plantations. Restor Ecol 15:156–167

Curtin RA, Squire RH, Mackowski CM (1991) Management of native hardwood forests in state forests of North Coast New South Wales. In: McKinnell FH, Hopkins ER, Fox JED (eds) Forest management in Australia. Surrey Beatty & Sons, Sydney, pp 77–106

de la Cretaz AL, Kelty MJ (2002) Development of tree regeneration in fern-dominated forest understories after reduction of deer browsing. Restor Ecol 10:416–426

Dudley N, Mansourian S, Vallauri D (2005) Forest landscape restoration in context. In: Dudley N, Mansourian S, Vallauri D (eds) Forest restoration in landscapes—beyond planting trees. Springer, New York, pp 3–7

Dutton I, Specht A, Stewart B (1989) Rehabilitation of a littoral rainforest. Aust Parks Recreation 25: 27–33

Fensham RJ, Fairfax RJ, Cannell RJ (1994) The invasion of *Lantana camara* L. in Forty Mile Scrub National Park, north Queensland. Aust J Ecol 19:297–305
Floyd AG (1966) Effect of fire upon weed seeds in the wet sclerophyll forests of northern New South Wales. Aust J Bot 14:243–256
Floyd AG (1991) Natural regeneration of rainforest in NSW. In: Phillips S (ed) Rainforest remnants—proceedings of a workshop on rainforest rehabilitation. NSW National Parks and Wildlife Service, Sydney, pp 20–25
Fox BJ, Fox MD (1986) Resilience of animal and plant communities to human disturbance. In: Dell B, Hopkins AJM, Lamont BB (eds) Resilience in Mediterranean-type ecosystems. Dr W Junk Publishers, Dordrecht, pp 39–64
Friedel MH (1991) Range condition assessment and the concept of thresholds: a viewpoint. J Range Manage 44:422–426
Gilmour AR, Gogel BG, Cullis BR, Welham SJ, Thompson R (2002) ASReml user guide. Release 1.0. VSN International Ltd., Hemel Hempstead, UK
Haggar J, Wightman K, Fisher R (1997) The potential of plantations to foster woody regeneration within a deforested landscape in lowland Costa Rica. For Ecol Manage 99:55–64
Harrington RA, Ewel JJ (1997) Invasibility of tree plantations by native and non-indigenous plant species in Hawaii. For Ecol Manage 99:153–162
Hartley MJ (2002) Rationale and methods for conserving biodiversity in plantation forests. For Ecol Manage 155:81–95
Hobbs RJ, Harris JA (2001) Restoration ecology: repairing the earth's ecosystems in the new millennium. Restor Ecol 9:239–246
Hobbs RJ, Norton DA (1996) Towards a conceptual framework for restoration ecology. Restor Ecol 4: 93–110
Holling CS (1973) Resilience and stability of ecological systems. Annu Rev Ecol Syst 4:1–23
Ihaka R, Gentleman R (1996) R: a language for data analysis and graphics. J Comput Graph Stat 5:299–314
Johnson JB, Omland KS (2004) Model selection in ecology and evolution. Trends Ecol Evol 19:101–108
Keenan R, Lamb D, Woldring O, Irvine T, Jensen R (1997) Restoration of plant biodiversity beneath tropical tree plantations in northern Australia. For Ecol Manage 99:117–131
King GC (1985) Natural regeneration in wet sclerophyll forest with an overstorey of *Eucalyptus microcorys*, *E. saligna* and *Lophostemon confertus*. Aust For 48:54–62
Kooyman RM (1996) Growing rainforest—rainforest restoration and regeneration. Greening Australia, Brisbane
Lake PS (2001) On the maturing of restoration: linking ecological research and restoration. Ecol Manage Restor 2:110–115
Lamb D (2005) Restoring tropical moist broad-leaf forests. In: Dudley N, Mansourian S, Vallauri D (eds) Forest restoration in landscapes—beyond planting trees. Springer, New York, pp 291–297
Lavorel S, Touzard B, Lebreton J, Clement B (1998) Identifying functional groups for response to disturbance in an abandoned pasture. Acta Oecol 19:227–240
Lindenmayer DB, Hobbs RJ, Salt D (2003) Plantation forests and biodiversity conservation. Aust For 66:62–66
New South Wales National Parks, Wildlife Service (1999) Bongil Bongil National Park plan of management. NSW National Parks and Wildlife Service, Sydney
NSW Scientific Committee (2004a) Subtropical coastal floodplain forest of the NSW north coast bioregion—endangered ecological community determination—final. NSW Department of Environment and Conservation, Sydney
NSW Scientific Committee (2004b) Littoral rainforest in the NSW north coast, Sydney basin and south east corner bioregions—endangered ecological community determination—final. NSW Department of Environment and Conservation, Sydney
Parrotta JA (1995) Influence of overstorey composition on understorey colonization by native species in plantations on a degraded tropical site. J Veg Sci 6:627–636
Parrotta JA (1999) Productivity, nutrient cycling, and succession in single- and mixed-species plantations of *Casuarina equisetifolia*, *Eucalyptus robusta*, and *Leucaena leucocephala* in Puerto Rico. For Ecol Manage 124:45–77
Parrotta JA, Knowles OH, Wunderle JM Jr (1997) Development of floristic diversity in 10-year-old restoration forests on a bauxite mined site in Amazonia. For Ecol Manage 99:21–42
Patterson HD, Thompson R (1971) Recovery of interblock information when block sizes are unequal. Biometrika 58:545–554
Pharo EJ, Beattie AJ (2001) Management forest types as a surrogate for vascular plant, bryophyte and lichen diversity. Aust J Bot 49:23–30

Pinheiro J, Bates D, DebRoy S, Sarkav D (2003) The nlme package—linear and nonlinear mixed effects models. http://cran.r-project.org/
Powers JS, Haggar JP, Fisher RF (1997) The effect of overstorey composition on understorey woody regeneration and species richness in 7-year-old plantations in Costa Rica. Forest Ecol Manag 99:43–54
Quinn GP, Keough MJ (2002) Experimental design and data analysis for biologists. Cambridge University Press, New York
Rayment GE, Higginson FR (1992) Australian laboratory handbook of soil and water chemical methods. Inkata Press, Sydney
Rubio A, Gavilan R, Escudero A (1999) Are soil characteristics and understorey composition controlled by forest management? For Ecol Manage 113:191–200
Scheffer M, Carpenter SR (2003) Catastrophic regime shifts in ecosystems: linking theory to observation. Trends Ecol Evol 18:648–656
State Forests of New South Wales (SFNSW) (1995) Proposed forestry operations in the Coffs Harbour and urunga management areas—environmental and fauna impact statement. State Forest of New South Wales. Coffs Harbour
State Forests of New South Wales (n.d.) Ecologically sustainable forest management—native silvicultural manual. State Forests of New South Wales, Sydney
ter Braak CJF (1986) Canonical correspondence analysis: a new eigenvector technique for multivariate direct gradient analysis. Ecology 67:1167–1179
ter Braak CJF, Smilauer P (1998) CANOCO version 4. CANOCO reference manual and user's guide to Canoco for Windows: software for canonical community ordination. Microcomputer Power, Ithaca, NY
ter Braak CJF, Verdonschot PFM (1995) Canonical correspondence analysis and related multivariate methods in aquatic ecology. Aquat Sci 57:255–289
Thomas SC, Halpern CB, Falk DA, Liguori DA, Austin KA (1999) Plant diversity in managed forests: understorey responses to thinning and fertilization. Ecol Appl 9:864–879
Totland Ø, Nyeko P, Bjerknes A, Hegland SJ, Nielsen A (2005) Does forest gap size affect population size, plant size, reproductive success and pollinator visitation in *Lantana camara*, a tropical invasive shrub? For Ecol Manage 215:329–338
Tucker NIJ, Murphy TM (1997) The effects of ecological rehabilitation on vegetation recruitment: some observations from the Wet Tropics of North Queensland. For Ecol Manage 99:133–152
Turton SM, Duff GA (1992) Light environments and floristic composition across an open forest-rainforest boundary in northeastern Queensland. Aust J Ecol 17:415–423
Venables WN, Ripley BD (2002) Modern applied statistics with S, 4th edn. Springer, Germany
Villa Castillo B, Reid N, Cummings J, Corral Rivas JJ, Smith M, Storrie G (2005) Restoration of *Eucalyptus saligna* and *Eucalyptus pilularis–Paspalum* plantations for biodiversity conservation in Bongil Bongil National Park, NSW, Australia. In: Proceedings of the 16th international conference of the Society for Ecological Restoration, 24–26 August 2004. Victoria, British Columbia, Canada, pp 1–13
Webb LJ, Tracey JG, Williams WT (1972) Regeneration and pattern in subtropical rainforest. J Ecol 60:675–695
Westoby MB, Walker B, Noy-Meir I (1989) Opportunistic management of rangelands not at equilibrium. J Range Manage 42:266–274
Wetzel S, Burgess D (2001) Understorey environment and vegetation response after partial cutting and site preparation in *Pinus strobus* L. stands. For Ecol Manage 151:43–59
Whisenant SG (1999) Repairing damaged wildlands—a process-orientated, landscape-scale approach. Cambridge University Press, United Kingdom
Yirdaw E, Luukkanen O (2003) Indigenous woody species diversity in *Eucalyptus globulus* Labill. ssp. *globulus* plantations in the Ethiopian highlands. Biodivers Conserv 12:567–582

MIX
Papier aus verantwortungsvollen Quellen
Paper from responsible sources
FSC® C105338

If you have any concerns about our products,
you can contact us on
ProductSafety@springernature.com

In case Publisher is established outside the EU,
the EU authorized representative is:
Springer Nature Customer Service Center GmbH
Europaplatz 3, 69115 Heidelberg, Germany

Printed by Libri Plureos GmbH
in Hamburg, Germany